Implementing Automation Initiatives in Companies to Create Better-Connected Experiences

Jorge Remondes
Instituto Superior de Entre Douro e Vouga, Portugal & ISCAP, Instituto Politécnico do Porto, Portugal

Sandrina Teixeira
ISCAP, Instituto Politécnico do Porto, Portugal

IGI Global
PUBLISHER of TIMELY KNOWLEDGE

A volume in the Advances in Marketing, Customer Relationship Management, and E-Services (AMCRMES) Book Series

Published in the United States of America by
IGI Global
Business Science Reference (an imprint of IGI Global)
701 E. Chocolate Avenue
Hershey PA, USA 17033
Tel: 717-533-8845
Fax: 717-533-8661
E-mail: cust@igi-global.com
Web site: http://www.igi-global.com

Library of Congress Cataloging-in-Publication Data

Names: Remondes, Jorge, 1971- editor. | Teixeira, Sandrina, 1975- editor.
Title: Implementing automation initiatives in companies to create better-connected experiences / Jorge Remondes, and Sandrina Teixeira, editors.
Description: Hershey, PA : Business Science Reference, [2023] | Includes bibliographical references and index. | Summary: "This book explores one of the most current and decisive topics for business and academia and that is automation and connectivity expected to help the decision-making process of companies and equip marketing managers and executives with the knowledge and skills to support them in the strategic and operational decision-making process"-- Provided by publisher.
Identifiers: LCCN 2022023435 (print) | LCCN 2022023436 (ebook) | ISBN 9781668455388 (hardcover) | ISBN 9781668455395 (paperback) | ISBN 9781668455401 (ebook)
Subjects: LCSH: Decision making--Data processing. | Strategic planning--Decision making. | Automation.
Classification: LCC HD30.28 .I473 2023 (print) | LCC HD30.28 (ebook) | DDC 658.4/012--dc23/eng/20220715
LC record available at https://lccn.loc.gov/2022023435
LC ebook record available at https://lccn.loc.gov/2022023436

This book is published in the IGI Global book series Advances in Marketing, Customer Relationship Management, and E-Services (AMCRMES) (ISSN: 2327-5502; eISSN: 2327-5529)

British Cataloguing in Publication Data
A Cataloguing in Publication record for this book is available from the British Library.

For electronic access to this publication, please contact: eresources@igi-global.com.

Advances in Marketing, Customer Relationship Management, and E-Services (AMCRMES) Book Series

Eldon Y. Li
National Chengchi University, Taiwan & California Polytechnic State University, USA

ISSN:2327-5502
EISSN:2327-5529

Mission

Business processes, services, and communications are important factors in the management of good customer relationship, which is the foundation of any well organized business. Technology continues to play a vital role in the organization and automation of business processes for marketing, sales, and customer service. These features aid in the attraction of new clients and maintaining existing relationships.

The Advances in Marketing, Customer Relationship Management, and E-Services (AMCRMES) Book Series addresses success factors for customer relationship management, marketing, and electronic services and its performance outcomes. This collection of reference source covers aspects of consumer behavior and marketing business strategies aiming towards researchers, scholars, and practitioners in the fields of marketing management.

Coverage

- Relationship Marketing
- B2B marketing
- Legal Considerations in E-Marketing
- Customer Retention
- Telemarketing
- Data mining and marketing
- Text Mining and Marketing
- Cases on Electronic Services
- CRM and customer trust
- Database marketing

IGI Global is currently accepting manuscripts for publication within this series. To submit a proposal for a volume in this series, please contact our Acquisition Editors at Acquisitions@igi-global.com or visit: http://www.igi-global.com/publish/.

The Advances in Marketing, Customer Relationship Management, and E-Services (AMCRMES) Book Series (ISSN 2327-5502) is published by IGI Global, 701 E. Chocolate Avenue, Hershey, PA 17033-1240, USA, www.igi-global.com. This series is composed of titles available for purchase individually; each title is edited to be contextually exclusive from any other title within the series. For pricing and ordering information please visit http://www.igi-global.com/book-series/advances-marketing-customer-relationship-management/37150. Postmaster: Send all address changes to above address.

Titles in this Series

For a list of additional titles in this series, please visit: http://www.igi-global.com/book-series/advances-marketing-customer-relationship-management/37150

Cases on Social Justice in China and Perspectives on Chinese Brands
Youssef El Haoussine (Beijing Normal University - Hong Kong Baptist University United International College, China) and Lulu Wang (Sinotrend Consulting, China)
Business Science Reference • © 2023 • 335pp • H/C (ISBN: 9781668449554) • US $215.00

Building a Brand Image Through Electronic Customer Relationship Management
Arshi Naim (King Kalid University, Saudi Arabia) and Sandeep Kumar Kautish (Lord Buddha Education Foundation, Asia Pacific University, India)
Business Science Reference • © 2022 • 315pp • H/C (ISBN: 9781668453865) • US $250.00

Social Customer Relationship Management (Social-CRM) in the Era of Web 4.0
Nedra Bahri Ammari (IHEC of Carthage, Tunisia)
Business Science Reference • © 2022 • 317pp • H/C (ISBN: 9781799895534) • US $250.00

Handbook of Research on Global Perspectives on International Advertising
Ipek Krom (Istanbul Esenyurt University, Turkey)
Business Science Reference • © 2022 • 397pp • H/C (ISBN: 9781799896722) • US $315.00

Handbook of Research on Consumer Behavior Change and Data Analytics in the Socio-Digital Era
Pantea Keikhosrokiani (School of Computer Sciences, Universiti Sains Malaysia, Malaysia)
Business Science Reference • © 2022 • 458pp • H/C (ISBN: 9781668441688) • US $315.00

Handbook of Research on Interdisciplinary Reflections of Contemporary Experiential Marketing Practices
Gökhan Akel (Antalya Akev University, Turkey)
Business Science Reference • © 2022 • 596pp • H/C (ISBN: 9781668443804) • US $315.00

Developing Relationships, Personalization, and Data Herald in Marketing 5.0
Jasmine Kaur (Chitkara Business School, Chitkara University, Punjab, India) Priya Jindal (Chitkara Business School, Chitkara University, Punjab, India) and Amandeep Singh (Chitkara Business School, Chitkara University, India)
Business Science Reference • © 2022 • 327pp • H/C (ISBN: 9781668444962) • US $250.00

IGI Global
PUBLISHER of TIMELY KNOWLEDGE

701 East Chocolate Avenue, Hershey, PA 17033, USA
Tel: 717-533-8845 x100 • Fax: 717-533-8661
E-Mail: cust@igi-global.com • www.igi-global.com

Table of Contents

Detailed Table of Contents

company, Consumers Trust, better known in Portugal as "O Portal da Queixa." Qualitative research will be undertaken using interview and content analysis. Its main objective is to understand and explain the causes and effects of this concept that, despite having originated the scientific research mentioned, in particular in these last two years, as a result of the COVID-19 pandemic, have not been explored enough and therefore are not yet consensual.

Section 2
Automation Initiatives and Better-Connected Experiences

The automatization, dematerialization, and innovative digital systems and technologies, named Industry 4.0, are transforming the economic and social system, encompassing opportunities and challenges for growth and international expansion. In Portugal, SMEs face great difficulties and challenges because innovative digital businesses and systems enhance globalization and intensify competition. Exploratory research was designed to explore how Industry 4.0 and digitalization will influence the organization, competitiveness, and internationalization process of the SMEs under study. Results indicate that Industry 4.0 appears to be critical because it positively affected the value chain and the organizational and competitiveness level. Therefore, the companies were better positioned in their competitive advantages. This work provides insights for the study and joint understanding of digitalization and business internationalization because it integrates and relates two autonomous, complex, and strategic fields of research, with no applied studies for the Portuguese SME background.

Omni-channel retailing integrates different contact points in social, virtual, and physical realms in customer journey. This chapter provides a framework to understand omni-channel customer behavior, first by focusing on customer journey in omni-channel retailing representing the actual step of multichannel and cross-channel retailing. To contribute to the conceptualization of omni-channel customer journey, some suggestions including customer experience with digital, social, and phygital dimensions are offered. Followingly, the critical antecedents of omni-channel customer experience are reviewed with the main theoretical models applied in the extant literature. To concretize the omni-channel consumption experience, well-structured scales of omni-channel experience are discussed. Lastly, insights for omni-channel customer journey are discussed with future research directions and managerial implications, which emphasize the importance of new approaches to omni-channel customer journey.

This chapter aims to understand the challenges in transitioning to the digital era of banking from the perspective of customers. This study uses an exploratory and quantitative approach. Quantitative data was collected from 329 bank clients, and then descriptive and correlation analysis was performed. The findings showed that performance expectancy, effort expectancy, and facilitating conditions have a positive relation, whereas social influence was found to have weak relation. Academically, this study presents valuable insights to mobile banking utilization and to the adaptation of the UTAUT model to banking sector. From a managerial perspective, this study provides insights into clients' challenges and suggestions to better integrate online platforms into branch service and to improve customer experience.

This study aims to identify the main applications of metaverse in business based on scientific publications. More specifically, the chapter aims to identify conceptual frameworks used in the literature. The authors conducted a systematic review based on the Web of Science and Scopus databases to achieve the objectives. This study followed a systematic literature review methodology consisting of protocol development, identified inclusion, and exclusion criteria for relevant publications; data extraction; and synthesis. The research resulted in 26 articles, which were identified and classified according to applications to the business environment, concluding that the main implications and applications of the metaverse in the business environment focus on an integrative framework that includes ethical issues, innovation management, marketing management, and knowledge management. This research has put down the foundation for establishing metaverse as a research topic in business research. With this study, the contribution lies both at the theoretical and practical levels.

With the evolution of technology and all the associated paradigms, the business reality had the need to adapt and incorporate all this evolution. Gamification strategies in social networks are used to optimize the use and involvement of users. Thus, in a business context, these strategies are used to increase user engagement in the company/brand's social networks with the aim of increasing and strengthening trust. This chapter's main objectives are to understand if internet users and social media users know the term gamification and if the associated strategies are perceptible in the social media context. Therefore, to

Preface

"Digital is changing the world quickly and irreversibly. As a result, companies must rethink their business model, from customer relations to production control to supplier communication in the digital age. Those that don't run the risk of failing and disappearing." (Meffert & Mendonça, 2017)

Welcome to *Implementing Automation Initiatives in Companies to Create Better-Connected Experiences*!

INTRODUCTION

Implementing Automation Initiatives in Companies to Create Better-Connected and Experiences is on the agenda in Europe and worldwide. If there is relevant content for companies and society, it is those related to automation and the improvement of the relationship between people and technology, which imply a digital transformation process that urgently needs to be intensified and deepened so that economies can grow and develop, providing a better quality of life for its populations. This science-based book aims to increase academia's knowledge of automation matters in companies and equip marketing managers and executives with the knowledge and skills to support them in strategic and operational decision-making.

The book explores one of the most current and decisive topics for business and academia. New content published on automation and connectivity will help the decision-making process of companies and the learning process of students and researchers. The digital transformation is underway, and the need for knowledge grows daily. Research in this area needs visibility and sharing as the market seeks digital transformation.

ORGANIZATION OF THE BOOK

The book is organized into two sections and 11 chapters: "Digital Transformation" and "Automation Initiatives and Better-Connected Experiences." A brief description of each of the chapters follows:

The two chapters composing the book's first section introduce the conceptualization of Digital Transformation and the results of studies applied in Portuguese companies.

The first chapter, "Digital Transformation in Organizations and its Impact on Knowledge Management: A Quantitative Study", by Manuel Sousa Pereira, António Cardoso, José Carlos Sá, Miguel Magalhães and Sílvia Faria, aims to understand the relevance of digital transformation and its impact on organizations in Portugal. This chapter intends to understand the explanatory factors of the impact of digital transformation and the contributions to the identification, contextualization and importance of better knowledge management. The study revealed that employees recognize a digital culture in organizations and are aware of the digital transformation process in the companies where they work. On the other hand, they confirm that the organization is committed to digital transformation, which already uses some digital technologies (mobile devices, smart sensors, cloud computing, big data and data analytics); recognizes the benefits of digital transformation for work productivity, productive efficiency and the optimization of work methods, with impacts on performance in knowledge management.

In the second chapter, "Digital Transformation and Business Models: Consumers Trust Case", Marco Lamas and Alexander do Amaral Pereira promote a generic theoretical framework for this very relevant topic, supported by a literature review of the research that has been developed over the years. It will also present the case study of a Portuguese company, Consumers Trust, better known in Portugal as "O Portal da Queixa".

Following is the book's second section, comprising nine chapters that propose different scopes of the Automation Initiatives and Better-Connected Experiences.

In the third chapter, "Relating Industry 4.0 and the SMEs Internationalization Process: A Case Study", Ana Rita Cardoso and Raquel Pereira present exploratory research designed to explore how Industry 4.0 and digitalization will influence the organization, competitiveness, and internationalization process of the SMEs under study. Results indicate that Industry 4.0 appears to be critical because it positively affects the value chain and the organizational and competitiveness level. Therefore, the companies were better positioned in their competitive advantages. Furthermore, this work provides insights for the study and joint understanding of digitalization and business internationalization because it integrates and relates two autonomous, complex, and strategic research fields, with no applied studies for the Portuguese SME's background.

The fourth chapter, "Understanding Consumer Behavior in a Hyper-Connected Omnichannel Retailing", by Ahmet Tugrul Tiger, provides a framework to understand omnichannel customer behavior, first by focusing on the customer journey in omnichannel retailing, representing the actual step of multichannel and cross-channel retailing. To contribute to the conceptualization of the omnichannel customer journey, some suggestions, including customer experience, are offered as the actual form of consumer behavior with digital, social and physical dimensions. Followingly, the critical antecedents of omnichannel customer experience are reviewed with the main theoretical models applied in the extant literature. To concretize the omnichannel consumption experience, well-structured scales of omnichannel experience are discussed. Lastly, insights for the omnichannel customer journey are discussed with future research directions and managerial implications, which emphasize the importance of new approaches to the omnichannel customer journey.

The fifth chapter, "Challenges in Adapting to Mobile Banking from the Perspective of the Customers", by Carolina Afonso and João Duarte Teixeira, aims to understand the challenges in transitioning to the digital banking era from the perspective of customers. This study uses an exploratory and quantitative approach. Quantitative data was collected from 329 bank clients, and descriptive and correlation analysis was performed. The findings showed that Performance Expectancy, Effort Expectancy and Facilitating Conditions have a positive relation, whereas Social Influence was found to have weak relation. Academically, this study presents valuable insights into mobile banking utilization and the adaptation of the UTAUT model to the banking sector. Furthermore, from a managerial perspective, this study provides insights into clients' challenges and suggestions to better integrate online platforms into branch service and improve customer experience.

The sixth chapter, "Metaverse Applications in Business: A Systematic Literature Review and Integrative Framework", by Julien Diogo and Pedro Mota Veiga, aims to identify the main applications of Metaverse in business based on scientific publications. More specifically, the chapter aims to identify conceptual frameworks used in the literature.

In the seventh chapter, "Gamification Strategies for Social Media", Jorge Esparteiro Garcia, Pedro Rodrigues, Jorge Simões and Manuel José Serra da Fonseca, the main objectives are to understand if Internet users and social media users know the term gamification and if the associated strategies are perceptible in the social media context. Therefore, an online survey was developed and applied to analyze better the users' behavior regarding these strategies and their knowledge about Gamification. After the survey, it was possible to conclude that 16.3% of the respondents already knew the term gamification. Furthermore, the survey made it possible to conclude that most respondents had already been in contact with Gamification strategies, even if 90.6% did not adhere to them frequently.

The eighth chapter, "AI vs. Human: Investigation of User Experience in Chat Support", by Chaitra Katti and Belem Barbosa, aims to examine user experience between AI and human-based chat support and understand whether AI-based chatbots necessitate replacing human interaction in customer support. The study's results indicated that perceived benefit and relevance among human-based chats are relatively high, and perceived risk was only limited to delayed response and negative attitude. However, chatbots evidenced their perceived risk was high anthropomorphic chatbots. Risk factors such as ethical concerns and lack of human touch were commonly seen in anthropomorphic chatbots. On the other hand, their perceived benefit and relevance were seen in chatbots acting as a substitute for search queries, web navigation or product demonstration.

The objective of the ninth chapter, "The Influence of Music on Online Advertising: The Case of YouTube", by Jorge Remondes and Sandrina Teixeira, is to understand how the use of a particular type of music may or may not capture the attention of the online user when viewing an ad and, perhaps, activate a click and an eventual conversion. An original advertisement was created to support this research and obtain results through quantitative research in the form of an online questionnaire application. How music can stimulate attention, memory, familiarity, the coherence of elements, rhythm and personal tastes was analyzed, as well as how these concepts interconnect with what the viewer visualizes in an online advertising campaign. The ad was published on Youtube because on this social network; there is only a short period of time of mandatory exposure of the ad to, consequently, lead to the conversion of the user to a landing page. It was concluded that the music of an online audiovisual campaign could induce the recipient's attention, confirming that the maximum optimization in an online advertising campaign occurs when the coherence of the music with the ad is prioritized.

The tenth chapter, "Newspaper Digital Content During and After the Silly Season: A Case Study", by Ana Maria Lima, Sandrina Teixeira and Belem Barbosa, aims to understand the distribution options of newspaper content online and offline. Adopting a case study approach, it analyses the content strategy of one of the most widely read newspapers in Portugal, Jornal de Notícias (JN). This chapter highlights the importance of aligning the content strategy and readers' preferences.

Finally, in the eleventh chapter, "Smart Cities: How Can Low-Density Territory Populations Be Smart Tourism Destinations? The Case of Guarda", by André Alves Toste, Dulcinea Catarina Moura and Lídia Aguiar, through the identification of the essential infrastructures, the importance is given by policymakers to the theme and understanding the perception of those who influence the destination, a proposal for a model for low-density smart tourism destinations was developed. It is concluded that low-density territories must satisfy all the related dimensions and conceive collective thinking that makes this goal possible – intelligent tourism destinations.

TARGET AUDIENCE

The target audience of this book will be composed of researchers, teachers, managers, executives and Master's and PhD students.

Good reading!

Jorge Remondes
Instituto Superior de Entre Douro e Vouga, Portugal & ISCAP, Instituto Politecnico do Porto, Portugal

Sandrina Teixeira
ISCAP, Instituto Politecnico do Porto, Portugal

REFERENCE

Meffert, J., & Mendonça, P. (2017). *Digital@Scale*. Planeta.

Acknowledgment

To IGI Global for accepting the proposal to organize this book "Implementing Automation Initiatives in Companies to Create Better-Connected Experiences";

To the professionals at IGI Global for the fundamental support provided throughout the process of managing and editing the book;

To all the authors who submitted chapter proposals;

To the authors of the eleven chapters published;

To the reviewers of the chapters for their excellent contribution to the final quality of the book;

To the future readers of this book.

Section 1
Digital Transformation

Chapter 1
Digital Transformation in Organizations and Its Impact on Knowledge Management: A Quantitative Study

Manuel Sousa Pereira
https://orcid.org/0000-0002-6238-181X
Instituto Politécnico de Viana do Castelo, Portugal

José Carlos Sá
https://orcid.org/0000-0002-2228-5348
School of Engineering, ISEP, Polytechnic of Porto, Portugal

António Cardoso
https://orcid.org/0000-0003-2545-0617
Fernando Pessoa University, Portugal

Miguel Magalhães
Universidade Portucalense, Portugal

Sílvia Faria
Universidade Portucalense, Portugal

ABSTRACT

This study aims to understand the relevance of digital transformation and its impact on organizations in Portugal. In this way, it is intended to evaluate the application of technologies, knowledge management, and incremental innovation in an organizational context. As a methodology, the authors intend to prepare a questionnaire in Google Docs, mainly using the Likert scale, which will be shared through LinkedIn. Having obtained 291 responses that will be analyzed in order to confirm or not the variables under analysis, this academic research intends to understand the explanatory factors of the impact of digital transformation and the contributions to the identification, contextualization, and importance for better knowledge management. The results proved to be relevant for the effective understanding of digital transformation in organizations in Portugal.

DOI: 10.4018/978-1-6684-5538-8.ch001

INTRODUCTION

This study arose from the need to understand to what extent digital transformation provides positive differentiation in organizations, as well as to understand the impact of digital transformation on organizations in Portugal. The central objective of this work is to analyze and understand the relevance of technology, digitization of processes aligned with the sharing of communication, promote success factors for organizations.

In this sense, the following research questions arise: are Portuguese companies carrying out the digital transformation? Is there a digital culture and a commitment to the digital transition in organizations? Does digital transformation contribute to improving knowledge management? Does knowledge management support and contribute to digital transformation?

In view of this problematization, the present research aims to analyze the digital transformation in organizations and evaluate its effect on Knowledge Management practices in organizations.

In specific terms, it is intended to know the digital culture in the organization, the knowledge about the digital transformation, the commitment to the digital transition, the adoption of digital technologies, the benefits of the digital transformation and the performance in the knowledge management.

This chapter is structured with abstract, introduction, literature review, data analysis, conclusions and literature review.

LITERATURE REVIEW

We can characterize digital transformation as the incorporation of technology in line with the organizational strategy, as mentioned by Rego et al. (2021) For Von Leipzig et al. (2017) digital transformation encompasses from technology to the strategy and culture of the organization, which results in new models of strategic management, thus making the analysis of cultural, methodological and technological aspects fundamental in the study of this process. More broadly, technology is also an engine of organizational development, or as Daehn and Costa says. Pereira, (2021) a broader concept that is applied (Krimpmann, 2015; Andriole, 2020; Imran; Shahzad; Butt; Kantola, 2021). Krimpmann (2015, p. 1209) defines it as the "sum of technologies that transform previously physical processes into processes that are partially or fully activated by technology".

The process of building a digital transformation strategy presupposes a predisposition and incentive to change, seeking to change the attitudes and behaviors of those responsible for the organization. digital transformation strategy, which can simplify the process and reduce obstacles. For this, there must be a person in charge in the company to outline this path of changes, who has previous experience as a facilitator of change and the leadership profile that the position requires (Christian & Thomas, 2015). However, for this to be a reality, it is necessary to provide a culture of change, internally promoting a set of practices, incentives and procedures that enhance this continuous improvement, involving the entire organization seeking success, as we can see in the words of Schwertner (2017). Successful organizations must leverage strategy, culture, and leadership to harness the digital transformation potential of the organization's business.

Digital transformation as an opportunity to create value and adapt to market needs and requirements, while simultaneously being the organization's ability to adapt to new times, as described by Bican & Brem (2020) [..]digital transformation is a profound change and accelerates business activities, processes, competencies and models to fully explore the changes and opportunities in digital technology and its impact on society in a strategic and prioritized way".

Digital transformation also contributes to the sharing and management of knowledge in an organizational context, as it provides a climate of participation, reflection and individual and collective analysis, as mentioned by Machado et al. (2022). From the 2000s to early 2018, related research has focused on social networks, autonomous media, and human-computer interactions, and this research shows that KM is strategic for organizations (Gaviria-Marin et al., 2018; Serenko, 2013; Tzortzaki & Mihiotis, 2014). Individual knowledge refers to an individual's ability to make judgments and decisions based on their knowledge. Or in other words, Kraus et al. (2022) [...] Digitization requires new forms of communication and Collaboration in the workplace and can be understood as the use of digital technologies and data (digitized and natively digital) to generate revenue, improve business and replace/transform business processes (not simply digitize them).

The digital transformation and its implementation in the organization's digital business model, contributes to efficiency, both in terms of speed of processes, in the relationship with stakeholders, as well as in the ability to meet the expectations of employees, as well as your consumers, customers or followers. In this sense, and according to Stjepić, Ivančić, & Vugec (2020) digital transformation can be carried out by either altering the existing business model with the help of digital technologies or creating a new and innovative one (Pejić Bach et al., 2018). Hence, if the organization chooses to carry out a digital transformation following the direction of changing the existing business model, it implies changing the existing business model and the organization's business processes. In concrete terms, and according to the authors Mohamed Hashim, Tlemsani, & Matthews, R. (2022), we found that from the point of view of universities, digital transformation strategies aim to (a) increase total revenue (b) increase productivity (c) generate value through innovative practices and (e) develop a reputation for brand and novelty (Matt et al., 2015; Shaughnessy, 2018).

With regard to information sharing in its relationship with digital transformation in a university context and taking into account its relevance of its role for the promotion of knowledge, Mohamed et al. (2022) mention that universities need to determine digital transformation management practices to manage complex transformation strategies (Angelopou-los, Christina & Yannis, 2019). That is, they need to formulate agile, realistic, and scalable digital transformation strategies that support the centered philosophy that integrates the entire university function.

Regarding its implementation and monitoring in practice, a constant reinvention of processes, techniques and practices in an organizational context is necessary, seeking to encourage, train employees in an innovative organizational culture, transforming threats into an opportunity for growth and co-creation of value, as describe AlNuaimi et al. (2022) although digital transformation offers many opportunities for companies of the day, many academics and professionals struggle to understand it, especially because it differs from the well-established concept of IT-enabled organizational transformation (Wessel et al., 2021). In other words, digital transformation requires organizational restructuring and has consequences for the metrics used to tune performance (Verhoef et al., 2021).

METHODOLOGY

The review of the literature previously carried out reveals a multiple influence and interdependence between digital transformation and knowledge management, because if digital transformation has an impact on knowledge management, this is crucial for the success of digital transformation (de Alvarenga, 2019; Diogo, Junior & dos Santos, 2019).

To answer the research questions and objectives, a quantitative study was developed, with a descriptive design (Pestana & Gageiro, 2014; Malhotra, 2019) based on a survey.

The questionnaire was developed based on studies identified in the literature (Gill & VanBoskirk 2016; Moreira el al, 2017; de Alvarenga, 2019; Diogo, Junior & dos Santos, 2019) and aimed to collect information about perceptions, opinions, workers' attitudes and behaviors regarding digital transformation and knowledge management in the organization, comprising 6 dimensions: (1) Digital culture in the organization (7 items); (2) Knowledge about Digital Transformation in the organization (6 items); (3) Commitment to Digital Transformation in the organization (7 items); (4) Adoption of Digital Technologies in the organization (3 items); (5) Benefits of Digital Transformation in the organization (11 items); (6) Performance in knowledge management (5 items).

In the end, a question was included to know the technologies already adopted in the organizations, and the respondents could mark all those that were already in use in the company where they worked. Included here are the main technologies associated with Industry 4.0 (Hanebrink et al, 2015; i-SCOOP, 2022): Bid data and Data analytics; Cloud computing; Cyber security; Drones; Mobile devices; AI - Artificial Intelligence; 3D printing; Systems integration; Iot - Internet of things; Common platforms; Virtual reality; RFID – Radio Frequency Identification; Autonomous robots; Smart sensors; Virtual simulation; and real-time location systems.

Before answering the statements presented to evaluate the dimensions presented above, some questions were asked to characterize the respondents: gender, age group, graduation (Basic education; High School; Graduation; Master degree; Doctorate), service time (< 3 years; 4 to 9 years; 10 to 14 years; 15 to 19 years; > 20 years) and activity sector (Teaching/training; Business; Banking; Industry; Technologies).

Regarding the propositions presented, respondents answered based on a 5-point Likert scale (Pestana & Gageiro, 2014; Malhotra, 2019), where "1" meant "totally disagree" and "5" "totally agree".

The survey was built on "google forms" was tested before (with 15 workers) it was given to ensure both validity and reliability (Pestana & Gageiro, 2014; Malhotra, 2019. Participants were told that their anonymity would be protected, and if they chose, they could withdraw their responses from the survey. After participants took the survey, they would be provided with a report of the most important results of the study. It was made available online between January and April 2022. This is a sample non-probabilistic for convenience (Malhotra, 2019). During the survey data collection, controls were built to ensure that each respondent only participated in the survey one time.

ANALYSIS OF RESULTS

291 questionnaires were collected and validated, which is considered a very satisfactory number that allows the analysis and statistical treatment of the data.

Data were imported from Google Forms for analysis in SPSS (Statistical Package For Social Science) version 25.0

To assess the reliability of the various dimensions under analysis, Cronbach's alpha value was calculated (Pestana & Gageiro, 2014).

For operational reasons, a univariate analysis of data will be carried out, using descriptive statistics (absolute and relative frequencies, means and standard deviation) (Hill & Hill, 2002) in order to answer the questions and objectives of the search.
The sample of respondents consisted of 62.2% male, with the remaining 37.8% being female (Table 1).

Table 1. Sample characterization. Source: Own Elaboration

Variables	Categories	N	%
Gender	Female	110	37,8
	Male	**181**	**62,2**
Age	21 – 30 years	44	15,1
	31 – 40 years	71	24,4
	41 – 50 years	**118**	**40,5**
	51 – 60 years	46	15,8
	61 – 70 years	12	4,1
Graduation	Doctorate	42	14,4
	Master Degree	86	29,6
	Graduation	**131**	**45,0**
	High school	18	6,2
	Basic education	14	4,8
Service time	=> 20 years	58	19,9
	15 to 19 years	31	10,7
	10 to 14 years	41	14,1
	4 to 9 years	67	23,0
	=< 3 year	**94**	**32,3**
Activity sector	Teaching/training	89	30,6
	Business	21	7,2
	Banking, security	11	3,8
	Industry	**94**	**32,2**
	Technologies	14	4,5
	Others	62	21,7

In terms of age, it appears that the majority of respondents are in the age group from "41 to 5 years old" (40.5%), followed by the age group from "31 to 40 years old" (24.4%). Only 4.1% were in the age group of "61 to 70 years old".

Regarding educational qualifications, most respondents had higher education, with 45% having a degree, 29.6% a master's degree and 14.4% a doctorate.

Despite the data showing that 32.3% of respondents have been with the company for "up to 3 years", 142 (48.8%) are in the ranks with 10 or more years in the company.

In terms of sector of activity, despite the great heterogeneity of responses obtained, it appears that 32.2% work in industry, 30.6% in the "education/training" sector, 7.2% in "Business", 4.5% in Technologies, and 3.8% in Banking/insurance.

In order to verify the internal consistency of the scale, Cronback's Alpha was analyzed on all items, having obtained a good internal consistency (Pestana & Gageiro, 2014): Cronback's Alpha of 0.972

The dimension "Digital culture in the organization" (Table 2) had good internal consistency (Cronback Alpha = 0.805). The agreement values on the 7 items that make up the scale are high (M=3.87; SD=0.977), with 35.9% of respondents agreeing with the seven propositions presented and 32.1% totally agreeing.

In more specific terms, 173 respondents (59.5%) fully agreed that the company's competitive strategy depends on digital (M=4.51; SD=0.682) and 127 (43.6%) agree that the organization invests in education and in the digital training of employees (M=4.15; SD=0.832).

Table 2. Digital culture in the organization

Alpha de Cronbach: 0,805	**Strongly disagree 1**	**Disagree 2**	**Undecided 3**	**Agree 4**	**Strongly agree 5**	**M**	**SD**
	F (%)	**F (%)**	**F (%)**	**F (%)**	**F (%)**		
V1. We believe that our competitive strategy depends on digital	1 (0,3%)	3 (1%)	16 (5,5%)	98 (30,7%)	173 (59,5%)	**4,51**	**,682**
V2. Top management promotes digital transformation	19 (6,5%)	40 (13,7%)	92 (31,6%)	84 (28,9%)	56 (19,2%)	3,41	1,139
V3. We have the right leaders to execute our digital strategy	17 (5,8%)	37 (12,7%)	62 (21,3%)	91 (31,3%)	84 (28,9%)	3,65	1,190
V4. The organization invests in targeted digital education and training of all employees	2 (0,7%)	9 (3,1%)	43 (14,8%)	127 (43,6%)	110 (37,8%)	**4,15**	**,832**
V5. We clearly communicate our digital vision both internally and externally	8 (2,7%)	18 (6,2)	66 (22,7%)	113 (38,8%)	86 (29,6)	3,86	1,001
V6. We dedicate the appropriate resources and means to the implementation of the digital strategy	9 (3,1%)	36 (12,4%)	78 (26,8%)	89 (30,6%)	79 (27,1%)	3,66	1,097
V7. Customer perceptions are considered in the organization's digital design and development	4 (1,4%)	12 (4,1%)	79 (27,1%)	131 (45%)	65 (22,3%)	3,83	,870
TOTAL (Average)	**60 (2,9%)**	**155 (7,6%)**	**436 (21,4%)**	**733 (35,9%)**	**653 (32,1%)**	**3,87**	**0,977**

The employees surveyed claim to have knowledge about the digital transformation (table 3) in the organization (M=4.03; SD=0.916), with the 6 propositions obtained that make up the scale obtaining agreement values above the arithmetic mean (M>2, 5). The "Knowledge about digital transformation" scale showed good internal consistency (Cronback Alpha = 0.754).

The vast majority of respondents (63.2%) fully agree that digital transformation is the future of organizational management (M=4.56; SD=0.620) and 127 respondents (43.6%) agree that they seek to understand the vision, the mission and strategies defined by the organization and apply them in daily activities (M=4.15; SD=0.832)

Table 3. Knowledge about Digital Transformation in the organization

Alpha de Cronbach: 0,754	**Strongly disagree 1**	**Disagree 2**	**Undecided 3**	**Agree 4**	**Strongly agree 5**	**M**	**SD**
	F (%)	**F (%)**	**F (%)**	**F (%)**	**F (%)**		
V8. I am aware of the goals of Digital Transformation in my organization	17 (5,8%)	37 (12,7%)	62 (21,3%)	91 (31,3%)	84 (28,9%)	3,65	1,190
V9. I try to understand the vision, mission and strategies defined in my organization and apply them in my daily activities	2 (0,7%)	9 (3,1%)	43 (14,8%)	127 (43,6%)	110 (37,8%)	**4,15**	**,832**
V10. In the process of Digital Transformation, I consider that I did not feel resistance to hange	1 (0,3%)	14 (4,8%)	52 (17,9%)	119 (40,9%)	105 (36,1%)	4,08	,872
V11. Digital Transformation has changed internal processes	2 (0,7)	11 (3,8%)	45 (15,5%)	141 (48,5%)	92 (31,6%)	4,07	,826
V12 Digital Transformation is the future of organizational management			20 (6,9%)	87 (29,9%)	184 (63,2%)	**4,56**	**,620**
V13. I am aware of the goals of Digital Transformation in my organization	17 (5,8%)	37 (12,7%)	62 (21,3%)	91 (31,3%)	84 (28,9%)	3,65	1,190
TOTAL (Average)	**39 (2,2%)**	**108 (6,19%)**	**284 (16,7%)**	**656 (37,5%)**	**659 (37,6)**	**4,03**	**0,916**

The scale of the "organization's commitment to digital transformation" achieved good internal consistency, with a Cronback Alpha of 0.882 recorded.

Likewise, respondents consider that the organization is committed to digital transformation (Table 4), with the 7 propositions that make up the scale obtaining agreement values above the arithmetic mean (M=3.9; SD= 1.006).

Table 4. Commitment to Digital Transformation in the organization

Alpha de Cronbach: 0,882	**Strongly disagree 1**	**Disagree 2**	**Undecided 3**	**Agree 4**	**Strongly agree 5**	**M**	**SD**
	F (%)	**F (%)**	**F (%)**	**F (%)**	**F (%)**		
V14 In my organization there are policies that prioritize the use of information technologies	3 (1%)	14 (4,8%)	38 (13,1)	121 (41,6%)	115 (39,5%)	**4,14**	**,892**
V15. My organization is prepared for the evolution of Digital Transformation.	12 (4,1%)	18 (6,2%)	65 (22,3%)	96 (33%)	100 (34,4%)	**3,87**	**1,083**
V16. Our example of organizational change can promote other Digital Transformation projects.	10 (3,4%)	26 (8,9%)	61 (21%)	102 (35,1%)	92 (31,6%)	3,82	1,080
V17. The leadership of organizations is prepared for digital change.	15 (5,2%)	33 (11,3%)	73 (25,1%)	97 (33,3%)	73 (25,1%)	3,62	1,131
V18.My direct managers are committed to digital change	6 (2,1%)	28 (9,6%)	71 (24,4%)	109 (37,5%)	77 (26,5%)	3,77	1,014
V19. Our managers alert us to what is important to know	7 (2,4%)	22 (7,6%)	72 (24,7%)	107 (36,8%)	83 (28,5%)	3,81	1,010
V20. I feel comfortable expressing my opinion and presenting my point of view to my colleagues and superiors. I feel that I will be heard.		8 (2,7%)	47 (16,2%)	91 (31,3%)	145 (49,8%)	**4,28**	**,832**
TOTAL (Average)	**53 (2,6%)**	**149 (7,3%)**	**427 (20,9%)**	**723 (35,5%)**	**685 (33,6%)**	**3,9**	**1,006**

Thus, 236 respondents (81.1%) fully agree and agree that the organization where they work has policies that prioritize the use of information technologies (M=4.14; SD=0.892). Furthermore, 196 respondents (67.4%) agree and fully agree that the organization is prepared for the evolution of digital transformation (M=3.87; SD=1.083).

Regarding the adoption of digital technologies in the organization, the scale obtained a Cronback Alpha of 0.657, which is considered satisfactory and acceptable (Table 5). The agreement values on the three items that make up the scale obtained agreement values above the arithmetic mean (M=3.79; SD=0.914).

Most respondents agree and fully agree that in their day-to-day work they use technologies and digital products, such as in processes, management and internal communication, meetings (M=3.96; SD=0.888) and that through technological innovations, manual operations were modified and made digital (M=3.83; SD=0.870).

Table 5. Adoption of Digital Technologies in the Organization

Alpha de Cronbach: ,657	**Strongly disagree 1**	**Disagree 2**	**Undecided 3**	**Agree 4**	**Strongly agree 5**	**M**	**SD**
	F (%)	**F (%)**	**F (%)**	**F (%)**	**F (%)**		
V21. In my day to day work I use technologies and digital products. In processes, management and internal communication, meetings...	3 (1%)	13 (4,5%)	63 (21,6%)	125 (43%)	87 (29,9%)	**3,96**	**,888**
V22. The processes in my service are fully digitized.	14 (4,8%)	31 (10,7%)	76 (26,1%)	106 (36,4%)	64 (22%)	3,60	1,089
V23. Through technological innovation, manual operations were modified and became digital	4 (1,4%)	12 (4,1%)	79 (27,1%)	131 (45%)	65 (22,3%)	**3,83**	**,870**
TOTAL (Average)	**21 (2,4%)**	**56 (6,4%)**	**218 (24,9%)**	**362 (41,4%)**	**216 (24,7%)**	**3,79**	**0,914**

The "Benefits of digital transformation in the organization" scale had good internal consistency: Cronback Alpha = 0.887. In global terms, as can be seen in table 6, the employees surveyed recognize the benefits of digital transformation in the organization (M=3.90; SD=0.911).

Most respondents (63.2%) fully agree that digital transformation reduces tasks performed manually (M=4.56; SD=0.620), contributes to the improvement of internal processes (M=4.31; SD =0.793), made the work simpler and more efficient (M=4.25; SD=0.706), contributed significantly to the reduction of paper use (M=4.15; SD=0.837) and allowed the optimization of work methodologies (M=4.05; SD=0.862).

The "Performance in knowledge management" scale had good internal consistency (Cronback Alpha=0.765).

As shown in the table below (table 7), respondents recognize that digital transformation is fundamental and has a major impact on knowledge management and organization performance (M=3.89; SD:0.896), with, on average, 67.3% agree and fully agree with the propositions presented.

Table 6. Benefits of Digital Transformation in the organization

Alpha de Cronbach: 0,887	Strongly disagree 1	Disagree 2	Undecided 3	Agree 4	Strongly agree 5	M	SD
	F (%)	F (%)	F (%)	F (%)	F (%)		
V24. Digital Transformation contributed to the improvement of internal processes.	2 (0,7%)	6 (2,1%)	30 (10,3%)	114 (39,2%)	139 (47,8%)	**4,31**	**,793**
V25. Digital Transformation has made services more transparent and secure.	14 (4,8%)	48 (16,5%)	101 (34,7%)	71 (24,4%)	57 (19,6%)	3,37	1,118
V26. Digital Transformation contributed significantly to the decrease in paper use.	3 (1%)	6 (2,1%)	47 (16,2%)	123 (42,3%)	112 (38,5%)	**4,15**	**,837**
V27. Digital Transformation has reduced the tasks performed manually.			20 (6,9%)	87 (29,9%)	184 (63,2%)	**4,56**	**,620**
V28. Digital Transformation has made jobs simpler and more efficient.		2 (0,7%)	39 (13,4%)	134 (46%)	116 (39,9%)	**4,25**	**,706**
V29. The Digital Transformation contributed to the increase in the speed of the execution of my tasks.	8 (2,7%)	22 (7,6%)	77 (26,5%)	111 (38,1%)	73 (25,1%)	3,75	1,004
V30. Digital Transformation contributed to the decrease in the number of tasks.	7 (2,4%)	13 (4,5%)	70 (24,1%)	117 (40,2%)	84 (28,9%)	3,89	,956
V31. I feel that with Digital Transformation I can be faster and more efficient in carrying out my tasks	12 (4,1%)	23 (7,9%)	101 (34,7%)	101 (34,7%)	54 (18,6%)	3,56	1,013
V32. Digital Transformation has increased knowledge sharing	9 (3,1%)	41 (14,1)	101 (34,7%)	96 (33%)	44 (15,1%)	3,43	1,009
V33. I consider the digital transformation to be a motivating change.	15 (5,2%)	31 (10,7%)	74 (25,4%)	102 (35,1%)	69 (23,7%)	3,62	1,112
V34. Technological change and innovation has the advantage of optimizing working methodologies	2 (0,7)	11 (3,8%)	56 (19,2%)	124 (42,6%)	98 (33,7%)	**4,05**	**,862**
TOTAL (Average)	**72 (2,2%)**	**203 (6,3%)**	**716 (22,3%)**	**1180 (36,8%)**	**1030 (32,1%)**	**3,90**	**0,911**

Specifically, 271 respondents (93.2%) agree or totally agree that digital transformation is essential for a better performance of the organization (M=4.51; SD=0.682). Likewise, 266 respondents (91.4%) agree and fully agree that they are aware of the importance of knowledge management and its impacts on digital transformation (M=4.49; SD=0.677).

The technologies currently adopted by organizations (Table 8) are mostly "mobile devices" (54%), Smart sensors (25.4%), Cloud computing (5.5%, Big Data and Data Analytics (4.8%) and Systems Integration (2.7%).

The study revealed that employees recognize that there is a digital culture in organizations and are aware of the digital transformation process in the companies where they work. They confirm that the organization is committed to digital transformation, which already use some digital technologies (mobile devices, smart sensors, cloud computing, big data and data analytics, and systems integration); recognize the benefits of digital transformation for work productivity, productive efficiency and the optimization of work methods, with impacts on performance in knowledge management.

Table 7. Performance in knowledge management

Alpha de Cronbach: 0,765	Strongly disagree 1	Disagree 2	Undecided 3	Agree 4	Strongly agree 5	M	SD
	F (%)	F (%)	F (%)	F (%)	F (%)		
V35. The implementation of the technological platform contributed to the increase in knowledge sharing among colleagues.	7 (2,4%)	47 (16,2%)	93 (32%)	87 (29,9%)	57 (19,6%)	3,48	1,055
V36. The knowledge gained during and after Digital Transformation can improve service delivery to citizens.	12 (4,1%)	47 (16,2%)	94 (32,3%)	96 (33%)	42 (14,4%)	3,37	1,047
V37. I believe that Digital Transformation has contributed to the improvement of Knowledge Management practices.	6 (2,1%)	33 (11,3%)	91 (31,3%)	95 (32,6%)	66 (22,7%)	3,63	1,021
V38. I am aware of the importance of Knowledge Management and its impacts on Digital Transformation.	1 (0,3%)		24 (8,2%)	96 (33%)	170 (58,4%)	**4,49**	**,677**
V39. Digital transformation is essential for a better performance of the organization.	1 (0,3%)	3 (1%)	16 (5,5%)	98 (33,7%)	173 (59,5%)	**4,51**	**,682**
TOTAL (Average)	**27 (1,8%)**	**130 (8,9%)**	**318 (21,8%)**	**472 (32,4%)**	**508 (34,9%)**	**3,89**	**0,896**

Table 8. Technologies currently used by the organization

Technologies	F	%
Big data e Data Analytics	14	4,8
Cloud Computing	16	5,5
Cyber Security	3	1,0
Drones	1	,3
Mobile devices	**157**	**54,0**
AI (Artificial Intelligence)	2	,7
3D printing	1	,3
Systems integration	8	2,7
IoT - Internet of Things	3	1,0
Common platforms	1	,3
Virtual reality	2	,7
RFID (Radio Frequency identification)	2	,7
Autonomous robots	1	,3
Smart sensors	**74**	**25,4**
Virtual simulation	1	,3
Real-time location systems	5	1,7
Total	**291**	**100,0**

CONCLUSIONS

The study carried out made it possible to verify that, from the perspective of the interviewed employees, the organizations where they work have a digital culture and are aware of the digital transformation process that is taking place, as stated by Von Leipzig et al. (2017) digital transformation encompasses technology as well as the organization's strategy and culture.

Therefore, they consider that the organization is prepared and committed to digital transformation and that the organization where they work has policies that prioritize the use of information technologies, as indicated by Kraus et al. (2022) referring to the use of digital technologies and data (digitized and natively digital) to generate revenue, improve business.

Likewise, employees consider that in their day-to-day work they use digital technologies and products, such as in processes, management and internal communication, meetings and that through technological innovations manual operations have been modified and become digital.

As a result, respondents widely recognize the benefits of digital transformation in the organization" and that it is critical and has a major impact on knowledge management and organizational performance.

Regarding the degree of implementation of digital technologies and tools of the so-called industry 4.0, it still seems to emerge in an incipient way, although the adoption of "mobile devices" and intelligent sensors stands out. Only a few organizations already use Cloud computing, Big Data and Data Analytics, Systems Integration.

However, despite this commitment and effort to advance the digital transformation in organizations, there is still a great effort to be made to make digitalization a reality. In fact, the most advanced technologies (Cyber security, Drones, 3D printing, IoT, RFID; Virtual reality; autonomous robots; virtual simulation) are still not used in most organizations.

By those reasons, this research provides a strong source of value for business, IT managers, and knowledge management, above all for organizations looking to make the digital transition. In terms of practical and managerial implications, the study highlights that managers and professionals should consider adopting digital technologies to improve their operations, production efficiency and working methods, given the impact that digital transformation has on business performance and knowledge management performance.

Despite the work developed and the reliability of the analyzed data, several research limitations guide future work. The main limitations of this study are a convenience sample that does not represent the reality of Portuguese organizations, as well as the absence of a qualitative study on the subject. The study includes employees from organizations of various dimensions (Small and Medium Enterprises, Big Corporations) sectors and industries (public, private, services, technologies, startups, fintech) where there are supposed to be differences in terms of digital transformation, but which this study, due to the size of the sample, was not possible to analyze. We believe that the COVID-19 pandemic caused a short-term acceleration in the adoption of various digital technologies and tools associated with Industry 4.0, as well as the birth of new companies and completely digital business models, but whose phenomenon was not taken into account in this search.

This study encourages further research opportunities. In future research, it would be important to explore dynamic enterprise architecture capabilities in order to analyze the alignment of business processes with knowledge management and digital technologies. Likewise, it could be interesting to study the relevance of digital transformation in organizational innovation and success, and the impact of public crisis (for example, COVID19) on digital transformation.

ACKNOWLEDGMENT

This research received no specific grant from any funding agency in the public, commercial, or not-for-profit sectors.

REFERENCES

AlNuaimi, B. K., Singh, S. K., Ren, S., Budhwar, P., & Vorobyev, D. (2022). Mastering digital transformation: The nexus between leadership, agility, and digital strategy. *Journal of Business Research, 145*, 636–648. doi:10.1016/j.jbusres.2022.03.038

Bican, P. M., & Brem, A. (2020). Digital business model, digital transformation, digital entrepreneurship: Is there a sustainable "digital?" *Sustainability, 12*(13), 5239. doi:10.3390u12135239

Daehn, C., Costa, A., & Pereira, R. (2021) Digital transformation and sustainability: Challenges and trends. In *ENGEMA 2021 - XXIII International Meeting on Business Management and the Environment*. FEA/USPAt.

de Alvarenga, A. (2019). *Digital Transformation in Public Administration: Case Study* [Master's Dissertation]. ISCTE – Lisbon University Institute, Department of Information Sciences and Technologies.

Diogo, R., Junior, A., & dos Santos, N. (2019). Digital transformation and knowledge management: contributions to the improvement of productive and organizational processes. P2P & Innovation, 5(2), 154-175. doi:10.21721/p2p.2019v5n2.p154-175

Gill, M., & VanBoskirk, S. (2016). *The Digital Maturity Model 4.0 Benchmarks: Digital Business Transformation Playbook*. Forrester Research, Inc. Available: https://dixital.cec.es/wp-content/uploads/presentacions/presentacion06.pdf

Hanebrink, J., Kunze, H. M., Mix, T., Löffler, J., Felten, D., & Leppien, K. (2015). *How to Navigate digitization ofmanufacturing sector*. McKinsey Industry 4.0 GlobalExpert Survey. Available online: https://www.mckinsey.com/~/media/McKinsey/Business%20Functions/Operations/Our%20Insights/Industry%2040%20How%20to%20navigate%20digitization%20of%20the%20manufacturing%20sector/Industry-40-How-to-navigate-digitization-of-the-manufacturing-sector.ashx

i-SCOOP. (2022). *Industry 4.0 and the fourth industrial revolution explained*. Available online: https://www.i-scoop.eu/industry-4-0/

Kraus, S., Durst, S., Ferreira, J. J., Veiga, P., Kailer, N., & Weinmann, A. (2022). Digital transformation in business and management research: An overview of the current status quo. *International Journal of Information Management, 63*, 102466.

Machado, A., Secinaro, S., Calandra, D., & Lanzalonga, F. (2022). Knowledge management and digital transformation for Industry 4.0: A structured literature review. *Knowledge Management Research and Practice*, 1–19. doi:10.1080/14778238.2021.2015261

Malhotra, N. (2019). *Marketing Research: An Applied Orientation*. Person Education.

Mohamed Hashim, M. A., Tlemsani, I., & Matthews, R. (2022). Higher education strategy in digital transformation. *Education and Information Technologies*, *27*(3), 3171–3195.

Moreira, M., Au-Yong-Oliveira, M., Goncalves, R., & Costa, C. (2017). *Digital transformation – Opportunities and threats for more consistent competitiveness*. Ed. Syllables & Challenges.

Pestana, M., & Gageiro, J. (2014). *Data Analysis for Social Sciences: Complementarity of SPSS*. Ed. Syllabus.

Rego, F. (2021) Digital transformation in business: a systematic study from the web of science and scopus, *11th International Symposium on technological innovation*, 11(1), 1610-1619. 10.51722/S2318-3403202100011364

Schwertner, K. (2017). Digital transformation of business. *Trakia Journal of Sciences*, *15*(Suppl. 1), 388–393. doi:10.15547/tjs.2017.s.01.065

Stjepić, A.M., Ivančić, L., & Vugec, D.S. (2020). Mastering digital transformation through business process management: Investigating alignments, goals, orchestration, and roles. *Journal of Entrepreneurship, Management and Innovation*, *16*(1), 41-74.

KEY TERMS AND DEFINITIONS

Co-Creation: Form of organizational innovation that involves different interlocutors (suppliers, distributors, retailers, customers, employees, and another's business partners) in order to enhance the collective creative potential, resulting into innovative solutions customized and valued by the market.

Commitment: Digital transformation represents an organizational change, which can be difficult to achieve if there is no determination and commitment from top management, as well as the involvement of all the structure and functions of the organization, and the participation of all employees. Organizational commitment means developing a corporate culture that promotes innovation and creativity within the company.

Digital Culture: Sets of fundamental skills, competences and knowledge relating to the use of digital technologies by businesses and citizens, as well as by society at large.

Digital Transformation: Process that contemplates the use of digital technologies to solve problems in different domains (economy, education, work, communication, production and communication system, business activity, etc.)

Incremental Innovation: Form of innovation that implements improvements in a process or in a technology that already exists (improving actual technologies), gradually allowing to achieve greater levels of efficiency and performance. This type of innovation differs from disruptive innovation that creates a new technology/solution (creating something new), making existing technologies/solutions (products and services) in the market obsolete, through disruptions and new consumption habits creation.

Knowledge Management: Set of technologies and processes whose objective is to support the creation, transfer and application of knowledge in organizations, in order to make large amounts of organizational information accessible, supporting decision making.

Technology Adoption: Represents the degree of incorporation of digital technologies in organizations in order to respond to today's competitive challenges.

Chapter 2
Digital Transformation and Business Models:
Consumers Trust Case

Marco Lamas
https://orcid.org/0000-0001-7365-2016
ISCAP, Instituto Politécnico do Porto, Portugal

Alecsander do Amaral Pereira
Instituto Superior de Entre Douro e Vouga, Portugal

ABSTRACT

The ongoing globalization process has been intensified with the constant evolution of technology, which leads companies to face digital transformation as both mandatory to guarantee their survival and essential to create value for their customers. Starting from digital transformation, this chapter's mission is to promote a generic theoretical framework in this very relevant topic, supported on a literature review of the research that has been developed over the years. It will also present the case study of a Portuguese company, Consumers Trust, better known in Portugal as "O Portal da Queixa." Qualitative research will be undertaken using interview and content analysis. Its main objective is to understand and explain the causes and effects of this concept that, despite having originated the scientific research mentioned, in particular in these last two years, as a result of the COVID-19 pandemic, have not been explored enough and therefore are not yet consensual.

INTRODUCTION

People and companies' routines changed throughout the various phases of human existence for several reasons, but this process underwent an exponential development after the industrial revolution, which took place in the second half of the 18th century, considered by some authors, as the first industrial revolution. The speed with which we began to witness new revolutions has increased considerably since then.

DOI: 10.4018/978-1-6684-5538-8.ch002

The progress of means of transport and communication, as well as the development of the oil, electric and steel industries, which took place in the period between 1850 and 1945, came to be known as the second revolution.

Beginning in 1950, the third industrial revolution was based on automation, due to the exploration and innovation of electronic means, which were the basis for the development of computing, robotics and the Internet, which caused a replacement of analog mechanics by digital force.

In order to stay ahead of their competitors and taking advantage of not only the spread of the Internet, but also the ease of use and acquisition, companies have invested in various digital technologies. The simple fact of acquiring technology alone cannot be considered digital transformation, this investment, which depending on the size of the company could easily reach millions of euros, were made, in many cases, in a disorderly way and without a predefined digital strategy.

According to the author of the book "The Fourth Industrial Revolution", Klaus Schwab, the fourth revolution, or the digital transformation itself, which began in 2011, is linked to a business model that uses, in order to improve productivity, technologies currently available. However, in 2020, the pandemic COVID-19, caused by the SARS-CoV-2 coronavirus, first identified in the city of Wuhan, China, in January 2020, accelerated the need for companies to implement digital transformation. Until then, there was a lot of theory, and often there was no consensus among authors on the definition of digital transformation, just as there were very few companies that had the entire process completed.

It is important to note that this chapter deals with the subject in a generic way, briefly explaining some of the key themes of digital transformation, such as the main technologies used, disruption, digital maturity and related business models. As it is important, we also briefly comment on the digital revolution and the history of the Internet, as well as success and failure cases. And finally, we present the Portal da Queixa case and its relation with the main topic on this chapter.

DIGITAL TRANSFORMATION

Long ago, it dates to the oldest time of humanity, that the human being generates and stores information (data). When these are used through electronic technology, they become known as "Digital", thus giving rise to the "Age of Digitization and Information". Transformation is an act, process or instance of transforming or being transformed. We can then argue that digital transformation will be the way we transform the use of this data used through electronic technology. However, the existing literature shows that there are different understandings and meanings about what digital transformation is. Perdomo (2019) claims that the term has been widely used in recent years, and summarized, as can be seen in table 1, the perspectives of some authors.

Table 1. Other definitions of digital transformation. Source: (Perdomo, 2019, p. 23)

BMWi (2015)	Digitization means the complete network of all sectors of the economy and society, as well as the ability to collect relevant information and analyze and translate that information into action. Changes bring advantages and opportunities but create entirely new challenges.
Bowersox et al. (2005)	Digital Business Transformation is a process of reinventing a business to digitize operations and formulate extended supply chain relationships. DBT's leadership challenge is to re-energize already successful businesses to capture the full potential of information technology across the supply chain.
Westerman et al. (2011)	Digital transformation – the use of technology to radically improve the performance or reach of companies – is becoming a hot topic for companies around the world. Executives across industries are using digital advancements such as analytics, mobility, social media and embedded smart devices, and enhancing the use of traditional technologies such as ERP to change customer relationships, internal processes and value propositions.
Mazzone (2014)	Digital transformation is the deliberate and continuous digital evolution of a company, business model, process or ideation methodology, both strategically and tactically.
PwC (2016)	Digital Transformation describes the fundamental transformation of the entire business world through the establishment of new Internet-based technologies with fundamental impact on society as a whole.
Beuée & Schaible (2015)	We understand digital transformation as a consistent network of all sectors of the economy and the adjustment of actors to the new realities of the digital economy. Decisions in networked systems include data exchange and analysis, calculation and evaluation of options, as well as initiating actions and introducing consequences.
Nwankpa e Roumani (2016)	Digital transformation refers to changes and transformations that are driven and built on a foundation of digital technologies.
Piccinini et al. (2015)	It refers to how advances in digital technology are reshaping a wide range of activities in society at large.

Fitzgerald et al. (2013) and Vaskal et al. (2021) warn of the speed with which digital transformation is becoming a very popular phenomenon, while Pereira (2021) observes that technology changes rapidly and, thus, influences digital transformation in such a way that it affects all companies, from those that are slow to adopt more advanced digital technologies, to those that, because they have managers with a more focused view on what new technologies are, use them with the purpose of obtaining all the potential that digital transformation can provide for the organization.

There are authors who understand that digital transformation is a social phenomenon, as is the case of Stolterman and Fors (2004). For other authors, such as Rogers and Sparviero (2011), Pînzaru and Mitan (2012) and Pardo and Etayo (2014), the best way to define digital transformation is to understand it as a cultural evolution.

It is important to go back a little in time, because, in fact, electronic technology generated a digital revolution that began with the invention of the transistor, in 1947. A brief history of this revolution can be seen in Figure 1.

Figure 1. Brief history of the digital revolution. Source: Adapted from UKRI - Science and Technology Facilities Council (2018)

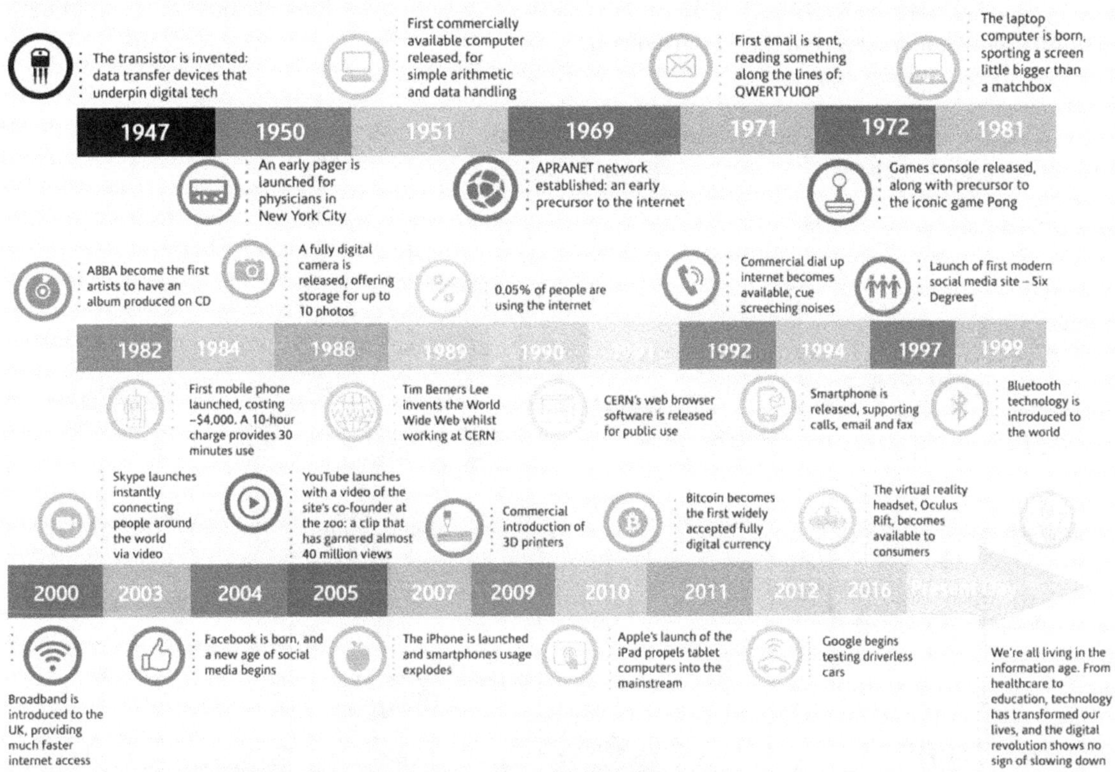

The transistor, by itself, would not change the world as we know it were it not for an article, whose title is "A Mathematical Theory of Communication", which turned out to be the basis of the information theory upon which the entire Internet was built. The article, published in 1948 by engineer and mathematician Claude Elwood Shannon, eventually gave him the nickname "The Father of Modern Digital Communications and Information Theory".

Shannon's (1948) article managed to solve a problem. Remote communication, that is, sending a message, was used through analog signals, which transformed it into varying pulses of voltage along an electrical conductor (a wire, for example). These pulses were measured at the other end, interpreted and converted into the message. For short distances, this process worked well, however, with every meter that the analog signal traveled along the conductor, the signal became weaker and contained noise. Of course the signal could be boosted at the beginning of the process, however the noise would also increase. Shannon defined the smallest possible pieces of information, which could not be further divided, as the most commonly used digital code in modern electronics. Shannon called them Binary Digit, being commonly used as we know it today, "bit". A bit has only one value (which can be 0 or 1). In 1956, during the design phase of the IBM Stretch, Werner Buchholz defined the term Byte for the unit of digital information that consists of eight bits.

For this reason, if it weren't for Shannon's article, the Internet could have not existed. And it is a fact that, on the ride with the Internet, the digital revolution was present in the four corners of the world and ingrained itself in society. Despite being "born" in 1969, as Siroshi (2015) observes, at the time the Internetcould not be accessed by the general public and its name was ARPANet, changing its name to INTERNET in 1982, as can be seen in Figure 2. In the same figure, we look at some of the applications and companies that emerged in the "Internet Age".

Figure 2. Age of the internet. Source: Adapted from Hyche, J. (2016, p. 13)

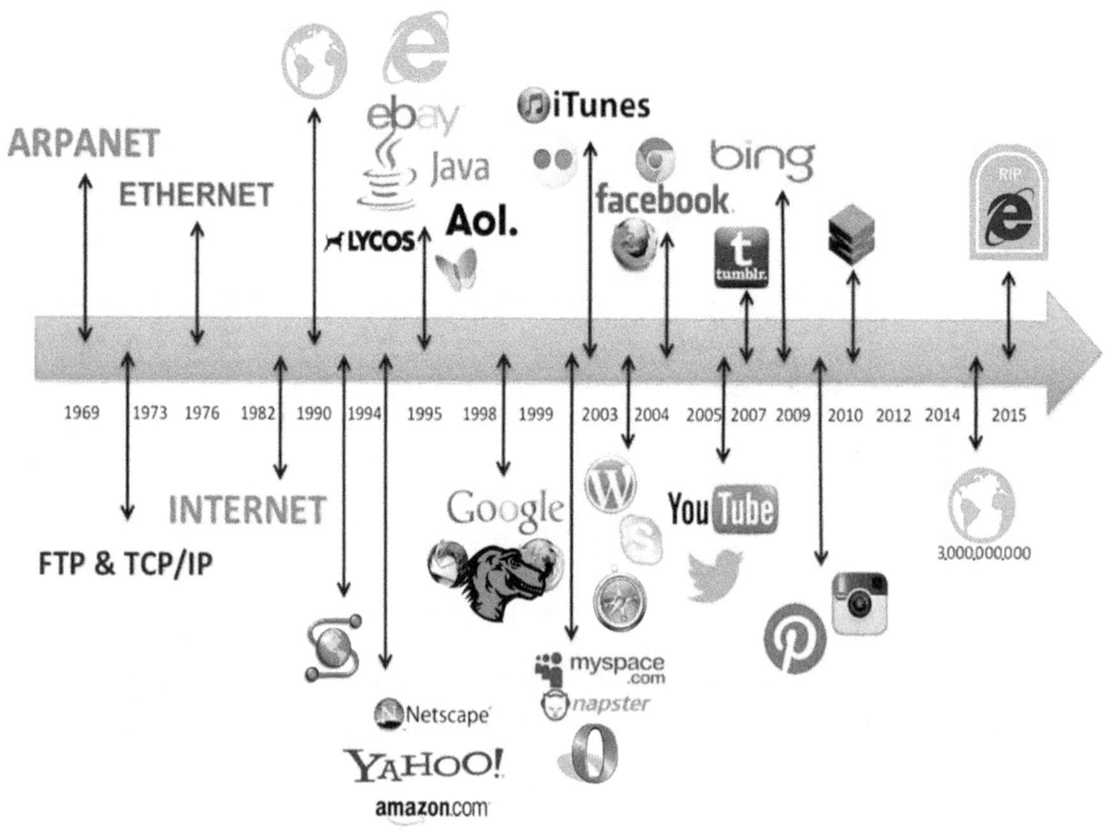

In fact, nowadays, the Internet is very important for the life of the human being, and it accentuates the strong inclination that most people have for taking advantage of the ease that technologies provide. Obvious examples are being able to call a taxi, order food or shop online, through a few touches on any mobile device. Companies such as Uber, Airbnb and Amazon were pioneers in offering that type of services. Consumers have now come to expect companies to follow this digitalization trend, otherwise, they will surely seek other suppliers, and this can lead to the disruption of a service, as we will see later. After all, no one wants to have to get up from their comfortable sofa to go to a restaurant get food anymore, do they?

TRANSFORMING TECHNOLOGIES

A business transformation is digital, as stated by Wade (2015), when it is developed based on digital technology.

The directions of this new digital market, fruits of the digital transformation that makes companies more competitive, are leading organizations to change the interaction, which until then were enough, in order to maintain an almost instantaneous contact with customers, and business partners, as reinforced by Fonseca (2015).

It is very important to know the technologies, because there are a lot of existing solutions. Let's address some of the most well-known and important:

- **Big data** – It is a strategy used for mining and manipulating data that cannot be handled by traditional techniques. Due to the large volume of data, of different sizes and formats, the name Big data was found in order to better describe this new situation. It is one of the most complex of digital transformation environments.
- According to Azevedo (2017), the main characteristics, which can be seen in Figure 3, are known as the 4 "Vs"[1], which are:
    - Volume (scale of data);
    - Variety (different form of data);
    - Velocity (analysis of streaming data);
    - Veracity (uncertainty of data).

Figure 3. 4 "Vs" of big data. Source: (Qasource, 2017)

- **Cloud Computing**- It is the distribution of computing services – servers, software, databases, storage, among many other services known as aaS (as a Service), over the Internet. According to the Synergy Research Group (2022) report, Amazon Web Services (AWS) is the leader in the cloud infrastructure market, followed by Microsoft, Google, Alibaba and IBM. In Figure 4, we can see the positions of each one since the fourth quarter of 2017.

There are several services available in the cloud, however, the main services, whose descriptions below we replicate exactly as described on the IBM website (2019), are:

§ **IaaS**, or infrastructure as a service, is on-demand access to cloud-hosted physical and virtual servers, storage and networking - the backend IT infrastructure for running applications and workloads in the cloud.

§ **PaaS**, or platform as a service, is on-demand access to a complete, ready-to-use, cloud-hosted platform for developing, running, maintaining and managing applications.

§ **SaaS**, or software as a service, is on-demand access to ready-to-use, cloud-hosted application software.

Figure 4. Quarterly Cloud Spending Jumps. Source: Synergy Research Group (2022)

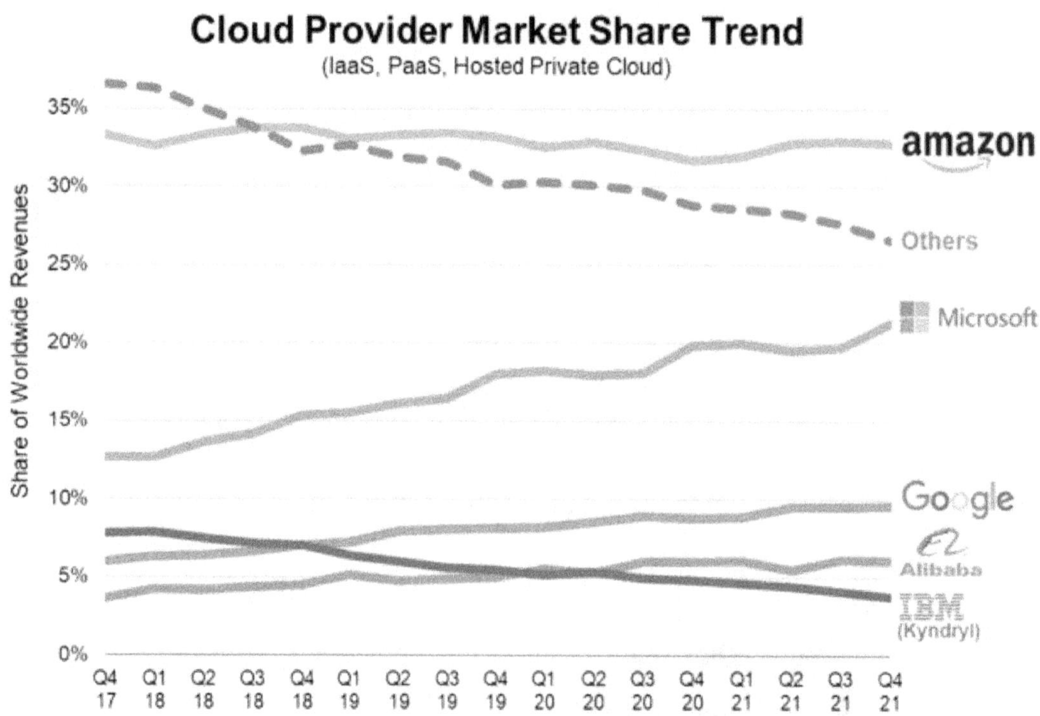

- **Artificial Intelligence**: It is the ability of a machine, and/or software, to artificially acquire intelligence and knowledge in order to develop the ability to make intelligent decisions, equivalent to human reasoning.

- **Internet of Things**: The phrase "Internet of Things", according to Ashton (2009), even with his own reservations, began to be used after a presentation he made at Procter & Gamble (P&G) in 1999. Known as IoT, the Internet of Things covers devices capable of connecting via the Internet and transmitting events and/or data. Ashton (2009) understands that the IoT has a potential, greater than the Internet, to change the world. Some of the IoT applications are present in:
  - Smart Homes;
  - Smart Buildings;
  - Smart Cities;
  - Smart Factories;
  - Smart Vehicles.

Is possible to mention other technologies:

- 5G;
- Blockchain;
- Autonomous Driving;
- Fintech;
- 3D Printing;
- Virtual Reality;
- Augmented Reality;
- Robotics.

IMPACT OF COVID-19 ON DIGITAL TRANSFORMATION

The COVID-19 pandemic has had an unprecedented impact on our way of life, causing people and organizations to change their way of life and the way they relate to their friends and customers. Stalmachova, Chinoracky & Strenitzerova. (2021) reinforce this statement in their study, when they mention that almost everyone started working at home with some type of digital technology.

Pereira (2021), states in his dissertation on digital transformation, which was developed in the midst of the pandemic, about the consequences caused by it:

We are all still feeling the consequences of the crisis caused by COVID-19 and also trying to predict this impact in the future. We are certain that this outbreak is a tragedy that will have far-reaching implications for humanity and the global economy. It has already changed the way we live, the way we work, and it has jeopardized some of the business initiatives of many companies. Schools and universities all over the world needed to adapt the way they teach, hospitals and medical centers needed to change the way they attended to users and essential services needed to change the way they attended to their clients, such as, for example, the catering, which started to make take-away sales in order to deliver meals without the customer consuming them in their establishment and, as in other cases, with as little contact as possible.

Although Filev (2020) argues that due to COVID-19, companies had to massively adopt teleworking, the idea of working from home using ICT did not emerge after the pandemic. This idea was proposed by companies based in California, such as Yahoo, under the name of "telecommuting" or "tele work", still in the 1990s, which may be the first generation of telecommuting, as Stalmachova, Chinoracky & Strenitzerova state. (2021).

The impact was felt in the closure of the activities of several companies and, Lamas and Lamas (2021), state that organizations that were not well prepared, with regard to information technology, more specifically SMEs, were the ones most likely to have gone bankrupt.

Despite unemployment in some areas, the COVID-19 disease has given a boost to digital transformation. Grove (2020), in addition to understanding that organizations have to focus on improving strategies that align technologies and processes, without forgetting to improve communication between people, stressed that strategies would change as a result of COVID-19. Nicolls (2020) warned that, due to the pandemic, many organizations have benefited from online businesses. Marous (2020) stated that it was a time when banking service organizations could have taken the opportunity to better develop digital platforms. Filev (2020) claimed that the areas of Cloud Computing, events, telecommuting and food and service orders would be the ones that would undergo the most radical changes. Cerqueira, Baranovskiy & Lamas (2021) claim that this change has affected, mainly, in an intense and profound way, the way of interaction between people and organizations (and even countries), where the screens of our personal computers, are no longer exclusive for our family and friends, and began to be "invaded" by co-workers, customers and suppliers. They also state that we are experiencing transformations that change our life habits, similar to so many others that we have read, which will be recorded in storybooks.

DIGITAL MATURITY

As we have seen, COVID-19 has brought new challenges to many organizations from the most different sectors of activity, which, in order to overcome them, and to not be extinguished, they had to go through a rapid change and adaptation to be able to sustain and maintain the business. In many of these cases, coupled with the global trend of digital technologies, one of the goals when adopting digital transformation in organizations was to seek a competitive digital environment. But to be able to achieve it, they need to reach digital maturity and that maturity doesn't come overnight. In the same way that a human being goes through phases, naturally, from childhood and puberty, until reaching adulthood, companies, their managers and employees, will, in an appropriate manner, go through different stages of digital maturity and, in each of them, they will learn to develop a working knowledge of digital trends.

It was in the 1970s that, according to Rafael et al. (2020), the concept of maturity model (MM) emerged for the first time, dedicated to software engineering. Since then, the model has evolved to be a widely used tool to improve business practices in several areas.

Fitzgerald et al. (2013), who already stated that organizations either implement digital transformation or run the risk of being overtaken by their competitors, classified, based on a survey carried out by the MIT Center for Digital Business and Capgemini Consulting, in 4 categories the level of digital maturity, whose base model is still used today. These categories and their definitions can be seen in Figure 5.

Figure 5. Classification of organizations. Source: Adapted from Fitzgerald et al. (2013)

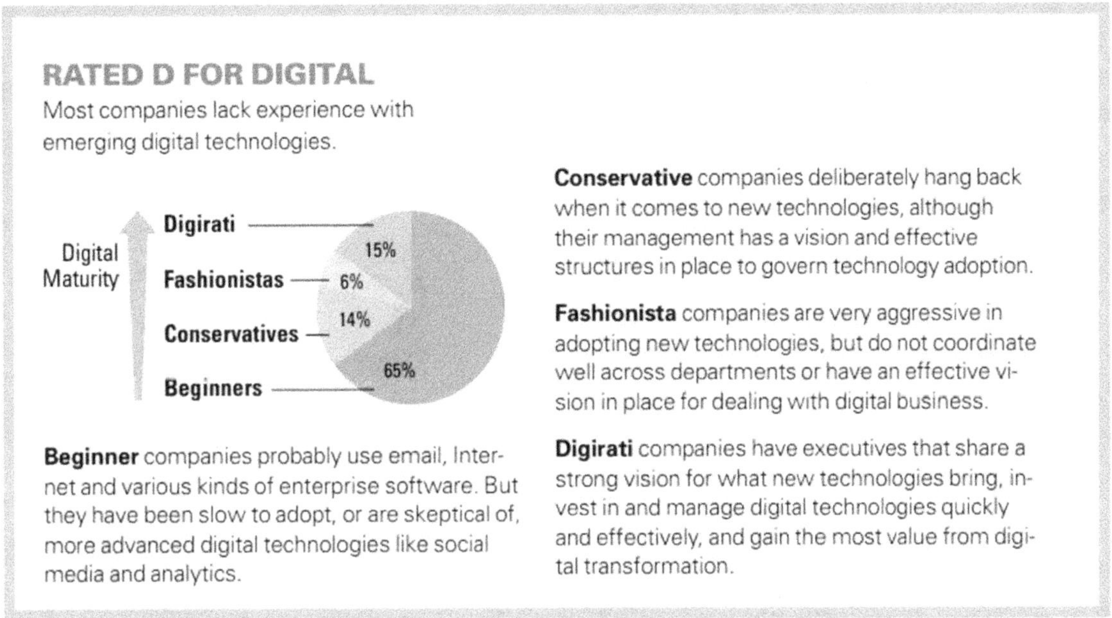

The organization's position at one of these levels of digital maturity changes depending on the level of leadership and the company's level of digital intensity. If a company has a low level of leadership and a low level of digital intensity, they are at the beginner's level. At the conservative level, there are companies that have a low level of digital intensity but that have a greater intensity in leadership, capable of being effective in managing the use of technology. A company is considered fashionista if it invests a lot in technology, being, therefore, an organization that has a high level of digital intensity, but because it has a low level of leadership intensity, it is unable to use technology in a way that involves the company as a whole. Finally, the company is considered digirati, that is, the elite of digitalization, when both the intensity of technology and leadership are high. When the company reaches this level, it has a horizontal governance with Innovative projects and has several digital initiatives implemented with measurable results.

There are several dimensions that can be analyzed to define the level of digital maturity, however, when checking some studies, the ones listed below are the ones that are most represented:

- Client
- Culture of the organization
- Strategy
- Leadership
- Operations
- People
- Technology

We can see in Figure 6, an example of the structure of the Digital Maturity Model (DMM) by Deloitte (2018), where digital maturity is evaluated through 5 of the dimensions indicated above, as well as the definition of each of these 5 dimensions. This Deloitte model (DMM) assesses digital maturity as a mean to provide the direction to achieve the expected results and also prevents duplication in the use of digital assessment tools.

Figure 6. Pan-organization. Source: (Deloitte, 2018)

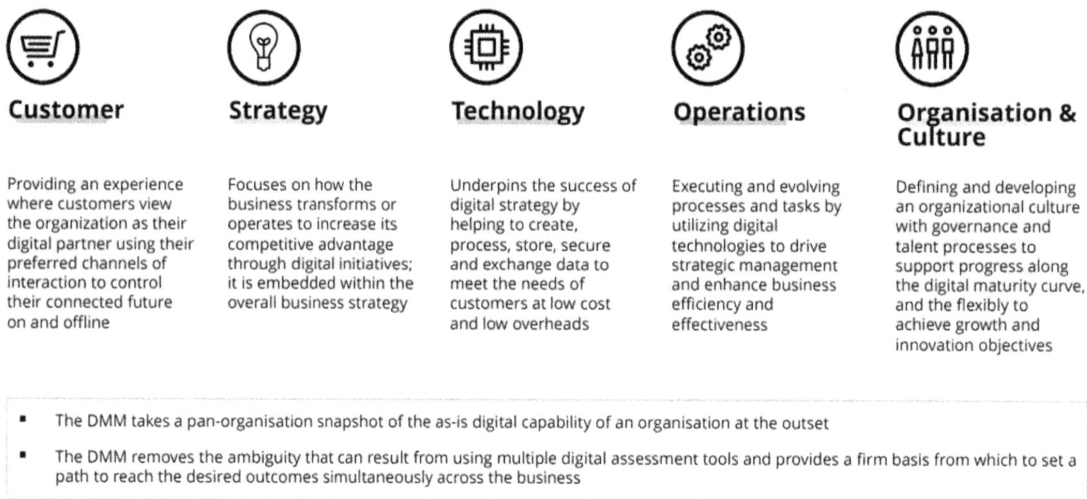

These 5 dimensions of Deloitte's DMM contain a total of 28 sub-dimensions, as can be seen in Figure 7, which in turn generate 179 individual criteria to obtain the final analysis.

Figure 7. Survey structure Deloitte. Source: (Deloitte, 2018)

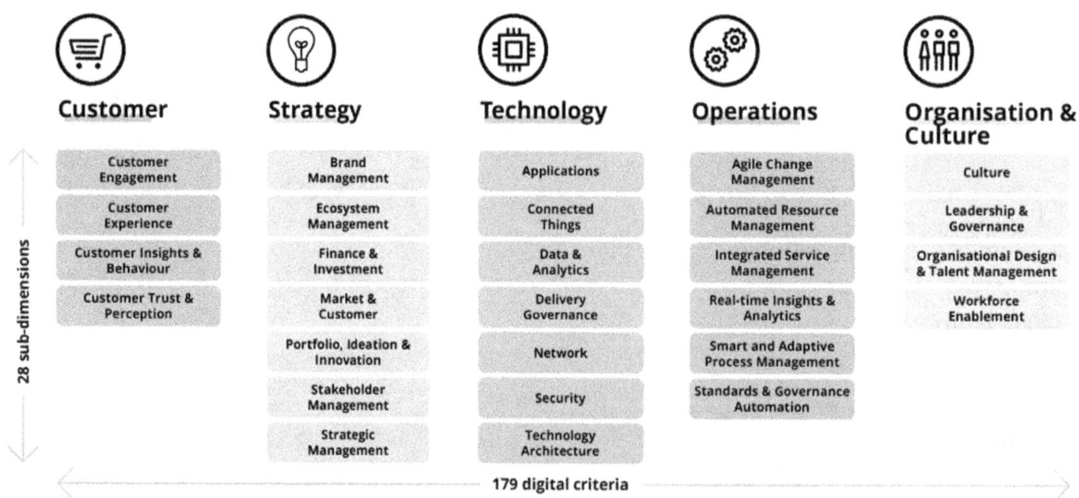

As technology constantly evolves, digital maturity is not a goal you reach and mark complete on your to-do list, but an ongoing process that must be adapted throughout the life of the organization. For example, an organization that at the beginning of this century had a static website, developed with basic tools, would have been considered digitally mature. However, if it did not upgrade to a dynamic and responsive website, with SEO and tools necessary to meet (and understand) the wants and needs of its market, its maturity level would drop radically.

IMPACT ON BUSINESSES

Despite the different definitions, there is a common understanding, that is, digital transformation has a strong influence on the company, as well as on the competition. The FujiFilm and Kodak examples, competitors in the field of photography, provide, as Wade (2015) explains, patterns of failure and success in the use of digital transformation. Reddy and Reinartz (2017) developed a broader approach. In addition to understanding that, practically, this transformation does not fail to reach a single moment in the daily lives of people and companies, and is literally happening, they claim that digital transformation has an innovative and, quite possibly, lasting impact, and thus changes the way we work and interact.

As doing something and doing the right thing are different concepts, applying digital transformation is not just applying a new digital technology through the simple acquisition of cutting-edge hardware and software, it is more than that. According to Pereira (2021), it is recommended that an organization understands why it is transforming and what must be transformed and has a guide on how to make the changes, so that it can then begin a process of digital transformation. Figure 8 depicts, even if in a humorous way, the need to know what must be transformed.

Figure 8. why transform? Source: (Wulff, M. & Morgenthaler, A., 2020)

To prepare a script of what should be transformed, Wade (2015) recommends that the organization introspects or the purpose of knowing:

- How the company makes money (business model);
- How the company is organized (the structure);
- Who works for the company (the employees);
- How the company does things (the processes);
- How information is managed (IT capacity);
- What products and/or services the company offers (the offers); and
- How the company interacts with its stakeholders (engagement model).

Cerqueira, Baranovskiy & Lamas (2021), mention that the globalization process has intensified with the galloping advance of technology, and even made a connection with the attention and importance, that companies should also have with internationalization.

And as there is a change, mainly in the paradigm, many of the basic principles of the strategy end up being changed based on new digital technologies. In Figure 9, Rogers (2016) presents, focusing on key concepts and strategic points, the changes in five domains as businesses move to digital.

Figure 9. the 5 domains. Source: (Rogers, 2016)

Domains	Strategic themes	Key concepts
Customers	*Harness customer networks*	• reinvented marketing funnel • path to purchase • core behaviors of customer networks
Competition	*Build platforms, not just products*	• platform business models • (in)direct network effects • (dis)intermediation • competitive value trains
Data	*Turn data into assets*	• templates of data value • drivers of big data • data-driven decision making
Innovation	*Innovate by rapid experimentation*	• divergent experimentation • convergent experimentation • minimum viable prototype • paths to scaling up
Value	*Adapt your value proposition*	• concepts of market value • paths out of a declining market • steps to value prop evolution

BUSINESS MODELS

It is important to note that an organization, when acquiring IT equipment, is not performing a digital transformation. Pereira (2021), states in his dissertation:

Digital transformation is not just about technology, but also about how a company can best compete using technology. And not only. The AEP (2019) indicates that, for the transformation to be successful, the organization needs to orchestrate a perfect tuning of the agents that make up the organization's ecosystem. It is important to define the strategy for the development of new skills, innovation and digital efficiency.

According to Birkinshaw & Ansari (2015) as the physical world increasingly becomes digital and software is able to analyze, control and interact with equipment and people, more companies are driven to create new business processes, new business models, and even new management models. Therefore, digital transformation only really exists if the companies' business model is changed. An evolution, transformation or even the creation of new business models is an integral part of digital transformation, with regard to companies, according to the authors Zhu et al. (2006), Rogers and Sparviero (2011) and Pardo and Etay (2014).

This digital evolution causes changes in customer consumption, and the increase in companies in the market leads to them having a very similar offer of products and services, leaving the customer with a much greater variety of offer. Customers select not necessarily by the product itself but by a much larger set of criteria, which represent the value that the company has to deliver (Bourbon & Lamas, 2021). A company must therefore, whenever a new idea, product or service is launched, define a value capture strategy, otherwise these can become just a passing fad and not a true case of success (Teece & Linden, 2017). This value capture strategy is associated with the concept of Business Models, a tool consensually recognized as fundamental to success in the business world.

There are many definitions, one of the simplest and most objective is that of Alexander Osterwalder and Yves Pigneur, who argue that a business model describes the logic of creation, delivery, and appropriation of value by an organization (Osterwalder & Pigneur, 2010). Based on this definition, Alexander Osterwalder initially developed the Business Model Canvas. This tool is a kind of screen, divided into 9 blocks (figure 10) closely linked with the creation of value (focus and central part of the screen).

Figure 10. Business Model Canvas. Source: (Osterwalder & Pigneur, 2010)

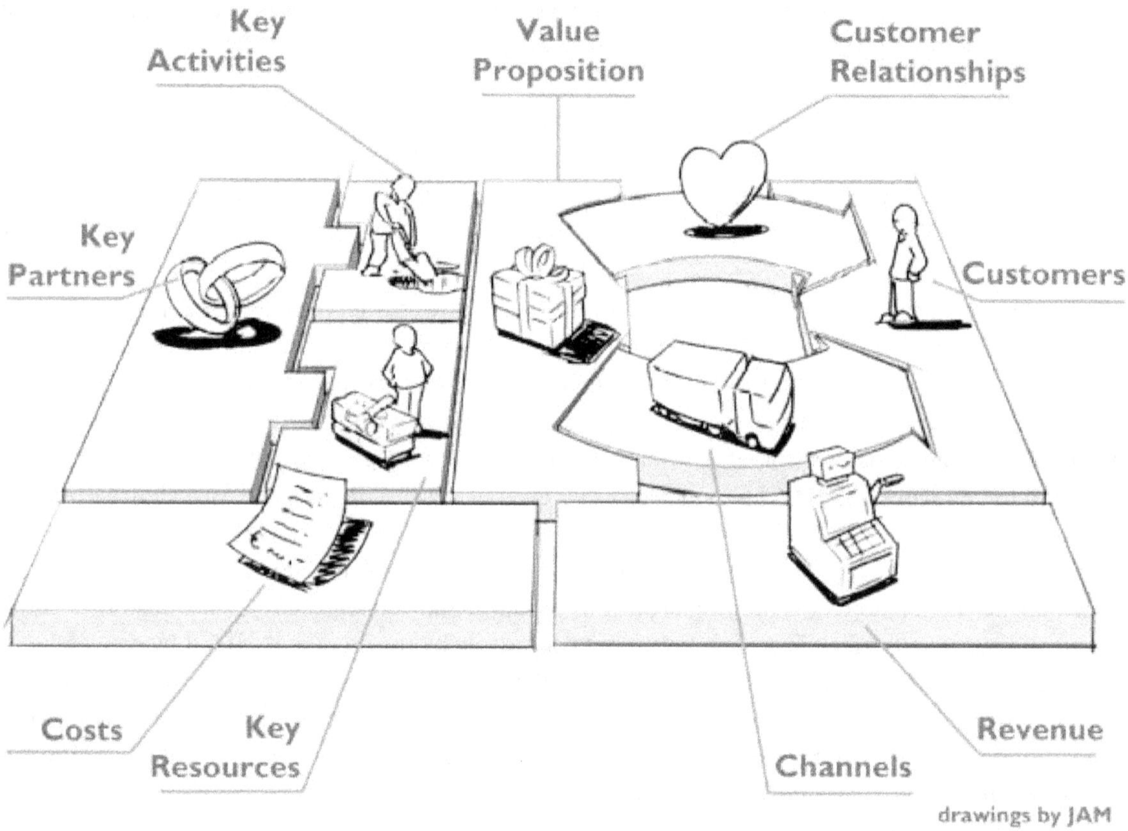

The remaining blocks are closely related, always based on the value proposition, and are related to the authors' definition of the concept. In the left part of the Canvas, there can be found the key partners, the key activities, and the key resources, which are the elements responsible for creating value. On the right, are the relationship with customers, distribution channels and customer segmentation, which are part of the value delivery process. Finally, at the bottom of the model are the structure of costs and revenues, which are related to the appropriation of value by companies (Bourbon & Lamas, 2021).

The success of a business, reinforces Teece (2010), depends as much on the business model and its implementation, as on the selection of technologies and the management of tangible assets and equipment. The business model indicates the way in which technological innovation and knowledge and the use of tangible and intangible assets are converted into profit. (Teece, 2010). Returning to the importance of value, we easily recognize that companies have to evolve, they cannot deliver value to customers as they used to do and must necessarily follow the changes that have taken place in recent decades, namely in the constant search for improvement in the quality of life that led to innovations in health, technology, economy and communication (Lamas, L. B. & Lamas, M., 2022).

The aforementioned changes and the development of technology have increased the ease and speed of contact between companies and consumers, wherever they are, even on the other side of the world

(Hagel III, 2016). Technology has also allowed us to move from physical to digital, revolutionizing the business world and the market itself (Baranovskiy, S., Cerqueira, C., Lamas, M., 2021), which forces companies to reinvent the way they do business, design their business models and deal with management, marketing, finance and, above all, managing people (Lamas, L. B. & Lamas, M., 2022). Vaskal et al. (2021), analyzed the development of digital transformation, in order to understand the impact of new technologies on business model innovation (BMI), a topic that we will emphasize next.

DISRUPTION

Imagine having to wake up early to get bread at a bakery near your house. You discover another bakery, far from your home, which allows you to place an order through an App and have the bread delivered to your home, warm, at the desired time. If you choose to receive bread at home, will you want to go back to the traditional model or will you adapt to the new model? Well, as we have seen, the Internet appeared in the mid-90s, and it was a tool used in disruption. One example, among many other cases, was in the music industry, when Shawn Fanning and his co-founder Sean Parker created Napster. And, as it evolved, getting faster and faster, even in the movie industry (which Netflix took advantage and Blockbuster did not). It's easy enough that that it makes it possible to communicate through several different media formats. Rogers and Sparviero (2011) claim that authors could, through the Internet, sell and deliver their work without the need for intermediaries.
According to Pereira (2021), some examples of disruptive companies are:

- **Airbnb**
    - *It doesn't have any rooms, but it's the biggest accommodation company in the world.*
- **Alibaba**
    - *It does not have stored goods, but it is one of the largest retail companies in the world.*
- **Facebook**
    - *It does not produce any content, but it is the largest media company in the world.*
- **Uber**
    - *Despite not owning any cars for the purpose, it is the largest taxi company in the world.*
- **YouTube**
    - *It doesn't have cameras or studio, but it's the biggest streaming company in the world.*

Disruption, even if not in an implicit way, is also observed by Cerqueira, Baranovskiy & Lamas (2021), when they point out that the development of technology allowed the contact between companies and consumers to became shorter, quicker and simpler, despite the physical barriers of distance.

Returning to the bakery case, the next step will be when the baker (who has the App) realizes that, to continue his business, he no longer needs a physical store to sell bread.

SUCCESSFUL VS FAILURE CASES

Large companies, traditional or digital, with brands known all over the world, ended their activities, while others adapted, survived, and even grew. In table 2, we cite some companies and the main causes responsible for success or failure.

Table 2. Main causes responsible for success or failure. Source: (Pereira, 2021)

COMPANY	SITUATION	MAIN CAUSES
Amazon	Success	Expanding the area of operation. Ability and willingness to go beyond the core business.
Domino's Pizza	Success	Developing digital capabilities and a close alignment with the company's business model.
Subway	Success	Bringing the customer closer to the brand, through the feeling that the product is personalized for him (innovation).
Kodak	Failure	It didn't innovate at the right time. It used the website to try to get more people to print digital photographs, rather than creating a new experience for customers.
Fujifilm	Success	Focusing on your core business. It combined its knowledge and, through SWOT analysis, joined existing forces with new digital capabilities.
Blockbuster	Failure	It didn't innovate at the right time. Didn't know how to use big data. It started the development of a rental service via the Internet but did not complete it. It didn't bother to understand who its customers were, and what they wanted.
Netflix	Success	Using big data, adapting to the rapid transformations of the digital age, customer focus.
Encyclopaedia Britannica	Success	It changed the way of bringing academic knowledge to people. It understood that there is a change in values in different generations. Realized that using the old distribution format, information could quickly become obsolete.

CONSUMERS TRUST

The selected case study is Consumers Trust, Lda (original name Megaklique, Lda), created in 2009, after its creator, Pedro Lourenço, detected a gap in the market, when he made a complaint in a supermarket. It concluded during the process that there was no agile process for complaints and their subsequent resolution and decided to develop a solution for the problem detected, thus the Portal da Queixa was born. "Today, it is the largest social network of consumers in the country and has reached a position of national reference, on the internet, in terms of consumption" (Portal da Queixa, 2022) and is already in the midst of internationalization with a presence in South Africa. Spain and France. And in fact, Consumers Trust was founded thanks to a cake with spoiled chocolate.
Pedro Lourenço, founder of Consumers Trust and CEO of Portal da Queixa, argues that

The focus on the internationalization of this unique project in Portugal and pioneer in Europe has always been our main objective, from the moment I decided to go ahead with the development of the platform of the Portal da Queixa, in 2009, after an unresolved episode of a bad bollycao I bought for my son, in a supermarket. Today, we are a proud team for feeling that we were the promoters of a new form of relationship between brands and consumers, changing the consumption paradigm and the vision of most Marketing and Customer Service managers. After a lot of experience gained in the successful case of Portal da Queixa in Portugal, we moved very recently to Spain, where we launched the 'beta' version of the Libro de Quejas platform online. This historic step allowed us to be the first social network of consumers in the world to be international and to take national creativity to other countries in Europe.

We will focus, as mentioned, only on the qualitative component used in the study, the interview technique. This is divided into three parts: the first addresses the origin of the company; the second focuses on the company's history and development; the third focuses on the business model, from its initial design to its evolution over the years. An interview was carried out online via Zoom, and the Founder and CEO, Pedro Barros Lourenço, was selected. With this interview we intended to get to know and analyze the target company, Consumers Trust, and in particular its business model. We now present the results we arrived at with these interviews.

Portal da Queixa was launched online in its 1st version in June 2009 (Joomla), with the aim of creating a community and in an attempt to classify the companies by sectors of activity and create a Ranking. Their base business model – advertising revenue, didn't work. According to Pedro Barros Lourenço, there were several reasons, but the most relevant, where the fact that it is an unknown brand in the market and its name "Portal da Queixa" that had associated to its negative connotation and that "frightened" the public.

The resilience associated with entrepreneurs manifested itself in this case and the project continued. The presence in December 2009 on the radio program Prova Oral (by the well-known Fernando Alvim) on Antena 3 (with an audience of 100 thousand people) worked almost as a validation of the project. It was also possible to perceive the differentiation of the project compared to what existed in the market.

In 2012, the portal was redesigned, but another setback arose, the main shareholder intended to close the company, so that this would not happen, one of the interns at the time, Vitor, took 10% of the company. The Portal da Queixa project had to be monetized to ensure its permanence in the market. The decision was to change the focus, to become a social network for consumers. They closed the portal and only reopened it in 2013, after a total rebrand. However, it was still all free, without generating revenue. Besides consumers, brands were also allowed to create an account. These changes allowed greater dynamism and interactivity. They maintained three advertising blocks as a revenue stream, which was still insufficient to guarantee sustainability.

Some competitors that appeared ended up leaving the market, failing to create a business model that would capture value, a problem that Portal da Queixa also had. Visibility was increasing and another important milestone took place, the participation in the television program, Savings Accounts in the newscast of the generalist television channel SIC. The project gained even more visibility and Pedro Barros Lourençao began to participate regularly with a space on a SIC program.

At the end of 2014, there was a first contact of the Brazilian project similar to Portal da Queixa, aiming to enter the European market, starting from Portugal, investing in the project and buying part of the share capital of Portal da Queixa. The negotiation took a long time and in February 2015 it was planned to close the deal. The promoter's lack of experience led him to invite the former majority partner to lead the final negotiation. The final amount requested (350 thousand euros) was very high and the negotiation did not end at that time. The representatives of the Brazilian project stayed to talk to the other investors and then give feedback. The opinion was that the process would move forward.

The negotiation also gave rise to the starting point for the design of a new business model based on software aimed at companies that allowed maintaining neutrality in the Portal da Queixa, while maintaining impartiality. "Time passed, 1 month, 2 months and there was no answer. They ended up transmitting in April that the business would not advance, the value was very high, and Brazil had entered a crisis.". It was more of a shock, a bucket of cold water, suddenly the promoters went from the dream to the stark reality. One more decision to make, keep the project or close it?

In April 2015, the 1st edition of Shark Tank was on the air, Pedro Lourenço, when he saw the program, identified Tim Vieira as the potential investor, he still thought about applying but chose to get in touch with Tim Vieira personally. The planned investment would be for the development of the business model. "On the 8th of May, less than 1 month after the bucket of cold water with the Brazilians, he had a meeting with Tim Vieira and they closed the deal, for 125 thousand euros and 50% of the capital was sold (company valuation of 250 thousand euros)."

This was the turning point for Portal da Queixa, the guarantee of continuity. Change of premises, hiring of personnel. Another update of the portal was carried out, launched on March 15, 2016, with yet another total rebrand and new business model. B2B business, companies would pay to have an account, have a page, respond directly to consumers (backoffice, management of their presence) and access to information about consumers and clear visibility. Segmentation, medium and large companies. It is a freemium model (complaints management is free), the premium includes statistics, business analytics, gamification, access to the possibility of having the "seal" of the Portal da Queixa. It was released with free and three more packages (standard, premium and premium plus).

Most of the customers are, as expected, medium and large companies. After the implementation of this new business model and according to customer feedback, a new "custom" plan was created, tailored to the needs of each customer.

In 2017 there was growth and in 2018 it continued, with companies already looking for Portal da Queixa. Public rankings were fundamental for companies to compete and want to be in the first places, investing for this and to obtain seals and be the "brand of the month". It was necessary to hire more people for customer management and end-user management.

There was a need, due to growth, to adapt the business model again, as many of the companies that required more resources were customers who had the cheapest package. It was necessary to bet on a negotiation at the time of renewal of contracts (12 months). The price was then defined not by the services but by the number of complaints, reassessed every 12 months based on history but also providing that if there was a "peak" the contract could be renegotiated halfway through the contract. More features were also added, mainly within business analytics and marketing. The integration of Artificial Intelligence is already being considered. News changes were made in the model, with only 2 packages, the basic and the custom, which will continue to evolve, always based on the feedback from end users and customer companies, incorporating the provision of support services for customer management and satisfaction. Advertising continues to be a source of revenue, but it only weighs 5%.

Internationalization started to make sense because the Portuguese market is very small. With the project going well in Portugal, as it did, international expansion was an expected next step, namely in Europe.

They decided to start with Spain, for reasons of geographic proximity and, consequently, the ease of travel, less costs and of course the Iberian market. The initial experience at the time of the go to market went very well, maintaining the focus on medium and large companies. Another reason for choosing Spain as the 1st internationalization market was the fact that they already had several Spanish companies as customers in Portugal. In October 2019 they took the 1st step, translating all the material into Spanish, working in Spain, and creating and registering the Consumers Trust brand for international expansion. The ever-present cultural issue, getting to know Spain better, they realized that Consumers Trust would not work and created the "Libro de Quejas", and with learning from this first experience, they started to create a specific brand for each country, South Africa "Complains book", France - "Reclame ici". Consumers Trust will be the umbrella brand. The Process is ongoing.

A second round of investments was necessary, with the entry of another partner (Business Angel), Dieter, with 30,000 euros to fund investments to enter the Spanish market. To speed up the internationalization process, they will start a new round of investment, which should be around two million euros.

The Consumer Trust is supported by 2Venture, which has already developed international expansion projects for several companies, provides support in the preparation of internationalization on a larger scale, namely in adapting the business model to make the expansion process less "personal" and less dependent on promoters, and allows automating the internationalization process and the respective negotiation. They also adapt the knowledge transfer and operationalization process, from a company with 12 years of experience to companies with 1 or 2 years. The process has to be prepared to be gradual.

A strategic road map has already been created (based on the existing history of the Portal da Queixa) to enter the market for Spain and France, which allows, based on the results and the subsequent necessary adaptations for the creation of a road map base to be used for the entry into upcoming markets. They closed 2020 with a turnover of 300 thousand euros (50% growth compared to 2019), they estimate for 2021, 600 thousand euros.

CONCLUSION

As digital transformation is an ongoing process that will continue on its way, some companies still create some resistance, many were forced to do it due to the COVID-19 pandemic while others embraced this transformation as a duty or even as a way to ensure their growth. They take advantage of the moment as an opportunity, realizing the positive implication that this has on their business model but also on their internal processes, allowing them to achieve better products and/or services and with that an increase in customers.

COVID-19 was not the main topic of this chapter, but it was important to include it, as it highlighted the importance of companies being proactive, as well as implying that information technology should not be seen as an expense and yes as an investment, but it is important to emphasize that this must be applied in a digital transformation in a conscious, planned and well-defined way, so that organizations are prepared for this transition.

In this chapter we analyzed a specific case, Consumers Trust, a company that embraced digital transformation since its inception and took advantage of the transformative technologies mentioned above, namely using SaaS, or software as a service, an on-demand access to ready-to-use. -use, cloud-hosted application software. Soon it will also introduce Artificial Intelligence, the ability that a machine, and/or software, artificially acquires intelligence and knowledge in order to develop the ability to make intelligent decisions.

The relationship between digital transformation and its business model was clear, which has evolved over the years based on experience and, more than that, the company's ability to learn from its surroundings and always being aware of what their customers and end users really want. It evolves from a business model based on advertising to a SaaS, maintaining the necessary independence in a portal that intends and even assumes itself as a social network for consumers.

Its internationalization process has highlighted the need for companies to adapt to different cultures and practices. Culture is thus undoubtedly a fundamental factor for the success or failure of companies in an international adventure. Consumers Trust has, despite being recent, demonstrated its digital maturity involving, as previously mentioned, the main areas of the business, namely: Customer; Organization's culture; Strategy; Leadership; Operations; People; and Technology.

As advocated by researchers, entrepreneurs and even governments, companies must face digital transformation as an opportunity and be an active part of a process that, having started, will intensify in the coming years. More than guaranteeing its growth, it is about guaranteeing its survival. To promote it, the Portuguese government created the Digital Maturity Seals (DMS), as one of the measures of the Action Plan for the Digital Transition. This initiative was developed in partnership with the National Press, the Portuguese Institute of Accreditation, and the Portuguese Institute of Quality, as well as with certifying entities. Any company, whether in the private or public sector, can apply for a digital maturity certification on each of the following domains: Sustainability, Cybersecurity, Privacy and Data Protection and Accessibility. Depending on its maturity level, different seals (Bronze, Silver, and Gold) can be granted.

REFERENCES

Ashton, K. (2009). That 'Internet of Things' Thing. *RFID Journal*. https://www.rfidjournal.com/articles/view?4986

Cerqueira, C., Lamas, M., Baranovskiy, S. (2021). Interculturality Role in Startups Internationalization: Eligent Club's Virtual Accelerator Program Case. *E-Revista de Estudos Interculturais do CEI–ISCAP*, *3*(9), 1–22. doi:10.34630/erei.v3i9.4217

Birkinshaw, J., & Ansari, S. (2015). Understanding management models: going beyond "what" and "why" to "how" work gets done in organizations. In Business model innovation: The organizational dimension (pp. 85-103). University Press.

Bourbon, R., & Lamas, M. (2021). Modelos de Negócio e Internacionalização: O caso da Jump Willy. *Portuguese Journal of Finance, Management and Accounting*, *7*(13), 47-69. http://u3isjournal.isvouga.pt/index.php/PJFM

Case, S. (2017). *A Terceira Onda da Internet*. HSM.

Certificado de Maturidade Digital. (2022). https://selosmaturidadedigital.incm.pt/SMD/

Deloitte. (2018). *Digital Maturity Model*. https://www2.deloitte.com/content/dam/Deloitte/global/Documents/Technology-Media-Telecommunications/deloitte-digital-maturity-model.pdf

Eden, R. E., Jones, A. B., Casey, V., & Draheim, M. (2019). Digital Transformation Requires Workforce Transformation. *MIS Quarterly Executive*, *18*(1), 1–17. doi:10.17705/2msqe.00005

Filev, A. (2020, March 30). COVID-19 Is A Before-And-After Moment In The Digital Transformation. *Forbes*. https://www.forbes.com/sites/andrewfilev/2020/03/30/covid-19-is-a-before-and-after-moment-in-the-digital-transformation/

Fitzgerald, M., Kruschwitz, N., Bonnet, D., & Welch, M. (2013). Embracing digital technology: A new strategic imperative. *MIT Sloan Management Review*. https://www.capgemini.com/se-en/wp-content/uploads/sites/29/2017/07/embracing_digital_technology_a_new_strategic_imperative.pdf

Fonseca, J. L. C. (2015). *Fatores de satisfação e fidelização no processo de compra online – Um estudo de caso na Universidade Lusófona do Porto* [Master Theses, Lusófona do Porto]. Repositório Científico Lusófona. http://recil.grupolusofona.pt/handle/10437/6956

Grove, W. (2020). How will COVID-19 Impact Digital Transformation? *Skybox Security*. https://blog.skyboxsecurity.com/how-will-covid-19-impact-digital-transformation/

Hagel, J. III. (2016). We need to foster our definition of Entrepreneurship. *Harvard Business Review*, (September), 1–5.

Hyche, J. (2016). *The Future of Branch Banking*. Louisiana Banker Association. https://www.lba.org/files/John%20Hyche,%20The%20Future%20of%20Branch%20%20Banking.pdf

IBM. (2019). *IaaS, PaaS and SaaS – IBM Cloud service models*. IBM. https://www.ibm.com/cloud/learn/iaas-paas-saas

Lamas, L. B., & Lamas, M. (2022). New insight into Business Innovation and Sustainable Development: The role of Education. In C. Sarmento (Ed.), *Concepts and Dialogues across Shifting Spaces in Intercultural Business* (pp. 15–31). Cambridge Scholars Publishing.

Marous, J. (2020). *COVID-19 Provides Opportunity For Digital Transformation*. The Financial Brand. https://thefinancialbrand.com/94700/covid-19-digital-banking-transformation-opportunities/

Nicolls, D. (2020). COVID-19: A Call for Digital Transformation. *Forbes*. https://www.forbes.com/sites/jumio/2020/04/01/covid-19-a-call-for-digital-transformation/?sh=79aa37434f11

Osterwalder, A., & Pigneur, Y. (2010). *Business Model Generation*. John Wiley & Sons, Inc.

Pardo, A., & Etayo, C. (2014). Movies and screens: The Spanish audience's choice. *Communicatio Socialis*, *27*(4), 131–145. https://dadun.unav.edu/bitstream/10171/37938/1/20141027124907.pdf

Perdomo, R. S. (2019). *Framework das capacidades dinâmicas para impulsionar a transformação digital: O Caso da Empresa de Construção Civil de Florianópolis* [Master Thesis]. Universidade Estadual de Santa Catarina, Brasil. https://sistemabu.udesc.br/pergamumweb/vinculos/000070/00007078.pdf

Pereira, A. (2021). *A Transformação Digital nas PME Portuguesas: Análise das Empresas da Região Centro* [Master Thesis]. Instituto Superior de Entre Douro e Vouga (ISVOUGA), Santa Maria da Feira, Portugal. https://comum.rcaap.pt/handle/10400.26/36356

Pînzaru, F., & Mitan, A. (2012). Social media and marketing of the "popcorn" music wave: The success of romanian commercial musicians analysed through their perceived image on facebook and youtube. *Economia e Sociologia*, *5*(2), 125–138. https://www.economics-sociology.eu/files/14_MOD_Paper_Pinzaru%20_Mitan_proofr_03.02.13E&S.pdf

Portal da Queixa. (2019). *Portal da Queixa internacionaliza-se e cria marca global: Consumers Trust.* https://portaldaqueixa.com/news/portal-da-queixa-internacionaliza-se-e-cria-marca-global-consumers-trust

Rafael, L. D., Jaione, G. E., Cristina, L., & Ibon, S. L. (2020). An Industry 4.0 maturity model for machine tool companies. *Technological Forecasting and Social Change, 159*(120203), 1–13. doi:10.1016/j.techfore.2020.120203

Raskino, M., & Waller, G. (2015). *Digital to the Core: Remastering Leadership for Your Industry, Your Enterprise, and Yourself.* Routledge.

Reddy, S., & Reinartz, W. (2017). Digital Transformation - From academic research to practical use. *GfK Verein, 9*(1), 11-17. https://www.nim.org/sites/default/files/medien/2327/dokumente/2017_gfk_mir_digital_transformation_eng_0.pdf

Research Councils UK. (2017). *Rise of the Machines.* Research Councils UK. https://blogs.rcuk.ac.uk/author/bena/page/2/

Rogers, D. L. (2016). *The digital transformation playbook.* Columbia University Press. doi:10.7312/roge17544

Rogers, J., & Sparviero, S. (2011). Same tune, different words: The creative destruction of the music. *The Observatory Journal, 5*(4), 1–30. https://pdfs.semanticscholar.org/bec0/5cc20dfc50c9c1eb5208a69eaff73c6c71fd.pdf

Shannon, C. E. (1948). A Mathematical Theory of Communication. *The Bell System Technical Journal, 15*(3), 379–423. doi:10.1002/j.1538-7305.1948.tb01338.x

Stalmachova, K., Chinoracky, R., & Strenitzerova, M. (2021). Changes in Business Models Caused by Digital Transformation and the COVID-19 Pandemic and Possibilities of Their Measurement—Case Study. *Sustainability, 14*(1), 127. Advance online publication. doi:10.3390u14010127

Stolterman, E., & Fors, A. C. (2004). Information Technology and the Good Life. International Federation for Information Processing Digital Library. *Information Systems Research, 143*, 687–692. https://www8.informatik.umu.se/~acroon/Publikationer%20Anna/Stolterman.pdf

Synergy Research Group. (2022). *As Quarterly Cloud Spending Jumps to Over $50B, Microsoft Looms Larger in Amazon's Rear Mirror.* https://www.srgresearch.com/articles/as-quarterly-cloud-spending-jumps-to-over-50b-microsoft-looms-larger-in-amazons-rear-mirror

Teece, D. J. (2010). Business Models, Business Strategy and Innovation. *Long Range Planning, 43*(2-3), 172–174. doi:10.1016/j.lrp.2009.07.003

Teece, D. J. (2020). Hand in Glove: Open Innovation and the Dynamic Capabilities Framework. *Strategic Management Review, 1*(2), 233–253. doi:10.1561/111.00000010

Teece, D. J., & Linden, G. (2017). Business models, value capture, and the digital enterprise. *Journal of Organization Design, 6*(1), 8. doi:10.118641469-017-0018-x

Vaskal, S., Massarol, M., Bagarottol, E. M., & Mas, F. D. (2021). The Digital Transformation of Business Model Innovation: A Structured Literature Review. *Frontiers in Psychology*, 1–18. https://www.frontiersin.org/articles/10.3389/fpsyg.2020.539363/full PMID:33584396

Wade, M. (2015). *Digital Business Transformation. A Conceptual Framework*. Global Center for Digital Business Transformations. https://www.imd.org/contentassets/d0a4d992d38a41ff85de509156475caa/framework

Westerman, G., Bonnet, D., & McAfee, A. (2015). *Liderando na Era Digital*. M. Books.

WulffM.MorgenthalerA. (2020). *Wumo*. https://wumo.com/wumo/2020/07/16

Zhu, K., Dong, S., Xu, S. X., & Kraemer, K. L. (2006). Innovation diffusion in global contexts: Determinants of post-adoption digital transformation of European companies. *European Journal of Information Systems*, *15*(6), 601–616. doi:10.1057/palgrave.ejis.3000650

ADDITIONAL READING

Andreini, D., Bettinelli, C., Foss, N. J., & Mismetti, M. (2021). Business model innovation: A review of the process-based literature. *The Journal of Management and Governance*. Advance online publication. doi:10.100710997-021-09590-w

Qasource. (2017). Big Data Testing Challenges and Solutions: TechnoCast - Spring 2017. *Qasource*. https://blog.qasource.com/hs-fs/hubfs/image/technocast/big-data-image.png

Schaub, K. (2017). *The Fourth Industrial Revolution*. Penguin Books Ltd.

KEY TERMS AND DEFINITIONS

aaS: It is the acronym for "as a service". It is a business model in the cloud that is the technological essence of the on-demand offer, as the customer chooses the service it wants and pays a price for it, as a subscription. The customer receives the data to access the service while the provider is responsible for providing and maintaining the necessary structure.

Artificial Intelligence: The concept concerns the use of technological solutions so that machines can simulate human capabilities linked to intelligence.

Business Model: Description of how a company develops its activities, what it offers (value proposition), to whom (segmentation) and what it represents in financial terms (revenue streams and cost structure). It's all about creating value for the company, customers, and society.

Digital Maturity: Digital maturity is closely linked to the level of improvement that companies reach, after adopting new digital technologies to create process automation, operations, and internal/external customer satisfaction.

Digital Transformation: Despite being a "digital" transformation, it is not only based on technological elements, since the true transformation is fundamentally human. It is a process of significant paradigm shift in the behavior and mentality of companies, with a structural change, in the search for efficiency and innovation, changing the way in which it operates and reaches its customers, using technology as a resource for your business solutions.

Digitalization: Digitalization is the transformation of something analog/physical to digital.

Disruption: New product or service that changes the market (or creates a new one) for customers, competitors, and market leaders.

Internet of Things: Well known by its acronym, IoT, it is the ability of some device, or "thing", to be able to collect and transmit data, through an internet connection and Bluetooth.

ENDNOTE

1 There are authors who define Big Data with 5 V's (Volume, Variety, Velocity, Veracity and Value), with 7 V's (Visualization and Variability) and even with 10 V's (Validity, Vulnerability and Volatility).

Section 2

Automation Initiatives and Better-Connected Experiences

Chapter 3
Relating Industry 4.0 and the SME Internationalization Process:
A Case Study

Ana Rita Cardoso
ISCAP, Instituto Politécnico do Porto, Portugal

Raquel Pereira
https://orcid.org/0000-0003-0990-4746
ISCAP, Instituto Politécnico do Porto, Portugal & CEOS. PP, Centre for Organisational and Social Studies, Portugal

ABSTRACT

The automatization, dematerialization, and innovative digital systems and technologies, named Industry 4.0, are transforming the economic and social system, encompassing opportunities and challenges for growth and international expansion. In Portugal, SMEs face great difficulties and challenges because innovative digital businesses and systems enhance globalization and intensify competition. Exploratory research was designed to explore how Industry 4.0 and digitalization will influence the organization, competitiveness, and internationalization process of the SMEs under study. Results indicate that Industry 4.0 appears to be critical because it positively affected the value chain and the organizational and competitiveness level. Therefore, the companies were better positioned in their competitive advantages. This work provides insights for the study and joint understanding of digitalization and business internationalization because it integrates and relates two autonomous, complex, and strategic fields of research, with no applied studies for the Portuguese SME background.

DOI: 10.4018/978-1-6684-5538-8.ch003

INTRODUCTION

The integration of digitalisation in production processes, industry and services has seen exponential growth, especially in recent decades, with the development of innovative digital systems and technologies, and the implementation of disruptive methods at all levels, which allow the interconnection of processes, machines, and people. This development has been called the fourth industrial revolution or industry 4.0 and it is characterised by the introduction of digital technologies that have a disruptive effect on companies' business models and the way they act and create customer value (Duričin & Herceg (2018); Salkin, aet al. (2018), Sjøbakk (2018), Xiaopu aet al. (2016)). The fourth industrial revolution is a revolution towards digitalisation, towards a knowledge-based, service-oriented society. According to Sukhodolov (2019), this revolution is evolving towards a self-organising and self-managing industrial model of fully automated production systems that instruct themselves autonomously and interactively, also leading to real-time data exchange, as well as to flexible production systems and progressively personalised production (Thoben aet al., 2017). Indeed, automatisation, dematerialisation and digitalisation are transforming the economic and social system, encompassing opportunities in terms of optimisation of resources, cost reduction and potential for growth and international expansion. This transformation is happening at a rapid pace because Information and Communication Technologies (ICT) are evolving at an extremely fast speed and, in the business context, where digitalisation and automation are the present and the future, companies, to be competitive, must be able to adapt quickly to this evolution. To improve their performance and achieve sustainable competitive advantages, companies are required to adopt digital transformation processes (Siachou aet al., 2020).

Altho there are challenges and barriers to the adoption and implementation of ICT, such as high investments, insufficient training and skills, lack of knowledge, implementation and maintenance costs, uncertainty about the return on investment, security, and reliability (Antunes aet al., 2019; Westerlund, 2020), it seems that the adoption of ICT and business digitalisation is an imperative path for companies in the current context and reality. ICT and its evolution integrate advantages in terms of productivity improvement, competitiveness, product or service quality, information sharing, multi-skilling, partnerships, remote work, and the marketing revolution (Westerlund, 2020). The implementation and adoption of these technologies and the Internet of Things (IoT), real-time information, Big Data, connectivity between people, objects, and systems, will lead to the personalisation of products and services on a large scale and therefore to a transformation of the value chain (Antunes aet al., 2019). Intelligent IoT systems enable the rapid production of new products, dynamic response to demand, and the real-time optimisation of production networks and supply chains (Antunes aet al., 2019). This context indicates that digitalisation is playing an increasingly key role in business and society, and all institutions are facing a fundamental need for radical changes in their strategy, structure, and corporate culture (Schmitt & Baldegger, 2020). Moreover, the implementation of ICT and digitalisation are transforming business models, and this is a key challenge in the context of international business and strategies (Chen & Kim, 2018; Watson aet al., 2018), and brings great opportunities, but also threats (Bankewitz aet al., 2016). According to Denicolai aet al. (2021), the contemporary economy stresses the relevance of digital transformation as a central driver towards innovation and business renewal, especially for Small and Medium-Sized Enterprises (SMEs). As SMEs make up the bulk of the world's businesses and the digital transformation is a key element in the progress and modernisation of companies and economies the phenomenon of digital transformation should be studied in the context of SMEs (Rupeika-Apoga aet al., 2022). But the question is: are they prepared?

SMEs are the core of the economic and productive system of European economies. They play the leading role as drivers of economic growth (Ardito aet al.,2021). In Portugal and most European countries, the business sector is made up of SMEs. These companies face great difficulties and challenges because ICT enhances globalisation, and intensifies competition between companies, which makes it the leading component for their expansion. The internationalisation process is affected by technology, allowing productivity, efficiency, and management improvements, and because it brings the world a "market without borders". Therefore, a question that emerges and to which an answer is required is: "What is the influence of Industry 4.0 in the process of Internationalisation of SMEs in Portugal?

Exploratory and descriptive research was designed, where the main objective is to analyse the importance of Industry 4.0 for the internationalisation of SMEs in Portugal. There is a large body of literature that analyses the internationalisation of companies and SMEs in Portugal. However, there are no research studies that specifically focus on the relevance of industry 4.0 and digitalisation to the internationalisation process of SMEs. Similarly, Hervé aet al., (2020) mention the lack of studies focussed on the integration and relevance of digital technologies in the internationalisation process of these companies. This seems to be the value-added and the highlight of this study. The research topic is complex, strategic, and dynamic. It aims to integrate and relate two autonomous research fields towards a collective understanding of the business digital internationalisation. Following a qualitative approach, thru a double case study, we intend to explore, inductively, how Industry 4.0, its technologies, and aggregated methods, will influence the organisational behaviour, competitiveness, and internationalisation process of the SMEs under study. The qualitative approach makes it possible to provide a thorough analysis goal of the research and theories used (Bergamaschi aet al., 2021). According to Yin (2015), the case-study method offers greater flexibility, allows an in-depth analysis of a complex research problem, and allows for a comparison between different cases. The remainder of this paper is organised as follows. Theoretical background and concepts and literature review useful to carry out the explorative research are provided in section two. Section three presents the methodological approach. Section four describes the major findings as well as the analysis and discussion of the results. The last section concludes with the synthesis and considerations of the results achieved, as well as the limitations and perspectives for future research.

BACKGROUND

Industry 4.0, Information Technologies and Digitalisation

The introduction of mechanical tools in manufacturing processes in late 18th century England, where machines and engines, powered by water, steam and coal, revolutionised the way goods were produced, and introduced society to the first industrial revolution (Bartodziej, 2017). From the mid-19th century, other energy sources based on electricity and oil, the division of labour and organisational innovations such as line production (Fordism) and scientific management of procedures (Taylorism) introduce a new industrial revolution, enabling mass production and the introduction of transport chains (Bartodziej, 2017). The third revolution is characterised by the implementation of electronic and information technologies for the automation of manufacturing processes, allowing the gradual replacement of human labour (Bahrin aet al, 2016). In the current economic system, characterised by demand and supply on a global scale, we are witnessing the 4th industrial revolution, in other words, the integration of Industry 4.0 into production systems, business models and everyday life.

Industry 4.0 is based on horizontal and vertical integration of production systems, driven by real-time data exchange and flexible manufacturing to enable customised production (Thoben aet al., (2017). Geissbauer aet al., (2016) add that Industry 4.0 focuses on the digitisation of all physical assets and integration into digital ecosystems with value chain partners. This new industrial revolution aims to interconnect products, production systems, supply chain and consumers thru the digitisation, automation, and artificial intelligence of production processes (Madsen, 2019; Pessôa & Becker, 2020). Industry 4.0 or the Fourth Industrial Revolution integrates and is driven by innovative Information Technologies (IT) and disruptive methods at all levels, which enable the total interconnection, in other words, of processes, machines and people. Authors such as Đuričin and Herceg (2018) Salkin, aet al. (2018), Sjøbakk (2018), Xiaopu aet al. (2016), state that Industry 4.0 is caused by digital technologies that have a disruptive effect on companies' business models and the way they act and create customer value. These authors consider that there are nine technological trends that integrate the primary components of Industry 4.0: The Industrial Internet of Things (IIoT) and Internet of Things (IoT); Big Data analytics; Cloud computing; Simulation; Augmented reality; Autonomous robots; Additive manufacturing; Cybersecurity; Horizontal and vertical integration of information systems.

ICT evolves at an extremely fast pace and, in the business context, where digitalisation and automation constitute the present and the future, companies to be competitive must be able to adapt quickly to evolution.

According to Antunes aet al. (2019), the interference of IoT in Industry 4.0 is extreme, with a constant interconnection of the digital and physical domains. The authors also refer that, real-time information, Big Data, the connexion between people, objects, and systems, will lead to the individualisation of products and services on a large scale and, therefore, to a control transformation of the value chain. Antunes aet al. (2019) argue that IoT also applies in Industry: intelligent IoT systems enable the fast production of new products, the dynamic response to demand and the real-time optimisation of production networks and supply chain, thru the management of networked machines using sensors and integrated control systems.

Organisational and Competitiveness Impacts

Digitalisation, technological advances, and digital information technologies bring and integrate great opportunities and advantages (Kraus aet al., 2019) being considered a resource that can be a sustainable competitive advantage (Porter, 2003; Zen & Faizin, 2021). However, to be able to benefit from these advantages and opportunities, companies will have to integrate significant changes in their activities, structure, and organisation (Kraus aet al., 2019; Quarato aet al. (2020)). Digital disruption is transforming the way companies organise themselves to create and deliver value, affecting the entire value chain from product design, production, logistics, marketing and sales, and after-sales service (Karlsson & Rundcrantz, 2017).

According to Porter (2003), the value chain of a company integrates the set of activities thru which a product or service is created and delivered to customers. Since each activity girds the creation, processing and communication of information, digital technologies execute and involve disruptive changes in the value chain (Porter & Heppelmann, 2015). Authors such as Rüßmann aet al. (2015), and Strange and Zucchella (2017), consider that the current digital transformation will allow production within value chains to become more efficient and flexible, enabling the customisation of production for niche markets. Rebelo (2020) adds that the existence of products or machines equipped with sensors allows access to data in real-time, enabling, for example, preventive maintenance and the coordination and synchronisation of information and product flows. In addition to the information produced by these sensors, the fact that companies can access data from a diversity of sources (such as search tools and social networks) with low storage costs (thru the existence of a cloud) support them, amongst other things, in decision making, purchasing and sales, production planning, and data management (Mittal aet al., 2018; Rüßmann aet al., 2015).

Also, robotisation and automation, arising from technological advances, are a source of organisational change, efficiency, and competitiveness. In addition to progressively cheaper, more functional, and higher performance, becoming increasingly accessible to SMEs (Rüßmann aet al., 2015), robotisation allows for the increasing quality control and efficiency of production, transport, and storage (Rebelo, 2020).

Other aspects with impacts on the activities of companies are the Internet and digital platforms that have enabled access to information and data in real-time, as well as simplifying communication and connectivity between buyers and suppliers at lower costs (Porter, 2003). Several factors affect the competitiveness of the company, but according to Malik (2019), two of these key factors are the value of customer opinions and product singularity.

The adoption of these platforms changes marketing and sales strategies, eliminate limitations and physical distances and allows companies to reach, more quickly and effectively, a wider range of customers (Nejadirani, aet al. 2011). On the other hand, the dematerialisation of production and distribution channels, arising from technological advances, allows firms to reduce their transaction costs in foreign markets (Coviello aet al. 2017). Moreover, by using the Internet as a sales channel, companies improve the efficiency of their production and distribution processes, enabling the automation of related internal production processes, and providing real-time information on inventories, production, sales, and distribution (Dethine aet al., 2020). Arising from artificial intelligence, IoT and digitalisation, the capacity to collect, process and disseminate data and information is immensely higher so they increase the capacity and speed of knowledge of markets and facilitate the customisation of products, being able to respond more easily and effectively to demand. In parallel, the emergence of communities and social networks enables companies to adapt their offer to local markets and, in some cases, diversify their activities (Hervé aet al., 2020).

With all the developments and innovations in terms of technologies and digitalisation, the value chains of companies are becoming global, focussing on core competencies and activities with the highest added value and subcontracting non-core activities. Therefore, according to Rebelo (2020), companies must know their capabilities to effectively manage their business models or value chains and remain competitive.

Internationalisation and Digitalisation

In a borderless world with extremely competitive and dynamic markets, digitalisation is becoming a dominant paradigm. Gartner (2018) defines digitalisation in a more business-oriented perspective, which means that they consider digitalisation to be the process of transformation into a digital business, by applying digital technologies to change a business model and provide new opportunities to generate value. The implementation of digital technologies is adapting new business models, and this is one of the most important challenges to be addressed to pursue international strategies (Chen & Kim, 2018; Watson aet al., 2018), which gives rise to great opportunities, but also threats (Bankewitz aet al., 2016).

Digitalisation has led to the dematerialisation of borders, bringing producers closer to consumers, reducing costs, influencing, according to Coviello aet al. (2017), the internationalisation process of companies in terms of access to resources, acquisition of knowledge and skills, as well as in terms of knowledge and learning of markets.

Traditional internationalisation models indicate that companies first develop their domestic markets, subsequently adopting internationalisation processes in stages, slowly and gradually. In this sequential and gradual model (Uppsala model), defined by Johanson and Vahlne (1977), the internationalisation process is an evolutionary process that begins with markets that are closer in geographical and cultural terms. Subsequently, the company progressively increases its involvement with the international market as its knowledge of the market increases. This progressive knowledge is the result of the experience acquired and helps the company to identify the risks and opportunities arising from a specific market (Johanson & Vahlnc, 1977). Despite being widely studied and coexisting for several decades, this model also suffered some criticism that eventually determined its revision and the emergence of alternative internationalisation models. One of these models is the theory of networks presented by Johanson and Mattsson (1988). This model has as a fundamental element the fact that markets are seen as networks of relationships between companies. For the authors, a network may consist of small groups of companies and even entire markets, which will be described as complex networks of relationships. This theory ends up being the development of the Uppsala model in that it considers that the experience acquired, and the market commitment now depends on the potential of the partners and their business network (Johanson & Vahlne, 2009). In a recent study, Child aet al., (2022) indicate that networking can help the internationalisation of SMEs and promote sales in foreign markets in several ways. Networks play a very important role for companies because they do not always have the essential resources to develop their internationalisation process. Establishing relationships with other business entities (manufacturers, deliverers, suppliers, resellers, government, customers, competitors, amongst others), these relationships will help companies overcome insufficient resources and protect their position in the market (Dominguinhos, 2000).

The growing increase and appearance of companies adopting fast internationalisation processes and the realisation that the traditional model of internationalisation by stages did not explain these new processes led to the development of new internationalisation models and perspectives. Altho there are several terms to refer to the phenomenon (such as Born Global, International New Ventures, Born International, Early International), the fundamental characteristic of these businesses is that they adopt rapid internationalisation processes (Zen & Faizin, 2021). The literature on the internationalisation of SMEs indicates that, due to the lack and limitations at the level of financial and human resources, exporting has been the most used mode by SMEs (Hollensen, 2011) and with the development of information technology, digital platforms and the internet, other internationalisation strategies relevant to SMEs have emerged (Hervé aet al. 2020).

The internet is considered a valuable tool for companies in that it can provide them with a faster global presence (that is, it allows the entry into several markets simultaneously), allows a quick connexion with all business partners, allows acquisition, processing, and transmission of information from new markets (Pereira & Pinto, 2017) and is a means to introduce products in global markets, facilitating consumers access to product information, prices, and advantages. Therefore, the internet, thru e-commerce and e-marketing, has streamlined the method of distribution and marketing for businesses by providing a diverse source of products for businesses and consumers (Caetora aet al., 2011). In recent decades, online sales and e-commerce have changed the way businesses are managed (Tiessen aet al., 2001). Indeed, modern technologies and the internet have allowed e-commerce to quickly become a new model of internationalisation and competitive strategy, particularly for SMEs (Gabrielsson & Gabrielsson, 2011). In the same sense, Zen and Faizin (2021) refer that the Internet is a facilitator and accelerator of SMEs exports and that digital transformation and digitalisation allow companies to develop digital business models that promote faster internationalisation, allowing to increase export performance and increase the value of companies. As Lee aet al., (2019), digital technologies and digitalisation reduce the barriers to access to information and communication across borders so leveraging digitalisation allows accelerating the internationalisation process of a company. Therefore, in this large virtual global market, companies must understand the advantages and relevance of digitisation and digital technologies because, and as Neubert (2018) states, the sooner a company realises the advantages of digitisation and digital technologies the faster it can improve its decision-making processes and accelerate the internationalisation process and improve the company's export performance (Zen & Faizin, 2021).

METHODOLOGY

This work aims to explain what Industry 4.0 consists of, how it is implemented and what implications it will have on SMEs in Portugal. Particularly, it will be sought to understand how Industry 4.0, the digital systems and technologies and aggregated methods will influence the operation of SMEs understudy, in other words, what impacts will be observed at an organisational level and on the competitiveness of SMEs and, finally, what the repercussions on their internationalisation process will be.

To pursue the objectives of the study, exploratory and descriptive research was designed, applying a comparative case study because it attempts to analyse and study the context and characteristics of specific phenomena. The qualitative method is exploratory in nature and helps us to understand, in a more detailed way, a subject or a problem. With this qualitative approach, it is intended to explore how Industry 4.0, its aggregated technologies, and methods, will influence the organisation, competitiveness, and internationalisation process of the SMEs under study. The qualitative approach allows the development of an in-depth analysis of the research and theories used (Bergamaschi aet al., 2021). The case study method can be described as an intensive and systematic investigation of an individual, group, community, or some other unit, in which the researcher examines in-depth data concerning several variables (Calanzaro, 1988), offers greater flexibility, allows an in-depth analysis of a complex research problem, and allows a comparison between different cases (Yin, 2015). When the context is complex and when it intersects a complex set of variables (Meirinhos & Osório, 2016), the case study is a powerful strategy that should collect and cross-cheque information from multiple sources, triangulate the information to answer the guiding questions and, finally, critically filter the problematic studied with the theoretical conceptual elements that underpin the study (Meirinhos & Osório, 2016). The disadvantages that are present in this methodological approach relate to the existing problems of objectivity, which may result from the researcher's lack of experience, lack of knowledge and lack of sensitivity (Sousa & Baptista, 2011). Another limitation is associated with the fact that the results are restricted to the context of analysis and cannot be generalised. To overcome these disadvantages the primary data collected were supplemented with secondary data collected from the companies' websites and reports, as well as other sources of information to proceed to the triangulation of these data with the primary data, analysing the results and their consistency and strengthening the knowledge of each company.

Given the objectives of the study, two SMEs were selected from the municipality of Santa Maria da Feira (Portugal), from two different activity sectors, one from the industrial sector and the other from the trade sector, which is internationalised. The cases under study were selected by convenience and not randomly, which is ultimately recommended by Eisenhardt (1989) to the extent that, for the author, the choice of cases without an objective is neither necessary nor desirable.

Data Collection

The Data were collected between May and July 2021, thru primary sources, semi-structured interviews with those responsible for internationalisation processes, and secondary data sources, namely websites and reports. According to Adhabi aet al., (2017), interviewing can be seen as a form of enquiry, where the researcher seeks to find out more about a subject dominated by the respondent, but it can also be understood as an interactive process where the interviewer asks questions to seek specific and relevant information to validate their research hypotheses. Adhabi aet al., (2017) argue that interviews, mainly semi-structured and unstructured, allow the researcher to intervene when necessary and ensure that the subject understands the topic or question under analysis. Furthermore, interviewers can use their interpersonal skills to explore significant issues raised by the participant, which are central to comprehensive data collection (Adhabi aet al, 2017).

The interviews were conducted face-to-face at the companies, from 25 September to 5 October and lasted approximately 30 minutes. The interview script included open questions, intending to allow the interviewees to develop aspects that they considered relevant to enable a clear understanding of the internationalisation processes and the relevance of Industry 4.0/ Digitisation. The interview script was divided into three parts. The first part included questions aimed at getting to know the fundamental characteristics of the company. The second part included a set of questions aimed at finding out about the internalisation process, reasons, barriers, strategies, markets, and methods, as well as suggestions that they would give to SMEs that now intend to start their internationalisation process. The third group of questions was used to find out about the implementation process of Industry 4.0 and digitalisation in the company, with the aim of understanding which Industry 4.0 tools have been implemented, the objectives to be achieved with this implementation, the advantages arising from it, as well as the impacts at organisational and competitiveness level and their relevance to the company's internationalisation process.

CASE STUDY

In this section, we start by briefly describing the cases under study. Subsequently, we present the analysis and discussion of the results, which attempt to highlight the convergences and divergences between the cases but, above all, to bring the results into line with the literature and theory reviewed.

Characterisation and Presentation of the Companies' Understudy

The "company A" was established in 1976, in Rio Meão (Portugal), and is dedicated to the activity of design and manufacture of high-quality electric pumps and systems for a vast variety of water applications. It positions itself amongst the best companies worldwide in the electric pumps and water systems market, thru its production quality, hydraulic efficiency, and reliability. The company offers its customers a complete range of products and solutions for both clean and wastewater, as well as high production flexibility with the ability to respond quickly to all its customers' needs. All its products are tested for electrical and hydraulic functioning and given its policy of monitoring and proximity to customers and partners, another advantage of the company is its ability to renew and adapt products following the standards and rules of some countries. It is in the process of expanding its operations worldwide to better meet customers' needs and demands. The company's growth and regular investments have allowed it to implement innovative research and development programmes, and automated processes have been developed to produce its components, so that they can be produced faster, more efficiently and at lower cost, and so that it can compete and expand its markets.

The "company B", founded in 1993, is based in S. Paio de Oleiros (Portugal), and its main activity is the recovery of textile waste. Aware of the growing use and consumption of energy, namely fossil resources, as well as the creation of waste, its founder realised the need for energy, waste and environmental management that would authenticate maximum efficiency with minimum costs, waste, and environmental impact. To reduce environmental impact, the company treats waste resulting from the manufacturing processes of various companies, particularly cotton waste, thus avoiding sending it to landfill sites. Thru this treatment it is possible to recycle textile waste, making it possible to reuse it as raw material, reducing the need for incineration and landfill. The company has been growing in the European Union markets. Keeping customers satisfied and offering a quality product is a fundamental policy to guaranty a relationship of trust and loyalty between the company and its customers.
Table 1 summarises other information on the companies under study.

Table 1. Companies' profile. Source: Own preparation based on the results of the interviews

Company	CASE A	CASE B
CAE and line of business	28130 – Manufacture of other pumps and compressors	38322 – Non-metallic waste recovery
Type of products/services sold	Electric pumps for clean and wastewater / electric motors	Textile waste
Genesis of the company	1976	1993
No. of employees: Before internationalisation Post internationalisation	16 44	16 16
Year of internationalisation	1997	1993
Responsible for internationalisation	Pedro Rodrigues	Joaquim José Carvalho Rola
Export volume on sales	55% (percentage referring to the year 2020)	79% (percentage referring to the year 2020)
Markets	Spain; Italy; France; Germany; England; Morocco; Indonesia; Cyprus; Australia; Ukraine; Luxembourg; Belgium; Poland; Netherlands; Slovakia; Austria; Jordan; Algeria.	Spain France

ANALYSIS AND DISCUSSION OF RESULTS

Characterisation of Internationalisation Processes

Table 2 presents, succinctly and comparatively, the internationalisation processes of the companies under study. The analysis and comparison of the two internationalisation processes indicate similar but also distinct aspects between them. This is, according to Yin (2015), one of the advantages of a multicase analysis compared to a single case study since it allows us to identify similar aspects, but also contrasting ones.

Table 2. Internationalisation process. Source: Own preparation based on the results of the interviews

Variables	CASE A	CASE B
Start	It took place in 1997, 21 years after the start of the activity and began with visits to fairs.	It occurred at the same time as its foundation thru a contact in Spain.
Factors/ motivations	- the need for growth that the company presented - the ease of geographical and cultural access - differentiation from the competition - access to technologies and the development of innovative capacity; - client follow-up; - the achievement of cost advantages; - the potential size of the external market.	- the inexistence of a domestic market in the textile recycling business, so that all the potential market for the company was the international market; - the easiness of geographic and cultural access - the clients follow up; the potential dimension of the external market; - the potential dimension of the external market.
Barriers	- strong competition in the destination market - lack of knowledge and experience facing external markets - lack of competitive advantage.	
Input modes	- Indirect export because the company had no experience in transporting products and, therefore, this method was the fastest and with the fewest risks. - E-commerce because it has a low investment and maintenance cost, allows access to a large public, is closer to the consumer and is also a very easy and safe way of doing business.	- Indirect exporting because some clients prefer to be in charge of the transport; - direct exporting because of the company's experience in transport, the reduction in intermediary costs, the internationalisation of the brand, a better profit margin in the medium and long term, greater autonomy over international negotiation strategies, greater control over internal planning, greater control over orders and sales and the possibility of having direct contact with customers.
First market	- Italy, due to an existing contact in the country, who liaised with the Italian company Matra and facilitated the entry into this country.	- Spain, thru an existing contact in that country.
Market in which it has faced the greatest difficulties	- Indonesia due to the existing bureaucracies in the country, causing difficulties both production and transport.	- Does not have. In the two markets where it is present (Spain and France), both were easy to enter due to the existence of contacts.
Strategy for the future	Continue to grow gradually, extending our presence to more countries, especially European countries, because they are easier to access, simpler to work with, less bureaucratic both in terms of production and transport, and present fewer legal problems compared to countries outside Europe.	Due to the limitations of the installed production capacity, it does not intend to expand its internationalisation markets.
Suggestions to other SMEs	- start the process in European countries because of less bureaucracy and legal problems and less cultural differences compared to countries outside Europe; - qualification and international experience of human resources; - structure and organisation of the company.	- start their process in the European market, in countries like Spain and France, because these are countries where entry is easier, and the barriers are lower compared to countries outside Europe.

Considering the literature on internationalisation models or theories, we can indicate that the case of company A is characterised by being a business that, after being well implemented in the domestic market, gradually evolves, and expands its activity to the international market. In this respect, it fits into the Uppsala model. Therefore, the company carried out its internationalisation process in a phased and incremental manner, as indicated by the Uppsala model. However, we found that the company did not follow all the sequential phases of the Johanson and Valhne (1977) internationalisation model because, despite its greater involvement and market commitment, exports remained in the entry mode, and did not evolve to the following phases.

In the case of company B, which started its internationalisation process in the same year as its foundation, it holds, in this sense, one of the main characteristics of a "Born Global", that is, the beginning of international activity occurs very quickly in its evolution (Rennie, 1993; Cavusgil aet al., 2008) and does not follow the stage model (Cannone & Ughetto, 2014).

Both cases mention that the process began due to the existence of a contact in the first market to be approached (Italy in company A and Spain in company B), which indicates the applicability of the revised Uppsala internationalisation process model (Johanson & Vahlne, 2009) and highlights the fundamental role of networks for international activity (Johanson and Mattsson, 1988).

As for the strategies and modes of entry into foreign markets, it should be remembered that this decision is one of the most complex in internationalisation (Morschett aet al., 2010). In the two cases under study, the most used strategy is exporting, confirming the perspective of Hollensen (2011) and Fernandes (2014) who indicate that this mode of entry is the simplest and most used by SMEs. In fact, both companies use indirect exporting, altho for distinct reasons. In case A, due to lack of experience in transporting the products and as a way to reduce risks (Hollensen, 2011). Company B indicates that this mode is used at the request of customers. Additionally, company A also uses e-commerce, enabled by ICT and digitalisation, mainly due to its advantages (Loane & Bell., 2006; Zen & Faizin, 2021; Hervé aet al., 2020). Online sales to external markets have become an important competitive strategy, particularly for SMEs (Gabrielsson & Gabrielsson, 2011; Coviello aet al., 2017; Hervé aet al., 2020). Company B also uses direct exporting as an entry mode because, as indicated by the literature (Hollensen, 2011; Simões, 2013) it allows greater access to the local market/destination and greater control of activities.

Regarding the motivations and barriers associated with the internationalisation process, we found quite similar aspects in both cases (Table 3), which are in line with the theory (Viana & Hortinha, 2005; Fletcher, 2001; Hollensen, 2011; Simões, 2013).

Concerning prospects, case B does not have the capacity to expand production and therefore does not intend to increase its international presence. Company A indicates that the objective is to continue with the current strategy: to gradually expand its presence in more countries and to European countries due to their greater proximity and ease of access.

Impacts of Industry 4.0 and Digitalisation

Table 3 summarises the responses of the companies' understudy to the questions aimed at understanding the impacts and relevance of Industry 4.0 and digitalisation in companies and their internationalisation processes. Case A presents a more advanced level of integration of Industry 4.0 components than case B, which may be associated with the company's own activity, industrial sector - "manufacturing of other pumps and compressors". Despite this, both pointed to common objectives to be achieved with the implementation of Industry 4.0, namely the improvement in service quality, improvement in working conditions, cost reduction, improved management efficiency, greater competitiveness, and support/ facilitation of the internationalisation process. Company A adds to this list increased productivity and improved product quality.

Both companies mentioned the implementation of Industry 4. 0/digitalisation in the company was truly relevant because it allowed production times and costs to be reduced, productivity and competitiveness to increase and information to be shared more easily. Case A also added that the implementation of Industry 4. 0/digitalisation allowed the company to increase production quality and safety. In this sense, we can mention that the companies' responses are in line with the perspective of several researchers such as Antunes aet al., (2019); Parida aet al. (2015), De Marchi aet al. (2018), Chen & Kim (2018), Watson aet al. (2018), Siachou aet al., (2020), Kongmanila and Takahashi (2009).

Table 3. Internationalisation process. Source: Own preparation based on the results of the interviews

<table>
<tr><th></th><th>CASE A</th><th>CASE B</th></tr>
<tr><td rowspan="2">Equipment/resources installed</td><td colspan="2">- production support machines;
- Internet;
- Cyber security systems;
- use of the Cloud.</td></tr>
<tr><td>- robots and software with simulators;
- the intranet and the industrial internet;
- augmented reality software.</td><td></td></tr>
<tr><td rowspan="2">Objectives with the implementation of Industry 4.0/digitalisation</td><td colspan="2">- the reduction of costs and the improvement of management efficiency;
- the improvement of service quality
- the improvement in working conditions;
- increasing competitiveness;
- support/facilitation of the internationalisation process;</td></tr>
<tr><td>- increased productivity;
- the improvement of product quality.</td><td>- reduction of production time;
- reduction and rationalisation of space.</td></tr>
<tr><td rowspan="2">Benefits gained by implementing Industry 4.0/digitisation</td><td colspan="2">- reduced production time
- greater flexibility of the production line;
- increased productivity;
- cost reduction and increased profitability;
- increased competitiveness;
- increased information sharing;</td></tr>
<tr><td>- increased production quality;
- increased safety.</td><td></td></tr>
<tr><td rowspan="2">Impacts at organisational level</td><td colspan="2">- positive impacts on the organisational level and value chain, more specifically on production/operations, marketing and sales, after-sales service, and the use of the Internet as a sales channel.</td></tr>
<tr><td>- positive impacts on the management of support activities such as infrastructure, human resources management, technology development, procurement/acquisition, and process automation.</td><td>- positive impacts on outbound logistics, mainly on customer service, product collection, storage, and distribution.</td></tr>
<tr><td rowspan="2">Impact on competitiveness</td><td colspan="2">- Online sales and e-commerce are essential to a company's competitive strategy and Industry 4.0/digitisation has boosted them, so the value of customer opinions and the uniqueness of the product have been positively affected by its implementation.</td></tr>
<tr><td>- allowed the company to increase its production;
- to have a faster response to orders.</td><td>- it allowed them to refine and improve their communication, exchange information and simplified international transactions;
- it was possible to create greater proximity with customers.</td></tr>
<tr><td rowspan="2">Impact on internationalisation</td><td colspan="2">- Industry 4.0/digitalisation is fundamental for the continuity of export/internationalisation of companies because:
- it allows and facilitates obtaining information, both at international market and partner level;
- it allows the reduction of risks throughout the process, the reduction of costs and facilitates the increase of financing capacity;
- digitalisation has become fundamental in maintaining and attracting new clients.</td></tr>
<tr><td>- facilitated access to new markets because it lowered several barriers that existed before the implementation of Industry 4.0;
- increased responsiveness and allowed the company to broaden horizons and enter markets considered more difficult and with more risks.</td><td>- digitalisation has substantially improved communication and contact with the customer;
- improved the company's ability to respond to orders.</td></tr>
</table>

Regarding impacts at the organisational level, both companies indicate positive impacts at the organisational level in the company. Case A mentioned that Industry 4.0 brought impacts in the areas of human resources, infrastructures, and technological development, purchasing and process automation. Case B indicated positive impacts at the customer service level, namely logistics, storage, and distribution. Additionally, both companies indicated positive impacts in production, marketing operations and sales.

At the level of company competitiveness, both companies report that the value of customer opinions and product uniqueness were positively affected by the implementation of industry 4. 0/digitalisation in the company, as stated by Malik (2019). Both case A and case B state that online sales and e-commerce are indispensable in a company's competitive strategy (as stated by Gabrielsson & Gabrielsson (2011); Cateora (2011); Hossain aet al., 2021; Zen & Faizin, 2021). Case B also indicated that the implementation of Industry 4.0 allowed the company to improve its communication, facilitated the exchange of information and simplified international transactions, also, it was possible to create greater proximity with customers.

As for the relevance of Industry 4.0/ digitalisation in the internationalisation process, given that the companies consider that information technology and digitalisation have facilitated the process of obtaining information, both at the level of the international market and at the level of partners, in reducing risks and costs, both companies state that it is currently essential to have Industry 4.0/ digitalisation implemented in the company in order to give continuity to exports and to carry out their internationalisation process, as well as to promote the expansion of the company in the international market. Company A further highlights the increase in productivity, whilst Company B also highlighted the possibility of reducing production time and space. In this sense, and as indicated by Sinkovics aet al. (2013), Falahat aet al. (2020), Coviello aet al. (2017), Chen and Kim (2018), Watson aet al. (2018), Lee aet al. (2019), we can state that also in the cases under study, Industry 4.0 and digitalisation prove to be fundamental for the continuity of exports and internationalisation of the companies. Additionally, both companies refer that the lack of access to technology at the time of their internationalisation (1993, in case A, and 1997, in case B) made the process limited and without much progression, both at the organisational and competitive levels. They also add that it is currently unthinkable, and even unfeasible, to start an internationalisation process without implementing Industry 4. 0/digitalisation in a company.

Finally, when asked whether digitalisation has changed the way of acting/performing the internationalisation process, the answers were positive, but the effects are not all coincident. Case B mentions that digitalisation has substantially improved communication as mentioned by Coviello aet al. (2017) and contact with customers (Hannibal & Knight, 2018; Strange & Zucchella, 2017), has improved the responsiveness of the company regarding orders (Jean aet al.,2010; Sinkovics aet al., 2013). As common effects, both companies mentioned that it was possible to increase the number of customers and markets (as indicated by Watson aet al., 2018; Mathews aet al., 2016; Bell & Loane, 2010; Foscht aet al., 2006). Company A also mentioned that Industry 4.0 and digitalisation allowed reducing several barriers that existed before the implementation of Industry 4.0 in the company, increased responsiveness (Rebelo, 2020), facilitated the entry into markets considered more difficult and with more risks, such as Eastern markets and, finally, digitalisation became fundamental in maintaining and acquiring new customers as noted by Falahat aet al. (2020).

It can thus be concluded that for the cases under study, the implementation of Industry 4.0 and digitalisation has brought positive impacts on the organisation, competitiveness, and internationalisation of companies, as noted by Neubert (2018) and Strange and Zucchella (2017).

FUTURE RESEARCH DIRECTIONS

As this research followed a qualitative approach, based on a case study, it has some limitations. Firstly, the small sample size and the limitations regarding the possibility of generalising the results, whose conclusions are limited to the context of analysis. In this sense, it would be important to extend the sample to a larger number of companies and a diversified analysis to different activity sectors. The literature also does not present international comparative studies, so this could be another topic of interest, in other words, future research could encompass companies from different countries to enable an international analysis and understand whether diverse cultural, geographical contexts and specific market factors influence strategic decisions regarding the implementation of Industry 4.0 and digitalisation. On the other hand, this study did not analyse the costs and risks inherent in the implementation of industry 4.0 and digitalisation in the process of internationalisation and international management, being another area of research to be explored.

CONCLUSION

Companies operate in global and borderless markets. With the implementation of industry 4.0, ICT and digitalisation, distance, geography, and other barriers assume less relevance in the international market entry decision. The literature review regarding industry 4.0/digitalisation indicates that industry 4.0/ digitalisation is a key competitive tool. The investment made in Industry 4.0 must be focussed on concrete objectives that enable the company to integrate its resources and business processes and increase its flexibility and agility, optimising the resources allocated to the control and management activity, providing the organisation with a complete focus on the development of its business and core competencies.

In a global market where they are subject to increasingly strong competition, SMEs feel the weight of their small size and reduced negotiating power. In this context, they must invest in their strengths and competitive advantages, namely in the capacity to innovate and adapt quickly to different contexts. The implementation of Industry 4.0, not only for its constant evolution but also for its growing competitiveness, assumes a fundamental relevance in companies, allowing SMEs to become more efficient at a productive and competitive level, facilitating their internationalisation process.

The results of this case study, applied to two Portuguese SMEs from different activity sectors, indicate that SMEs that started their internationalisation process, already in the nineties of the twentieth century, managed to do it without Industry 4.0, thru the contact network they obtained in the country of entry. With the technological evolutions and innovations that occurred, companies felt the extreme need to implement Industry 4.0, not only for their growth but also for their survival in the international market. The implementation of industry 4.0/digitalisation in these SMEs caused impacts at the organisational level and at the level of competitiveness, the value chain was positively affected by the implementation of industry 4.0 and, consequently, companies were better positioned to obtain a competitive advantage.

This work presents theoretical contributions to a theme that is still little explored. The research sought to analyse and understand the relevance and impacts of Industry 4.0 and digitalisation on the competitiveness and internationalisation process of Portuguese SMEs. In this sense, it highlights and relates several issues and aspects between two autonomous and complex but strategic research fields for SMEs. In practical terms, the results reveal that e-commerce and other platforms made possible by ICT and digitalisation are useful to create new sales channels for companies' products at the international level. On the other hand, the results of this study highlight the importance of implementing Industry 4.0 and digitalisation in Portuguese SMEs, being fundamental for them to understand its advantages and benefits and be able to enhance the process and speed of internationalisation.

REFERENCES

Antunes, J., Pinto, A., Reis, P., & Henriques, C. (2019). A Indústria 4.0 e o seu Impacto no Tecido Económico Empresarial Português: O Caso da Região Dão-Lafões [Industry 4.0 and its Impact on the Portuguese Business Economic Fabric: The Case of the Dão-Lafões Region]. *Fronteiras: Journal of Social, Technological and Environmental Science, 8*, 263-291. https://repositorio.ipv.pt/handle/10400.19/5654

Ardito, L., Simon, R., Vito, A., & Bernardo, B. (2021). The Duality of Digital and Environmental Orientations in the Context of SMEs: Implications for Innovation Performance. *Journal of Business Research, 123*, 44–56. doi:10.1016/j.jbusres.2020.09.022

Bankewitz, M., Åberg, C., & Teuchert, C. (2016). Digitalization and boards of directors: A new era of corporate governance? *Business and Management Research, 5*(2), 58–69. doi:10.5430/bmr.v5n2p58

Catanzaro, M. (1988). Using Qualitative Analytical Techniques. In N. Woods & M. Catanzaro (Eds.), *Nursing Research: Theory and Practice* (pp. 437–456). Mosby Incorporated.

Cateora, P., Gilly, M., & Graham, J. (2011). *International Marketing* (15th ed.). McGraw-Hill Companies, Inc.

Cavusgil, T., Knight, G., & Riesenberger, R. (2008). *International Business: Strategy, Management, and the New Realities*. Pearson Prentice Hall.

Chen, C., & Kim, C. (Eds.). (2018). *The digitization of business in China: Exploring the transformation from manufacturing to a digital service hub*. Springer International Publishing. doi:10.1007/978-3-319-79048-0

Child, J., Karmowska, J., & Shenkar, O. (2022). The role of context in SME internationalization – A review. *Journal of World Business, 57*(1), 101267. doi:10.1016/j.jwb.2021.101267

De Marchi, V., Di Maria, E., & Gereffi, G. (2018). *Local clusters in global value chain: Linking actors and territories through manufacturing and innovation*. Routledge. doi:10.4324/9781315182049

Denicolai, S., Zucchella, A., & Magnani, G. (2021). Internationalization, digitalization, and sustainability: Are SMEs ready? A survey on synergies and substituting effects among growth paths. *Technological Forecasting and Social Change, 166*(12), 120650. doi:10.1016/j.techfore.2021.120650

Dethine, B., Enjolras, M., & Monticolo, D. (2020). Digitalization and SME's export Management: Impacts on Resources and Capabilities. *Technology Innovation Management Review, 10*(4), 18–34. doi:10.22215/timreview/1344

Dominguinhos, P. (2000). *A Internacionalização das Empresas Portuguesas: Balanço da Evidência Empírica* [The Internationalization of Portuguese Companies: Balance of Empirical Evidence]. ESCE-IPS.

Duričin, D., & Herceg, I. V. (2018). Industry 4.0 and Paradigm Change in Economics and Business Management. In AMP 2018, LNME (pp. 37–56). Academic Press.

Eisenhardt, K. (1989). Building theories from case study research. *Academy of Management Review*, *14*(4), 532–550. doi:10.2307/258557

European Commission. (2021). *Annual Report on European SMEs 2020/2021*. https://ec.europa.eu/docsroom/documents/46062

Fletcher, R. (2001). A holistic approach to internationalisation. *International Business Review*, *10*(1), 25–49. doi:10.1016/S0969-5931(00)00039-1

Gabrielsson, M., & Gabrielsson, P. (2011). Internet-based sales channel strategies of born global firms. *International Business Review*, *20*(1), 88–99. doi:10.1016/j.ibusrev.2010.05.001

Hervé, A., Schmitt, C., & Baldegger, R. (2020). Internationalization and Digitalization: Applying digital technologies to the internationalization process of small and medium sized enterprises. *Technology Innovation Management Review*, *10*(7), 28–40. doi:10.22215/timreview/1373

Hollensen, S. (2011). *Global Marketing: A Decision Oriented Approach* (5th ed.). Pearson Education Limited.

Hossain, M., Azam, M. S., & Quaddus, M. (2021). Small firm entry to e-marketplace for market expansion and internationalization: A theoretical perspective. *Journal of International Entrepreneurship*, *19*(4), 560–590. doi:10.100710843-021-00297-5

Johanson, J., & Mattson, L. (1988). *Internationalization in Industrial Systems: A Network Aproach*. Croom Helm.

Johanson, J., & Vahlne, J. (1977). The Internationalization Process of the Firm - A model of knowledge development and increasing foreign market commitment. *Journal of International Business Studies*, *8*(1), 23–32. doi:10.1057/palgrave.jibs.8490676

Johanson, J., & Vahlne, J. (2009). The Uppsala Internationalization Process Model Revisited: From liability of foreignness to liability of outsidership. *Journal of International Business Studies*, *40*(9), 1411–1431. doi:10.1057/jibs.2009.24

Karlsson, E., & Rundcrantz, S. (2017). *'Surfing on the wave of Digitalisation' - Digital technology facilitating an SME's internationalisation process in the Fashion Industry - A single case study of Trendy AB* (Master Degree). University of Gothenburg. Retrieved from http://hdl.handle.net/2077/53149

Kongmanila, X., & Takahashi, Y. (2009). Innovation, export performance and profitability of Lao garment exporters. *International Journal of Economics and Management*, *3*(2), 225–236. http://www.ijem.upm.edu.my/vol3no2/bab01.pdf

Lee, Y., Falahat, M., & Sia, B. (2019). Impact of Digitalization on the Speed of Internationalization. *International Business Research*, *12*(4), 1. doi:10.5539/ibr.v12n4p1

Loane, S., & Bell, J. (2006). Rapid internationalisation among entrepreneurial firms in Australia, Canada, Ireland and New Zealand: An extension to the network approach. *International Marketing Review, 23*(5), 467–485. doi:10.1108/02651330610703409

Madsen, D. O. (2019). The Emergence and Rise of Industry 4.0 Viewed through the Lens of Management Fashion Theory. *Administrative Sciences, 9*(3), 71. doi:10.3390/admsci9030071

Meirinhos, M., & Osório, A. (2016). O estudo de caso como estratégia de investigação em educação. *Eduser - Revista De Educação, 2*(2). https://www.eduser.ipb.pt/index.php/eduser/article/view/24

Mittal, S., Romero, D., & Wuest, T. (2018). Towards a smart manufacturing toolkit for SMEs. In *IFIP International Conference on Product Lifecycle Management* (pp. 476–487). Springer. doi:10.1007/978-3-030-01614-2_44

Nejadirani, F., Behravesh, M., & Rasouli, R. (2011). Developing countries and electronic commerce the case of SMEs. *World Applied Sciences Journal, 15*(5), 756–764. https://idosi.org/wasj/wasj15(5)11/22.pdf

Neubert, M. (2018). The Impact of Digitalization on the Speed of Internationalization of Lean Global Startups. *Technology Innovation Management Review, 8*(5), 44–54. doi:10.22215/timreview/1158

OECD. (2019). *OECD SME and Entrepreneurship Outlook 2019*. OECD Publishing. https://www.oecd-ilibrary.org/industry-and-services/oecd-sme-and-entrepreneurship-outlook-2019_34907e9c-en

Parida, V., Westerberg, M. & Frishammar, J. (2012). Inbound open innovation activities in high-tech SMEs: the impact on innovation performance. *Journal of Small Business Management, 50*(2), 283-309. doi:10.1111/j.1540-627X.2012.00354.x

Pereira, P., & Becker, J. (2020). Smart design engineering: A literature review of the impact of the 4th industrial revolution on product design and development. *Research in Engineering Design, 31*(2), 175–195. doi:10.100700163-020-00330-z

Pereira, R., & Pinto, M. (2018). Estratégias de internacionalização – Estudo de caso para empresas portuguesas. *Empresas E Empreendedores Na CPLP, 3*(2), 9–33. doi:10.29073/e3.v3i2.45

Porter, M. E. (2001). Strategy and the Internet. *Harvard Business Review*. https://hbr.org/2001/03/strategy-and-the-internet PMID:11246925

Porter, M. E., & Heppelmann, J. E. (2015). How smart, connected products are transforming companies. *Harvard Business Review, 93*(10), 96–114. https://hbr.org/2015/10/how-smart-connected-products-are-transforming-companies

Rebelo, A. (2020). *The impact of digitalization on SMEs international strategies: A Bibliometric Analysis* [Master Thesis]. Faculdade de Economia Universidade do Porto.

Rupeika-Apoga, R., Larisa, B., & Kristine, P. (2022). Digital Transformation of Small and Medium Enterprises: Aspects of Public Support. *Journal of Risk and Financial Management, 15*(45), 45. Advance online publication. doi:10.3390/jrfm15020045

Rüßmann, M., Lorenz, M., Gerbert, P., Waldner, M., Justus, J., Engel, P., & Harnisch, M. (2015). *Industry 4.0: The future of productivity and growth in manufacturing industries*. Boston Consulting Group. https://image-src.bcg.com/Images/Industry_40_Future_of_Productivity_April_2015_tcm9-61694.pdf

Salkin, C., Oner, M., Ustundag, A., & Cevikcan, E. (2018). A Conceptual Framework for Industry 4.0. In A. Ustundag & E. Cevikcan (Eds.), *Industry 4.0: Managing the Digital Transformation* (pp. 3–23). Springer. doi:10.1007/978-3-319-57870-5_1

Schmitt, C., & Baldegger, R. (2020). Editorial: Digitalization and Internationalization. *Technology Innovation Management Review*, *10*(4), 3–4. doi:10.22215/timreview/1342

Siachou, E., Vrontis, D., & Trichina, R. (2021). Can traditional organizations be digitally transformed by themselves? The moderating role of absorptive capacity and strategic interdependence. *Journal of Business Research*, *124*, 408–421. doi:10.1016/j.jbusres.2020.11.011

Siachou, E., Vrontis, D., & Trichina, R. (2021). Can traditional organizations be digitally transformed by themselves? The moderating role of absorptive capacity and strategic interdependence. *Journal of Business Research*, *124*, 408–421. doi:10.1016/j.jbusres.2020.11.011

Simões, V. C. (1997). *Estratégias de Internacionalização das Empresas Portuguesas em Comércio e Investimento Internacional*. ICEP - Investimentos, Comércio e Turismo de Portugal.

Sjøbakk, B. (2018). The Strategic Landscape of Industry 4.0. In I. Moon, G. M. Lee, J. Park, D. Kiritsis, & G. von Cieminski (Eds.), *Advances in Production Management Systems: Smart Manufacturing for Industry 4.0* (pp. 122–127). Springer. doi:10.1007/978-3-319-99707-0_16

Strange, R., & Zucchella, A. (2017). Industry 4.0, global value chains and international business. *Multinational Business Review*, *25*(3), 174–184. doi:10.1108/MBR-05-2017-0028

Sukhodolov, Y. A. (2019). The Notion, essence and peculiarities of Industry 4.0 as a sphere of industry. In *Industry 4.0: industrial revolution of the 21st century*. Springer. doi:10.1007/978-3-319-94310-7_1

Thoben, K.-D., Wiesner, S., & Wuest, T. (2017). "Industrie 4.0" and Smart Manufacturing – A Review of Research Issues and Application Examples. *International Journal of Automotive Technology*, *11*(1), 4–19. doi:10.20965/ijat.2017.p0004

Tiessen, J. H., Wright, R. W., & Turner, I. (2001). A model of e-commerce use by internationalizing SMEs. *Journal of International Management*, *7*(3), 211–233. doi:10.1016/S1075-4253(01)00045-X

Viana, C., & Hortinha, J. (2005). *Marketing Internacional, 2ª Edição*. Edições Sílabo.

Watson, G. IV, Weaven, S., Perkins, H., Sardana, D., & Palmatier, R. (2018). International Market Entry Strategies: Relational, Digital, and Hybrid Approaches. *Journal of International Marketing*, *26*(1), 30–60. doi:10.1509/jim.17.0034

Westerlund, M. (2020). Digitalization, Internationalization and Scaling of Online SMEs. *Technology Innovation Management Review*, *10*(4), 48–57. doi:10.22215/timreview/1346

Xiaopu, S., Runtong, Z., Xiaomin, Z., & Quan, Z. (2016). Design theory, modelling and the application for the Internet of Things service. *Enterprise Information Systems*, *10*(3), 249–267. doi:10.1080/1751 7575.2015.1075592

Yin, R. K. (2015). Estudo de caso. Planejamento e métodos. Tradução de Daniel Grassi (5th ed.). Bookman.

Zen, F., & Faizin, M. A. (2021). An Overview of Digitalization and Internationalization Msmes: A Conceptual Paper. Proceedings of the BISTIC Business Innovation Sustainability and Technology International Conference (BISTIC 2021). *Advances in Economics, Business and Management Research, 193*, 195-200. https://www.atlantis-press.com/proceedings/bistic-21/125963905

ADDITIONAL READING

Ballestar, M. T., Diaz-Chao, A., Sainz, J., & Torrent-Sellens, J. (2020). Knowledge, robots and productivity in SMEs: Explaining the second digital wave. *Journal of Business Research*, *108*, 119–131. doi:10.1016/j.jbusres.2019.11.017

Banalieva R., Dhanaraj C. (2019). Internalization theory for the digital economy. *Journal of International Business Studies*, *50*(8), 1372-1387. doi:10.1057/s41267-019-00243-7

Bouwman, H., Nikou, S., Molina-Castillo, F. J., & de Reuver, M. (2018). The impact of digitalization on business models. *Digital Policy. Regulation & Governance*, *20*(2), 105–124. doi:10.1108/DPRG-07-2017-0039

Cassetta, E., Monarca, U., Dileo, I., Di Berardino, C., & Pini, M. (2020). The relationship between digital technologies and internationalisation. Evidence from Italian SMEs. *Industry and Innovation*, *27*(4), 311–339. doi:10.1080/13662716.2019.1696182

Etemad, H., Wilkinson, I., & Dana, P. (2010). Internetization as the necessary condition for internationalization in the newly emerging economy. *Journal of International Entrepreneurship*, *8*(4), 319–342. doi:10.100710843-010-0062-x

Hervé, A., Schmitt, C., & Baldegger, R. (2020). Digitalization, Entrepreneurial Orientation and Internationalization of Micro-, Small- and Medium-Sized Enterprises. *Technology Innovation Management Review*, *10*(4), 5–17. doi:10.22215/timreview/1343

Ramdani, B., Raja, S., & Kayumova, M. (2022). Digital innovation in SMEs: A systematic review, synthesis and research agenda. *Information Technology for Development*, *28*(1), 56–80. doi:10.1080/0 2681102.2021.1893148

Thrassou, A., Uzunboylu, N., Vrontis, D., & Christofi, M. (2020). Digitalization of SMEs: A Review of Opportunities and Challenges. In A. Thrassou, D. Vrontis, Y. Weber, S. M. R. Shams, & E. Tsoukatos (Eds.), *The Changing Role of SMEs in Global Business*. Palgrave Macmillan. doi:10.1007/978-3-030-45835-5_9

Wittkop, A., Zulauf, K., & Wagner, R. (2018). How Digitalization Changes the Internationalization of Entrepreneurial Firms: Theoretical Considerations and Empirical Evidence. *Management Dynamics in the Knowledge Economy*, *6*(2), 193–207. doi:10.25019/MDKE/6.1.01

KEY TERMS AND DEFINITIONS

Additive Manufacturing: Additive manufacturing comprises a set of technologies that ensure the production of small batches of products with a high degree of customisation. Reduction of waste, faster time to market due to timely prototyping, greater production flexibility and fewer indispensable tools are the main advantages of this technology.

Augmented Reality: Augmented reality (AR) authorises the invention of a virtual environment in which humans can interact with machines by leveraging devices capable of recreating the workspace. Interesting applications of AR are related to worker training and support in manual production activities.

Autonomous Robots: The evolution of traditional robots have developed the path towards new collaborative robot solutions (i.e., CoBots) that can operate in conjunction with humans safely and efficiently. Human-robot interaction can enable higher productivity. Similarly, artificial intelligence built into robots can provide that they learn from human activities, improving their autonomy and flexibility.

Big Data Analysis: Big data is qualified by volume, variety, and speed (the 3Vs) and calls for new data processing and analysis techniques. High-capacity computers and communication networks make it possible to store substantial amounts of information quickly. The visualisation, analysis and sharing of data are the support for making decisions based on this valuable information with greater precision and confidence.

Cloud Computing: Cloud computing is associated with ICT infrastructure that enables ubiquitous access to data from different devices, i.e., it is a computer system in which data is stored on specialised servers and information, services and programmes are accessed remotely via the internet. The cloud can be discussed as a service and support for collaborative design, distributed manufacturing, contribution to innovation, data mining, semantic web technology and virtualisation.

Cyber Security: To ensure the protection of the large amount of data collected, stored, and communicated thru the IIoT, cyber security strategies are one of the key challenges for the future.

Digital Transformation: Is a process whereby companies use digital technologies and tools to change their structure, processes, and business strategy and thus generate value and competitive advantage for the company.

Horizontal and Vertical Integration of the System: Horizontal integration indicates the creation of a global value network thru the integration and optimisation of the flow of information and goods between companies, suppliers, and clients. Vertical integration, in turn, is the integration of functions and departments from different hierarchical levels of a company, building a constant flow of data and information.

Industrial Internet of Things (IIoT): Enables communication between all devices inside and outside the factory. IIoT is an open, non-deterministic network in which self-organised intelligent institutions and virtual objects are interoperable and efficient to operate autonomously in pursuit of their own (or shared) goals depending on context, circumstances, or environments.

Internet of Things (IoT): The IoT (Internet of Things) is concept/project that enables all devices and systems to operate together, acquiring context information in real time, as well as reaching feedback from other working systems and finally analysing the gathered data. Therefore, this concept makes it possible to interconnect everything by turning objects into dynamic elements of an integrated network, whose plants will use this information intelligently. This form of digital connexion is expected to stimulate the emergence of new differentiated products and services.

Simulation: Simulation tools can be used widely across the value chain, from product design to operations management. Modelling and simulation tools are essential for the development of digital engineering and virtual reproduction of products and processes to identify potential problems early, avoiding waste of costs and resources in production.

Chapter 4
Understanding Consumer Behavior in the Hyper-Connected Omnichannel Retailing

Ahmet Tuğrul Tuğer
Piri Reis University, Turkey

ABSTRACT

Omni-channel retailing integrates different contact points in social, virtual, and physical realms in customer journey. This chapter provides a framework to understand omni-channel customer behavior, first by focusing on customer journey in omni-channel retailing representing the actual step of multichannel and cross-channel retailing. To contribute to the conceptualization of omni-channel customer journey, some suggestions including customer experience with digital, social, and phygital dimensions are offered. Followingly, the critical antecedents of omni-channel customer experience are reviewed with the main theoretical models applied in the extant literature. To concretize the omni-channel consumption experience, well-structured scales of omni-channel experience are discussed. Lastly, insights for omni-channel customer journey are discussed with future research directions and managerial implications, which emphasize the importance of new approaches to omni-channel customer journey.

INTRODUCTION

Technology transformed retailing and interactions between retailers and consumers in the last two decades. New technologies changed how consumers look for information, evaluate products and services, make purchase decisions, and share their experiences in an interactive way along their consumption journey. The actual shopping environment is not limited to independent retail channels, the lines between different channels has been eliminated in order to provide a seamless consumption experience (Rigby, 2011). Connected customers can interact with offline stores, online stores and other customers with the support of the new technologies (Inman & Nikolova, 2017). Digital, physical, human touchpoints are available

DOI: 10.4018/978-1-6684-5538-8.ch004

for customers who need more information, evaluate service/product offerings, purchase the offerings and request post-purchase support, (Lemon & Verhoef, 2016; Quach et al., 2020). In this hyper-connected environment, the main objective is to offer a consistent and seamless customer experience throughout different retailing channels (Chang & Li,2022). With the emphasis on the consistency and integration among various channels, omni-channel retailing, described by Beck & Rygl (2015) as different activities to sell products or services through all extensive channels supported by full integration, shares common points with omni-channel marketing which is a rather new concept (Nguyen et al., 2022).

Meanwhile, based on conventional marketing management perspective, omni-channel retailing can be assumed as a subset of omni-channel marketing, which can have a larger scope in terms of research and applications (Nguyen et al., 2022). Berman & Thelen (2018) describes omni-channel marketing as the fully integrated and progressed version of multichannel and cross-channel marketing and it includes customer shopping histories across channels, requires more promotions, prices and special customer loyalty program across channels. With a more inclusive approach, omni-channel marketing can be described as "the synergistic management of all customer touchpoints and channels both internal and external to the firm to ensure that the customer experience across channels as well as firm-side marketing activity, including marketing mix and marketing communication (owned, paid and earned), is optimized for both firms and their customers" (Cui et al., 2021, p.104). The scope of omni-channel marketing is expanding with different combinations of customer behaviors throughout their customer journey and different services such as e-government services (Schenk et al., 2021). Governments, producers in different industries (e.g. packaged goods, fashion), service providers (e.g. banks, restaurants) are actively implementing omni-channel marketing plans (Berman & Thelen, 2018; Cui et al., 2021).

The omni-channel concept was firstly applied and studied in retailing industry (e.g. Rigby et al., 2011) and the current state of research and applications is larger than omni-channel marketing area (Nguyen et al., 2022). Actually, the number of papers and investment decisions in omni-channel retailing is globally increasing in the post pandemic era. Some retailers decided to invest in building their online channels such as mobile applications, online platforms whereas some customers who were reluctant about omni-channel retailing before the pandemic adopted new omnichannel services for the first time (Salvietti et al., 2021). The pandemic period increased the technology acceptance and adoption of customers and retailers. Actually, omni-channel retailers face the challenge of integration to provide consistent service and to ensure connectivity among different offline and online channels (Gasparin et al., 2022). The customers can combine connected offline and online channels to search, evaluate, purchase and suggest and/or inform against a product or a service in this seamlessly connected environment (Chang & Li, 2022). Henceforth, an increasing number of recent studies reveal different details about the interactions between channels and customers in omni-channel retailing. Yet, the number of studies that study these probably intersecting details about the customer behaviors in omni-channel environment is limited (Mishra et al., 2021b) and there is no consensus on the conceptualization of omni-channel customer journey (Tuenrat et al., 2021b). In order to increase the limited theoretical understanding of omni-channel customer behavior and also to contribute to the conceptualization of omni-channel customer journey, this chapter provides an up-to-date and structured overview of omni-channel customer studies.

The objective of this chapter is fourfold. Firstly, it aims to provide a clean-cut separation among different retailing environments supported with an evolution perspective. Secondly, with the support of this categorization and increasing number of recent papers on omni-channel customer behaviors, this paper aspires to offers a review to understand the components of the omni-channel customer journey. Thirdly, this study aims to pinpoint at the critical role of digital, physical and phygital aspects of customer experience in different touchpoints of omni-channel customer journey which cannot be evaluated with a linear approach distinguishing pre-purchase, purchase and post-purchase stages. Fourthly, this chapter intends to provide insights for academics, researchers, businesspeople interested in omni-channel retailing literature and applications. More specifically, four research sub-questions are addressed: (a) What are the distinctive features of single channel, multichannel, cross-channel, omni-channel retailing? (b) What are the enablers of omni-channel customer behavior? (c) What are theoretical tools and academic measurement tools to evaluate omni-channel customer behavior? (d) What are the main research directions in omni-channel customer behavior?

To search for insights about main objectives and questions on omni-channel customer behavior, different factors affecting customer behavior are reviewed. Secondly, the new interaction possibilities between different types of retail channels and customers with the support of new technologies within the omni-channel marketing are studied. Accordingly, digital, physical and phygital dimensions of customer behavior, described as omni-channel customer experience, are assessed. Customer experience in several touchpoints that interact with customers throughout the omni-channel customer journey and different types of touchpoints in omni-channel retailing shopping experience are clearly defined. Key antecedents and measurement of omni-channel customer experience are reviewed with different theoretical applications. Finally, a review of the current state of omni-channel shopping experience is offered to draw conclusions for future research and applications.

METHODOLOGY

To generate and up-to-date overview of extant research on omni-channel retailing, omni-channel customer experience and omni-channel customer journey, a systematic literature review (SLR) is carried out in related topics. The literature review method structured is the Preferred Reporting Items for Systematic Review and Meta-Analysis (PRISMA) Statement (Moher et al., 2009). Data for the SLR was extracted and gathered from Scopus, one of the most wide-ranging sources of indexed academic publications and in order to reduce possible data homogenization with multiple database search, Scopus has been the single source for data extraction. To search database, a set of keywords have been identified: "omni-channel", "omnichannel", "omni-channel retailing", "omnichannel retailing", "omni-channel customer", "omnichannel customer", "omni-channel customer experience", "omnichannel customer experience", "omni-channel customer journey", "omnichannel customer experience", "omni-channel marketing", "omnichannel marketing", "webrooming", "showrooming", "omni-channel touchpoints". A search combining these keywords in the title, abstract and keywords is executed. The articles published after 2011 are taken into account. After excluding proceedings, book chapters, books, and materials not published in English, the search provided 565 articles.

To ensure that all papers are can be evaluated with less subjective views, three principles of inclusion and exclusion defined by Liao et al. (2017) are adopted in this study (See Table 1):

Following these criteria, the PRISMA flow chart that reports the different phases of the systematic literature review has been followed (Moher et al., 2009) (See Figure1). The exception was the lack of quantitative and qualitative analysis of included papers. The criteria and the research sub questions stated in the introduction are taken into account to minimize the subjectivity in exclusion and inclusion of the papers extracted from Scopus database.

Table 1. Inclusion and Exclusion Criteria and Their Explications

Inclusion / Exclusion	Criteria	Criteria Explanation
Exclusion	Search Engine Reason (SER) Without full-text (WF)	A paper has only its title, abstract, and keywords in English but not its full-text A paper without full text to be assessed
	Non-related (NR)	NR-1: A paper is not an academic article. For example, editorial materials, conference reviews, contents, or forewords NR-2: The definition about 'omni-channel" is not related to customer journey and customer experience
	Loosely Related (LR)	A paper doesn't focus on the review, survey, discussion, or problem solving of omni-channel customer behavior. In which, LR-1: Omni-channel is only used as an example fact LR-2: Omni-channel is only used as a part of its future research direction, future perspective or future requirement LR-3: Omni-channel is only used as a cited expression LR-4: Omni-channel logistics, operations is the main subject
Inclusion	Partially Related (PR)	PR-1: A research about the omni-channel retailing without mentioning omnichannel customer behavior PR-2: Omni-channel is only used to support the description of some challenges, issues, or trends that a paper intends to deal with PR-3: Omni-channel is one of several objects that to be reviewed, surveyed, or discussed
	Closely Related (CR)	The research efforts of a paper are explicitly and specifically dedicated to Omni-channel retailing and customer behavior, customer experience

Figure 1. The PRISMA Diagram

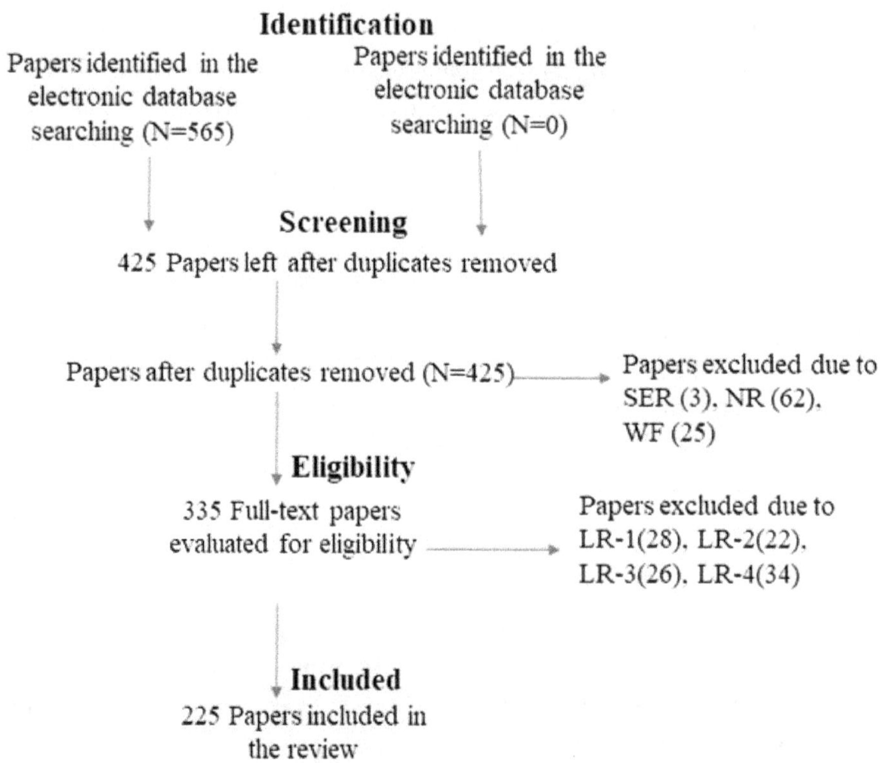

OMNI-CHANNEL RETAILING

The traditional retailing services in physical stores have gone through a quick development with the explosion of channels such as home shopping networks, catalogues, web stores, call centers in the last decades (Neslin et al; 2006). The channels are defined as interaction point between the company and the customer (Neslin et al; 2006). The increase of independent interaction points has created the need to manage these channels efficiently and emphasized the value that the customer can get from the interactions in different contact points. Consequently, Neslin et al. (2006, p.96) proposed a customer-centric definition of multichannel customer management referred as "the design, deployment, coordination, and evaluation of channels to enhance customer value through effective customer acquisition, retention, and development".

The multi-channel retailing can be accepted as the first sign of interactivity of retailers that were previously focused on maximizing value from the transactions with customers in single channels. Channels usually include the store, the Web, catalog, direct sales force, call center and other retail partners (Neslin & Shankar, 2009). The multi-channel retailing management focuses on offering shopping experiences in different contact points in pre-purchase, purchase and post-purchase stages. The customer uses different sources to search information, purchase, connect with retailer and other customers and to purchase the product or service (Hamouda, 2019). The customer can use instantly multiple channels and touchpoints (e.g. physical stores, website, mobile, social media, etc.) to complete even a single transaction (e.g. price/product information search, evaluation, quick purchase) (Sousa & Voss, 2006). In this regard, the customers can collect information in one channel and can purchase products and services in another channel (Verhoef et al; 2007). Differently from independent, single channel context, consumers' variety seeking behavior can increase competition amid different channels of the same retailer (Neslin & Shankar, 2009). Followingly, there might be channel migration among the different channels of the multi-channel retailer, or channel lock-in depending on channel interactivity, attributes and customer priorities in the multi-channel environment (Saghiri et al., 2017).

In order to fill the gaps or disruptions in the seamless shopping experience, the need to connect different channels is both expressed in multi-channel retailing (e.g., Neslin et al., 2006) and also in cross-channel retailing research (e.g., Beck & Rygl, 2015). Even though borders between multi-channel and cross-channel are unclear, based on the extant literature in cross-channel retailing, there are more interactions between channels and therefore, the increased consistency and integration of channels in terms of information, communication increases customer experience (Piotrowicz & Cuthbertson, 2014). Cross-channel retailing still relies on the customer's active involvement in the integration and interactivity of channels (Saghiri et al.; 2017).

The limited integrated consumption journey offered by multi-channel and cross-channel retailing can be further improved by omni-channel retailing (Verhoef et al., 2015; Beck & Rygl, 2015; Huré et al, 2017). The introduction of channels such as mobile applications, online platforms, IoT, smart devices, web portals in retail channels increased the channel options of the customers. In this context, as an evolution of cross-channel and multi-channel marketing, omni-channel retailing orchestrates different physical, digital and social retail channels to provide a seamless experience to the customer (Akter et al; 2021). Omni-channel retail management can be defined as, "the synergetic management of the numerous available channels and customer touchpoints, in such a way that the customer experience across channels and the performance over channels is optimized" (Verhoef et al. 2015, p. 176). The key particularity of the omni-channel retailing is that the organization is focused on the flow of customer's shopping experience, with a perspective to provide the consumer a complete experience (Juaneda-Ayensa et al., 2016). The distinction of multi-channel, cross-channel and omni-channel retailing can be described based on the enhanced interactivity and integration among different channels (See Table 2).

Table 2. Evolution of Channel Management in Retailing. Source: (Adapted from Verhoef et al; 2015; Beck and Rygl,2015; Mosquera et al., 2017)

	Multi-channel	Cross-channel	Omni-channel
Channel interactivity	Customers cannot start any interaction; there isn't integration of channels	Customers can start partial interaction, there is partial integration of channels	Customers can start full interaction, there is full integration of channels
Separation/connection of retail channels	Independent single channels with no connection	Connection across several channels but not across all channels	Fully integrated channels offering seamless shopping experience
Customer experience	Transaction/sales and experiences in independent channels	Transactions and experiences in connected channels	All channels work coherently to provide a unique and seamless experience

Based on the distinctions on the Table2 and current literature, the interconnectivity between touch-points is expected to be at the highest level in omni-channel retailing. Interconnectivity includes several components such as interconnection of service content and information which would enable real-time retailing services (Chang & Li, 2022). Omni-channel retailing service content interconnection describes the integration of marketing communication, order management and real-time inventory information available for customers (Chang & Li, 2022). The omni-channel retailing services such as "Buy online and pickup in store", "click and reserve", "ship from store" have different operational features but their integration has a critical role in customer loyalty (Mishra et al., 2021). The omnichannel retailers' services influence customer satisfaction and loyalty by increasing service and value, while also impacting both order fulfillment and returns management (Jones et al, 2021) (See Table 3).

Table 3. Omni-channel Retailing Service. Source: (Adapted from Murfield et al.,2017; Jones et al., 2021)

	Activities of Customers
Buy online and pickup in store (BOPS) Click and Collect (C&C)	Customers order and purchase products online (websites or mobile applications) and collect their order at the offline store
Click and Reserve (C&R)	Customers reserve products online (websites or mobile applications) and purchase the reserved products at the physical store
Ship from store (SFS)	The customer order is fulfilled from actual inventory at a physical store rather than from a distribution center
Next day delivery (NDD)	Quick fulfillment of customer order with the actual inventory
Online instore availability (OIA)	The customer can obtain the real-time and online information about the availability of a product in a physical store
In-store returns (ISR)	Customers who purchased goods in online stores return goods to a local store rather than shipping the product back
Free return shipping (FRS)	In addition to return product for free in-store, customers can send back the goods from their house

Different service features require a constant forward and reverse flow of information and goods in omni-channel retailing and in line with interconnectivity, the channel integration is a highly distinctive component of omni-channel retailing to enable real-time services (Jones et al., 2021). Omnichannel retailers are engaged to provide a perfect coordination and synergy across offline and online channels (Verhoef et al., 2015; Saghiri et al., 2017; Chang & Li, 2022; Neslin, 2022). In this sense, the concept of integration quality (IQ) initially promoted in the multichannel context (Sousa and Voss, 2006) seems to be more crucial within the omnichannel environment. In fact, IQ has been recognized as an important enabler of a unique omnichannel customer experience (Hossain et al., 2017; Shen et al., 2018). Actually, there is a long way to implement channel integration and connectivity in different industries (Neslin, 2022). In omni-channel retailing's objectives, the integration of all channels and touchpoints, abundant number of channels, various types of channels (e.g. physical, web, mobile, platforms, social media), high level of integration, real-time and synergetic management of all channels, seamless customer experience, control of channels, data sharing across all channels are critical points to realize in the long term (Hajdas et al., 2022).

In omni-channel retailing environment, differently from multichannel and cross-channel retailing context, customers move freely across multiple channels and touchpoints (Tuenrat et al., 2021). Customers control over their journeys in a powerful way (Van Nguyen et al., 2022). They can interact with a brand, retailer or marketer however and whenever they want (Hübner et al., 2016). The proliferation of channels and devices (e.g. smartphones, social media and voice-assistants) has allowed customers to connect with retailers via numerous touchpoints and personalize their journeys to satisfy their needs and preferences (Grewal et al., 2016; Rodríguez-Torrico et al., 2017). In this context with numerous touchpoints, customers are segmented in terms of channel preference and may use multiple channels throughout their consumption journey (Van Nguyen et al., 2022). Such a changing retail environment has resulted in more complex customer channel behaviour and journey patterns which were not possible in the past (Verhoef et al., 2015). A customer may start by product/service search for through voice-assistants or mobile apps and can end up purchasing it through the smartphone and also, they may search for more information and assessments on different platforms and social media (Chaudhary et al., 2021). With the proliferation of platforms and diversified devices, the customers can easily migrate across the channels and also use them instantaneously. In this context, omni-channel offers an integrated, consistent and seamless customer experience that associates the physical store with independent digital stores and platforms, with the purpose of providing exceptional customer experiences across all touchpoints (Chaudhary et al., 2021). This possibility of various combination of channels provides unpredictable, non-linear omnichannel customer journey.

Understanding Customer Journey in Omni-Channel Retailing

The customer journey is interpreted as a funnel, a linear track with search/purchase/aftersales steps, in the conventional customer decision-making perspective (Lemon & Verhoef, 2016). The omni-channel journey includes real and virtual settings and a variety of phases, relationships, touchpoints and channels (Mele et al., 2021). In omni-channel customer journey, the customer has the freedom to choose her/his own shopping track with the support of digital (e.g. kiosks, tablets, mobile apps, digital platforms, online, in-store technology), human (e.g. frontline service and customer service employees) and physical touchpoints (e.g. stores) (Grewal et al., 2020; Tuenrat et al., 2021). The combination of various different touchpoints in the pre-purchase/purchase/post-purchase can change in each transaction depending on consumer's free will (Alexander & Kent, 2020). Henceforth, the conventional linear funnel perspective may not be adequate to interpret omni-channel customer journey (Grewal&Roggeveen,2020). For example, in pre-purchase, product comparisons and instore availability can be conducted on mobile applications; in the purchase stage, consumers can buy online and pickup in store, click and reserve, make online payment; and the post-purchase stage concentrates on consumption experience, satisfaction, word-of-mouth, repurchase, engagement, as well as return and complaints. Therefore, to understand customer journey, there is a necessity to describe different customer behaviors and their various dimensions.

Omni-Channel Customer Behavior and Experience

It is essential for academics and professionals to describe customer behavior in connected omni-channel retailing environment, and above all, to understand the antecedents and outcomes of omni-channel consumer behaviors (Saghiri et al., 2017) such as satisfaction, positive and/ or negative word of mouth, engagement, loyalty (Lee et al., 2019) and webrooming, showrooming, cross-shopping (Flavian et al., 2020; 2021; Arora et al., 2020; 2022). Customer experience is referred as 'the cognitive, emotional, physical, sensorial, and social elements that mark the customer's direct or indirect interaction with market actors' (De Keyser, et al. 2015, p.14). Lemon and Verhoef (2016) conceptualized and developed customer experience and customer journey in omnichannel shopping context where the customer has the possibility of simultaneous migration throughout different channels of the same and/or other retailer(s). Based on the channel integration and interconnectivity, omni-channel retailers' priority is to offer seamless, consistent and combined customer experiences in different channels and stages (Le & Nguyen-Le, 2021). In search for the seamless and consistent customer experience, customers can choose different omni-channel behaviors, such as research shopping, cross-shopping, showrooming and webrooming.

Research Shopping

In the multi-channel perspective, the customers can collect information in one channel and can purchase products and services in another channel, called as, research shopping (Verhoef et al; 2007). Differently from single channels, consumers' variety seeking behavior can be supported in different channels. Therewithal, there might be channel migration, or channel lock-in, known as retaining the customer at one channel, depending on channel interactivity, attributes and customer priorities in the multi-channel environment. The importance of establishing interactive connections has been recognized to offer a seamless experience (Neslin & Shankar, 2009).

In multi-channel retailing, the interactions between channels and consumers are assumed to be very limited (Verhoef et al., 2007). Meanwhile, in this less-connected environment, there is a flow of customers between the channels of competitors. In channel switching or channel migration, customers collect information from the online channel of Company X but migrate to another online channel of Company Z to make purchases and in free-riding behavior, customers gather information from online channel of Company X, although they purchase from the offline channel of Company Z (Chou et al., 2016). In the connected environment of omni-channel retailing, the complete channel integration allows the customers to combine channels and behaviors to optimize customer experience. In multi-channel environment, the attractiveness of other service providers can increase channel switch intention. After searching product information via an online channel, consumers may not order instantly but purchase through another company with an online channel because they can obtain more information, higher value and satisfaction, reduced risk and a better customer experience by the end of their research shopping journey (Chou et al., 2016).

Showrooming

In the omni-channel retailing context, research shopping offered new cross-shopping possibilities. As a free-riding behavior, in showrooming, customers change channels and they benefit from the services provided by a retailer in the physical stores (seeing and testing the product, and customized service from the sales employees or kiosks), but the stores do not gain profit in return if customers do not purchase from the physical store or their online store. (Viejo-Fernandez et al., 2020). Showrooming, described as collecting data and examining products in physical stores and then purchasing online became a popular practice of hybrid shoppers who recognized the benefits offered by different channels (Flavian et al., 2020; Arora et al., 2022). With the introduction of smart devices such as tablets, smartphones, showrooming became a seamless and real-time consumption experience. In mobile showrooming supported with smart devices, customers visit the offline store to evaluate products and use a mobile device (e.g. smartphone, tablet) to acquire extra online information while they are in a physical store (Chimborazo-Azogue et al., 2022). Actually, showrooming represents customers' aptitude to firstly assess products/services and then use mobile technology to compare the products/services among eventual number of channels (Fiestas &Tuzovic, 2021).

As an important way to reduce risk in shopping, showrooming can be the last step of omni-channel retailing. For example, Kim & Park (2019) showed that consumer innovativeness supported showrooming behaviors. On the other side, showrooming can be a mediator in the omni-channel customer journey. Kang (2018) found that omnichannel customers' webrooming and showrooming behaviors positively influenced user generated content generation in social media so that customers can influence other customers with their comments. Critical antecedents of showrooming are price consciousness, perceived control, product quality, perceived search and online purchase benefits and attitude toward showrooming, multichannel search, perceived benefits of offline purchase (Sahu et al., 2021). Additionally, in-store services and technology are critical enablers of showrooming (Wang & Wang, 2022).

Webrooming

As an extended version of cross-channel behavior, webrooming behavior is defined as consumers' visits to an online store for information before purchasing offline (Flavian et al., 2016). Webrooming behavior tends to moderate the risk of online purchase; as available information supports the customers to make better-informed decisions (Flavian et al. 2016; 2020). Similar to showrooming, in the omnichannel shopping journey, webrooming can have different roles. The consumers can narrow down the alternatives before they decide to move to a physical store to complete purchasing process (Sahu et al., 2021). Flavian et al. (2016) found that found that webrooming improves purchase intention, search-process satisfaction, and choice confidence. In their comparative study of webrooming and showrooming, Flavian et al. (2020) showed that webrooming had a more positive impact than showrooming on smart shopping feelings, as it helped to perceive that they were saving time and/or effort and making the right choices to a greater extent than showrooming.

In addition to showrooming key antecedents, three information-processing motivations, (acquiring information, evaluating price and product and empowerment), and two uncertainty reduction motivations (perceived risk and confidence) are shown as critical drivers of webrooming (Flavian et al., 2016, Jing, 2018, Ax, 2019). Kleinlercher et al. (2020) have researched a large set of factors classified into four main categories: psychographic factors (customer involvement, pre-purchase certainty), shopping motivation (sales advice, having fun, convenience and time savings), channel-related factors (in-store shopping, online shopping experiences, price, assortment attractiveness of the physical store compared with online) and product-related factors. The combined and hybrid patterns of consumption within omni-channel environment can also blur the lines between showrooming and webrooming in the omni-channel journey.

Different Aspects of Omni-Channel Customer Experience

The omni-channel customer experience has different dimensions that can be concretely explained and measured with different approaches. Recent studies offer different tools and models such as journey mapping to illustrate how customers interact with touchpoints and contexts in omni-channel retailing (Alexander & Kent, 2020b; Mele et al; 2021 and Schneider, 2010). For instance, Le and Nguyen-Le (2020) defined the omnichannel customer experience as satisfaction with the experience and positive emotions, while Gao et al. (2021) described it as a two-dimensional construct composed of cognitive and affective experience. Shi et al. (2020) theorized it as a five-dimensional construct including connectivity, integration, consistency, flexibility, and personalization. Quach et al. (2020) operationalized omnichannel customer experience as flow and perceived privacy risk. Despite the existence of different conceptualizations in customer experience, the integration and evolution of different channels offer the chance to review the omni-channel customer experience with social, digital and phygital dimensions. As a critical part of customer journey, controlled (e.g. retailer's online websites, offline store) and uncontrolled (e.g. product/brand comparison websites) touchpoints should be interpreted as social, physical and digital dimensions of the omni-channel customer experience.

Social Dimension

Integrated marketing communications has a critical contribution in cross-channel synchronization (Manser Payne et al., 2017) and it can be assumed as the social dimension of omni-channel retailing. In the social dimension of omnichannel retailing, the consumers can interact with the retailers and other consumers before, during and after purchase decision through mobile, local and social applications (Hüseyinoğlu et al., 2018). In omni-channel fashion retailing, Kang (2019) suggested that SoLoMo consumers' perceptions of the value of showrooming and webrooming were antecedents of their omnichannel shopping intention, which increased their intention to share product reviews. In mobile showrooming context, Chimborazo-Azogue et al. (2022) found that mobile showrooming intention had a positive impact on user generated content (e.g. comments) which can be a part of another omnichannel customer's journey. Palazon et al. (2022) point at the significance of social touchpoints (controlled, owned social media, website; non-controlled touchpoints such as comparison websites, user generated content) to influence omnichannel customer behaviors. Based on these recent findings, social dimension of the customer experience should be assumed as part of omni-channel customer journey.

Digital Dimension

With the support of new technologies such as Internet of Things (IoT), Augmented Reality (AR), Virtual Reality (VR), Mixed Reality (MR), Artificial Intelligence (AI) based virtual assistants, chatbots, and robots are entirely changing the customer experience (Hoyer et al., 2020). In omni-channel retailing, in offline and online stores, the support of these technologies provide unique customer experience in different touchpoints (e.g. mobile apps, kiosks, virtual reality supported mobile applications). An increasing number of retailers are investing in digital touchpoints such as smart mirrors, interactive storefront windows to provide a seamless experience that relies on the interconnectivity of touchpoints (Vannucci & Pantano, 2020). In-store technologies, defined as the shopping process facilitating devices at various points in the physical store, empower consumer-channel interaction in digital dimension of omni-channel environment (Alexander & Kent, 2020). Fiestas & Tuzovic (2021) found that customers assisted by mobile digital touchpoints (applications) and showing showrooming tendencies, mobile-assisted showroomers followed a hybrid (mobile and physical) product evaluation phase during their offline store visit. Hilken et al. (2018), Farah et al. (2019) underlined the contribution of augmented reality, virtual reality, mixed reality to provide customers with a seamless omnichannel journey. The interconnectivity of various digital and other touchpoints remains as a critical challenge for omni-channel retailers due to introduction of new digital touchpoints (Larke et al., 2018).

Phygital Dimension

Phygital represents the new technological concept and evolution of retailing which associates and bonds the physical and digital components of the service to offer customers with unique, different, and perpetual experience at each point of service or sales (Banik, 2021). Phygital describes an environment that covers to different sectors ranging from education, banking and insurance, and airlines to health care, hospitality and restaurants, and entertainment (Batat, 2022). Omnichannel environments links customers with retailers through different channels using digital with customer journey perspective, whereas phygital, as a holistic approach, can be used in different marketing mix tools (Batat et al., 2022). In omnichannel customer journey, combining the physical and digital into a unique object of 'phygital' while starting customer communication provides multiple benefits to both retailers and customers (Mishra et al., 2021a)., Customers can reach additional information easily by relying on phygital dimension, thus increasing their consciousness of a retailer's products and services and this dimension supports better consumption decisions (Mishra et al, 2021a). Therefore, phygital dimension of customer experience became a crucial dimension of omni-channel customer journey.

Omni-Channel Customer Journey

Taking into account various (social, digital, phygital and also human) dimensions of omni-channel customer behavior, customer journey can be described as a process or system of touchpoints that a customer goes through to reach or purchase a product or service (Følstad and Kvale, 2018). Customers select touchpoints depending on their orders, a linear funnel is not realistic for omni-channel customers who can take on a longer or shorter journey (Mele et al., 2021). Existing research on the omnichannel customer journey has mainly focused on customer's interaction choice across channels (Barwitz and Maas, 2018), customer experience and satisfaction (Alexander and Kent, 2020; Tueanrat et al., 2021a), multichannel behaviors, and channel preferences (Nguyen et al., 2022). There are different conceptualizations and measurement tools in omni-channel customer journey and experience but the non-linearity of omni-channel customers' behaviors and different aspects (digital, social, phygital) of customer experience are also subject of research (e.g. Batat et al., 2021; Mele et al., 2021; Mishra et al., 2021a). Yet, actually, it may not be adequate to suggest a unique definition about a rather new concept omnichannel customer journey, which is studied with different perspectives.

Meanwhile, based on the reviewed studies, some critical issues of customer journey can support the conceptualization of omni-channel customer journey in the future research. In order to offer a seamless omni-channel customer experience, coordination and integration of social, digital, human, phygital touchpoints is a critical issue (Palazon et al., 2022). Furthermore, the combination of various omni-channel behaviors such as webrooming, showrooming, hybrid shopping, cross-shopping, research shopping should be assumed as part of this non-linear journey. Lastly, the customer's freedom to choose a short or a long journey with the combination of different behaviors and touchpoints and/or a unique behavior and touchpoint may be included within the future frameworks of omni-channel customer journey.

Key Factors in Omni-Channel Customer Behavior and Experience

Omni-channel marketing is a dynamically evolving research field. With different approaches and theoretical models used to define customer behavior and customer experience in omnichannel retailing, a review of key factors influencing customers' acceptance, intention to use omnichannel retailing services, showrooming, webrooming is offered in Table 4. This part aims to integrate common key factors into two main categories according to their direct and indirect impact on omni-channel customer behavior: individual factors, such as technology readiness, perceived value; and situational variables, such as social surroundings, channel integration.

- **Individual Factors:** Individual factors, the distinguishing characteristics of individuals which determine their consumption patterns in omni-channel retailing, include the following:
    - **Technology Readiness** is an individual's tendency to adopt and use new technologies in daily life and at work. Technology readiness is showed to influence omnichannel customer experience (Hickman et al., 2020).
    - **Perceived Value:** As a combination offline, online and mobile shopping value offers a review customer experience (Huré et al, 2017). In omni-channel banking (Hamouda, 2019) and in online, physical and mobile customer experience (Hickman et al, 2020), perceived value act as a critical antecedent.
    - **Perceived Customization:** Customization is described as the high level of individualization of services in omnichannel retailing (Chatterjee et al., 2021). Perceived customization defines to the extent to which the reactions of information systems are perceived as suitable or personally related to a customer's behaviors (Hickman et al., 2020). Tyrvainen et al. (2020) found that omnichannel personalization, customization has a positive impact on emotional and cognitive customer experience. Chatterjee et al. (2021) found that perceived customization had a significant impact on perceived benefit which can increase the intention to use omnichannel for shopping.

- **Brand Familiarity:** Brand familiarity is described as the amount of brand-related direct or indirect experiences that have been accumulated by the consumer (Hickman et al., 2020). Brand familiarity is found to have a positive impact on omnichannel experience (Hickman et al., 2020).
- **Perceived Risk**: Perceived risk is the complete evaluation of ambiguity and possibly contrary consequences during the consumption stages (Shi et al., 2020). Silva et al. (2018) found that perceived risk has a negative direct impact on intention to use an omnichannel retailing. Shi et al. (2020) found that customers' perceived risk had a negative impact on omnichannel shopping intention. Chatterjee et al. (2021) showed that perceived risk, as a combination of data privacy and data security factors, can decrease the intention to use omnichannel retailing. Dahana et al. (2018) found that perceived risk has a positive impact on showrooming behaviors as these behaviors can reduce risk.
- **Perceived Benefit**: Perceived benefit represents the concrete gain of the customer. Chatterjee et al., (2021) found that, as a second order factor of choice breadth, content consistency and customization, perceived benefit influences positively the intention to use omnichannel retailing for shopping.
- **Hedonic Motivations:** Hedonic motivations are associated with emotions such as fun, pleasurable, and enjoyable (Juaneda-Ayensa et al., 2016). Tyrvainen et al. (2020) found that hedonic motivation had a significant impact in creating omnichannel customer experiences.
- **Personal Innovativeness:** Personal innovativeness is described as "the willingness of an individual to try out any new information technology" (Agarwal & Prasad, 1998:206). Personal innovativeness contributes to adoption of omnichannel retailing services (Juaneda-Ayensa et al., 2016).
- **Perceived Trust:** Perceived trust is defined as consumers' subjective credence that the service provider or producer will accomplish its responsibilities (Kazancoglu & Aydin, 2018). In the omnichannel shopping environment, data privacy is also thought as a critical tool to diminish perceived risk and increase perceived trust (Chatterjee et al., 2021).
- **Perceived Compatibility:** Perceived compatibility represents the degree to which an experience is perceived as constant with potential users' current values, beliefs and actual and past experiences (Shi et al., 2020). In a study concerning Portuguese consumers, perceived compatibility is found to have a positive direct impact on omnichannel behavioral intention (Silva et al., 2018). Similarly, as a second factor of connectivity, integration and consistency, perceived compatibility had a positive impact on omni-channel shopping intention (Shi et al., 2020).

Table 4. Key Factors in Omni-channel Customer Behavior and Experience

Dimension	Concept	Author(s)
Individual Factors (Individual-led variables)	Technology readiness	Hickman et al., 2020; Mimoun et al., 2021
	Perceived value	Huré et al., 2017; Hamouda,2019; Kang,2019; Hickman et al. 2020
	Perceived customization	Hickman et al. 2020 ; Tyrvainen et al, 2020; Chatterjee et al., 2021
	Brand familiarity	Kim & Park, 2019; Hickman et al., 2020
	Perceived risk	Silva et al., 2018; Dahana et al., 2018; Flavian et al., 2019; Quach et al., 2020 ; Chatterjee et al.,2021
	Perceived benefit	Arora & Sahney, 2017 ;Sahu et al., 2021; Chatterjee et al.,2021
	Hedonic motivations	Juaneda-Ayensa et al., 2016; Kazancoglu&Aydin,2018; Kleinlercher et al., 2020
	Personal innovativeness	Juaneda-Ayensa et al., 2016; Silva et al., 2018; Kim&Park, 2019; Shi et al., 2020
	Perceived trust	Kazancoglu&Aydin,2018; Chatterjee et al.,2021
	Perceived compatibility	Silva et al., 2018; Shi et al., 2020
Contextual Factors (Retailer-led variables)	Channel integration	Zhang et al., 2018 ; Fang et al., 2021; Le & Nguyen, 2021; Gao et al, 2021 ; Mishra et al., 2021
	Service quality	Kleinlercher et al., 2020; Le & Nguyen, 2021; Park & Kim,2022
	Channel consistency	Quach et al., 2020 ; Akter et al., 2021; Gao & Huang,2021; Chatterjee et al., 2021
	Synchronization	Rodriguez-Torrico et al., 2020; Chang&Li, 2022

- **Contextual Factors:** Omni-channel retailing environment symbolizes a high level of interconnectivity between channels. Channel-related factors would influence customer behavior, customer experience as well as customers' intention to use omnichannel services as indicated here below:
    - **Integration of channel** includes the integrated promotion, product and price, flow of information and transaction, customer service across different channels in a consistent way (Le & Nguyen, 2021; Gao et al, 2021). Lee et al. (2019) found that channel integration quality with breadth of channel-service choice, transparency, process and content consistency positively influenced customer engagement in omni-channel retailing. In omni-channel banking, Hamouda (2019) showed that the perceived quality of omni-channel service integration increased significantly customer's satisfaction with the service. In a study related to the phygital experience in omni-channel retailing, Mishra et al. (2021) showed that cross-channel integration influences positively the customer retention.
    - **Service Quality:** Service quality concerns the level of service provided in different parts of customer journey (Pantano & Viassone, 2015; Le & Nguyen, 2021; Park & Kim,2022). As part of omni-channel service quality, offline or in-store service quality has been shown to influence positively omni-channel customer satisfaction (Park & Kim, 2022).
    - **Channel Consistency:** Channel consistency is based on the fact that information, content provided across different channels of the company needs to be reliable (Akter et al., 2021; Gao & Huang,2021; Chatterjee et al., 2021). Content consistency and process consistency are key components of channel consistency and they significantly contribute to omni-channel customer experience.

- **Synchronization**: In line with channel integration, synchronization describes interconnectivity between offline and online channels (Rodriguez-Torrico et al., 2020) Chang & Li (2022) showed that the scope of the interconnectivity between touchpoints through the complete customer journey is a critical component of seamless omni-channel customer experience.

Theoretical Models on Customer Behavior in Omni-channel Retailing

Customer behaviors may rely on different individual and contextual factors and the extant literature in different parts of customer journey. The impact of these factors is followed via controlled experiments, surveys or experimental designs. In order to explain causal relationships between different antecedents and outcomes in omnichannel customer behavior, following theoretical models provide a framework for actual and future researches:

- **Stimulus-Organism-Response (S-O-R) Theory:** Based on a framework developed by Mehrabian & Russell (1974), S-O-R Theory proposes that it is the stimuli (S) in an environment which evokes variations in persons' internal states (O), which also produces a behavioral reaction (R) as approach or avoidance behavior. As a practical model to discover the determinants and consequences of customer behaviors in omnichannel retailing, S-O-R model is frequently applied. In a social lab, Pantano & Viassone (2015) used S-O-R to explain the impact of simultaneously managed channels' impact on customers' reactions and concluded that participants of the experiment reacted positively to the interactive shopping atmosphere. Arora et al. (2020) used this model to increase insights about showrooming and showed that situational conditions such as assortment issues, poor offline store service, past showrooming experiences can increase showrooming intentions. Mishra et al. (2021a) applied S-O-R model to explore phygital customer experience in omnichannel retailing and concluded that with as a reaction to cross-channel integration, customer empowerment and retention increased. Ürgüplü & Yumurtacı Hüseyinoğlu (2021) used S-O-R framework to showed the impact of omni-channel capability, as a stimulus and customer empowerment, as organism was significantly positive on customer satisfaction, as a response.
- **Social Exchange Theory (SET):** Social exchange theory suggests that based on equity and reciprocity principles, individuals make rational decisions to engage in social relationships via cost and benefit analysis and it is a theory applied in different company-customer interactions (Cropanzano & Mitchell, 2005). Ferm & Thaichon (2021) applied cost benefit analysis SET and found that online interaction propensity and trust had a significant impact on the attitudinal loyalty toward social media participation in online banking services. In a study made concerning the customers of Apple/Kroger physical and/or virtual stores in U.S.A., Lee et al. (2019) used SET to show that omnichannel integration quality influences directly customer engagement. In their paper aiming to explore the interaction between perceived integration quality of channels and customer reactions in Taiwan and in China, Chen et al. (2022) used SET to hint at the impact of omni-channel service quality on customer experience and customer empowerment.
- **Technology Acceptance Model (TAM):** The TAM focuses on the impact of external factors on the beliefs, attitudes and intentions of customers to accept a new technology (Davis et al., 1989). In this model, perceived usefulness and perceived ease of use directly influence behavioral intention, which has a direct impact on the adoption of new technology. In omni-channel retailing context, Silva et al. (2018) show that perceived usefulness and ease of use influence significantly

future use intention of omni-channel retailing. In an extended TAM framework, Kaabachi et al. (2021) suggest that perceived usefulness and ease of use has a positive impact on omni-channel banking adoption. Similarly, with the use of TAM framework, Herrero-Crespo et al. (2022) found that the increase in perceived usefulness in fashion omni-channel retailing environment had a positive impact on webrooming and showrooming behaviors.

- **UTAUT Model:** The unified theory of acceptance and use of technology (UTAUT) is a technology acceptance model developed by Venkatesh et al. (2003) and UTAUT aims to describe users' intentions to use a technological system and following usage behavior. The theory has four key constructs: performance expectancy, effort expectancy, social influence, and enabling conditions. The first three are direct determinants of usage intention and usage behavior, and enabling conditions is a direct determinant of user behavior. Joo & Lee (2016) used UTAUT model to explain omni-channel customer behavior. Mimoun et al. (2022) applied UTAUT framework and found that the introduction of 3D virtual reality application in omnichannel shopping had a positive impact on performance and effort expectancy perceptions.
- **Innovation Diffusion Theory (IDT):** According to Innovation Diffusion Theory, "an innovation is communicated through certain channels over time among the members of a social system" (Rogers,1995, p.5). In this theory, an innovation characterizes a concept accepted as unique by an individual or a group (Shi et al., 2020). The main assumption of IDT is that individuals' innovation-decision process contains three phases: antecedents, process and outcomes (Cao and Li, 2018; Shi et al., 2020). Shi et al. (2020) applied IDT to develop an omni-channel customer experience with connectivity, integration, consistency, flexibility, personalization dimensions and found that these dimensions have a significant impact on perceived compatibility and perceived risk.

Measurement of Customer Experience in Omni-channel Retailing

The customer experience in omni-channel retailing includes different physical, social and phygital touchpoints during the shopping journeys. Differently from multichannel, single channel or cross-channel retailing, it is not easy to measure the customer experience as omni-channel customers can easily move across channels (Verhoef et al.,2015). Meanwhile, research into customer experience measurement is also providing new scales in line with new additional studies focused on omni-channel customer behaviors. Some of them are used as antecedents of outcomes such as satisfaction, loyalty, word of mouth, trust in retailer (e.g. Shi et al,2020; Rahman et al., 2022), whereas other scales are used for specific contexts such as seamless experience (Chang & Li, 2022); mobile customer experience (Lawry & Bhappu, 2021) and omni-channel retailer brand experience (Frasquet-Deltoro et al., 2021).

- **Omni-channel Customer Experience Scale (OCE):** To provide a complete assessment of the critical factors relating to omni-channel experience, Shi et al. (2020) proposed an omni-channel customer experience scale with 5 first-order factors: connectivity (4 items), integration (3 items), consistency (3 items), flexibility (4 items), personalization (5 items). Omnichannel experience construct dimensions are found to be significant antecedents of perceived compatibility and perceived risk, which correspondingly impact customers' purchase intention (Shi et al., 2020). As a main antecedent construct, the impact of this scale on different omnichannel related constructs can be further assessed: omni-channel customer satisfaction, customer loyalty, customer engagement and word-of-mouth.

- **Omni-channel Retail Brand Experience Scale (ORBE):** Frasquet-Deltoro et al. (2021) developed the omni-channel retail brand experience (ORBE) scale in omni-channel retailing to evaluate a holistic brand experience in different physical, digital touchpoints of a single brand retailer. As a second order construct, ORBE comprises 8 dimensions: Sensory dimension (2 items), affective dimension (2 items), intellectual dimension (2 items), behavioral dimension (2 items), lifestyle dimension (3 items), pragmatic dimension (3 items), relational dimension (3 items) and social dimension (2 items). ORBE offers an evaluation of brand experience for a unique brand retailer and this scale can be further applied to assess to assess the brand loyalty and engagement of a unique brand retailer as well as multi-brand retailer (Frasquet-Deltoro et al., 2021).
- **The Mobile in-Store Index (MIX):** Developed by Lawry & Bhappu (2021), MIX is an effective scale to assess consumer engagement in omni-channel retailing. The MIX can be used in combination with validated measures of consumer motivation for using mobile devices in webrooming behaviors, for finding new sources of value co-creation in physical stores, thus offering new insights on increasing consumer engagement in omni-channel retailing (Lawry & Bhappu, 2021). Based on Activity Theory Framework, The MIX consists of 6 validated items: "Review special coupons/promotions sent to their email or phone account; Identify their location on a store map in order to find products; Obtain product information and reviews by scanning labels/tags; Build a gift/shopping list by scanning labels/tags. Order out-of-stock products that are shipped to their home for free; Pay for products in order to skip the checkout line" (Lawry&Bhappu,2021). MIX would provide insights for customer value co-creation in phygital omni-channel customer experience in future applications.
- **Seamless Experience Scale (SE):** In order to understand and measure the seamless experience across different touchpoints of the omnichannel shopping journey, Chang & Li (2022) have recently conceptualized and developed a scale of seamless experience (SE) for the overall evaluation of customer experiences. Through mixed-method approach, the SE construct is developed as a second-order construct including six dimensions: availability of links (3 items), consistency of sales strategies (3 items), information visibility (3 items), simplicity of payment (3 items), flexibility of fulfillment (3 items), convenience of sharing (3 items). The overall SE scale is found to have a direct impact on customers' repurchase intention, word-of-mouth (Chang & Li, 2022). As a scale developed from customer's perspective, SE offers an original common ground to evaluate omnichannel customer experience on individual basis. As a recently developed scale, there is a need to adopt it in different omni-channel retailing contexts such as banking, luxury.
- **Perceived Omni-channel Customer Experience Scale (OCX):** Rahman et al. (2022) propose a new measurement tool that can be thought as a comprehensive scale of omni-channel customer experience. With a mixed-method research design and studies in eight phases, the authors obtain a second-order omni-channel customer experience construct with nine first-order omni-channel evaluation dimensions: social communications (3 items), value (4 items), personalization (4 items), customer service (4 items), consistency (4 items), information safety (4 items), delivery (4 items), product returns (4 items), loyalty programs (4 items). Furthermore, the positive relationship of omni-channel customer experience with customer satisfaction, customers' loyalty intentions, word-of-mouth intentions, share of wallet and trust in omni-channel retailer is also validated (Rahman et al., 2022). As a new construct, it can be used in conjunction with customer engagement scale in the future research.

SOLUTIONS AND RECOMMENDATIONS

The findings in the extant literature in omni-channel customer behavior have significant implications for managers and researchers who aim to understand and implement effective omni-channel marketing strategies in different contexts. Creating awareness about omnichannel retailing of different private and public services, products while offering a consistent and seamless customer experience is critical for the success of omnichannel retailing. With the introduction new customer facing technologies supported by AR, VR, MR, AI in online and offline environments, the customer experience can be enhanced. Meanwhile, as the literature on omni-channel retailing is rather new and also the corporations need more time and investment to establish their hyper-connected omni-channel environment, there are lots of new links to discover between customer behaviors and retailer's activities and investments.

First of all, retailer marketing managers should acknowledge that the customer journey may not have a repetitive and linear cycle so that they set up their omni-channel strategies based on their target market's realities. Secondly, even though, omni-channel retailing priorities may change upon the market realities, businesspeople should recognize the importance of perceived trust, risk, value and channel integration as crucial factors for omni-channel retailing of different products and services. Thirdly, showrooming and webrooming behaviors are critical part of omni-channel journey and the support of social, human, digital and phygital touchpoints would play a critical role in a customer's final decision. For example, as suggested by Fiestas & Tuzovic (2021), mobile assisted showroomers may avoid interacting with frontline sales employees, as a human touchpoint in the physical store. Therefore, businesspeople should be cautious about the new patterns of their customers' omni-channel journey with the support of observations, data analysis and journey mapping. As a newly growing area of research and application, phygital dimension of customer experience can provide new sources of sustainable competitive advantage for the companies involved in omni-channel retailing. Furthermore, as Mimoun et al. (2022) suggest, even if retailers implement an integrated omnichannel strategy, customers continue to be the queens or kings as they have the freedom to adopt a new omni-channel retail technology or touchpoint. Hence, managers might test new technologies with a representative group of their customers before investment decisions, if possible. Social dimension of customer experience, (e.g. positive/negative word of mouth, user generated content) is a non-controllable touchpoint in any omni-channel retailer organization, therefore, as in a classical retailer case, customer service is a key factor to increase customer satisfaction in omni-channel context. Finally, in line with the user generated content and social touchpoint enabling the active involvement of customers, customer value co-creation would offer a critical competitive advantage to omni-channel retailers. An agile omni-channel retailer would enhance the possibility to have valuable market insights and build long-term relationships with the customers who would be part of value co-creation. The role of external and internal partners (loyal and/or actual customers) to support and enhance omni-channel interactivity would be critical to offer real-time ubiquitous customer experience.

FUTURE RESEARCH DIRECTIONS

In omni-channel environment, there is a need for a more consolidative understanding of the actual complexity of different interactions between different touchpoints along the non-linear omni-channel customer journey (Barwitz & Maas, 2018). Some components of omni-channel customer journey are beyond the control of marketers and retailers. Hence, to better evaluate the impact of non-controllable dimensions of customer experience on omni-channel shopping journey and to detect possible gaps between different touchpoints, service blueprinting literature and applications would be supportive in different retailing contexts (Lemon & Verhoef, 2016). The channel integration along the omnichannel customer journey remains as the critical point and there would be need to make longitudinal studies to create journey maps (Mele et al., 2021; Gasparin et al., 2022). Moreover, phygital omni-channel experience requires more research and application to review the integration and interaction of physical, digital and social touchpoints of omnichannel shopping journey. At the beginning, the first-time user customers' acceptance and adoption of phygital touchpoints can be evaluated. Followingly, there would be a need to examine the actual customers' perception of a newly introduced phygital touchpoint. Future research could examine how in-store, mobile and virtual omni-channel technology may impact both the utilitarian and the hedonic aspects of the customer experience (Mimoun et al., 2022).

In webrooming and showrooming behaviors, future research is necessary to evaluate its influence on customer engagement and loyalty. Additionally, there is a need to examine the impact of omni-channel customer services such as buy online pickup in store, buy online return in-store on omni-channel customer engagement. Despite the differences between research context of the studies included in this chapter, the subjectivity of customers, perceived risk, integration remains as a critical antecedent of omni-channel behaviors. There is a need to determine cultural factors that can be prevalent in different industries and countries omni-channel research to evaluate the impact of the subjectivity. Additionally, future field studies involving consumers would be fruitful in order to assess their perceptions of the omni-channel journey. With the introduction of new digital and phygital touchpoints in the short term and long term, the need to re-examine omni-channel customer experience would be dominant. Hence the impact of additional and existing touchpoints on customer experience should be examined in different omni-channel contexts.

CONCLUSION

This chapter aimed to examine the current state of omni-channel customer journey with the main objective and research sub-questions stated in introduction. As the actual stage of an evolution that starts from single channel shopping, and continues multi-channel and cross-channel shopping, omni-channel customer behavior requires new perspectives to understand shopping patterns. Actually, there are different perspectives to explain omni-channel customer journey and there is a need to execute further research to evaluate its complete framework. Based on the extant research and definitions, in order to contribute to future conceptualization of omni-channel customer journey, non-linearity of processes, human, digital, social and phygital touchpoints, customer freedom, control and power remain as critical components of omni-channel customer experiences. An overview of the critical factors influencing omnichannel customer behavior is offered based on the frequently used measurement scales. The findings show that to improve omni-channel customer journey there is a necessity to interpret customers' perceptions in individual and contextual antecedents of different omni-channel customer behaviors. Furthermore, channel consistency and integration are crucial to manage customers' perceptions about the omni-channel capability of a retailer. To better interpret the connections between antecedents and outcomes in omni-channel marketing context, the frequently applied theories are also reviewed with their distinctive results. Finally, the actual results of research of omni-channel customer journey are outlined and recommendations for business people and academicians are also shared. In conclusion, the chapter has provided a wide-ranging review of omni-channel customer behaviors from multiple approaches.

REFERENCES

Ailawadi, K. L., & Farris, P. W. (2017). Managing Multi- and Omni-Channel Distribution: Metrics and Research Directions. *Journal of Retailing*, *93*(1), 120–135. doi:10.1016/j.jretai.2016.12.003

Akter, S., Hossain, T. M. T., & Strong, C. (2021). What Omnichannel Really Means? *Journal of Strategic Marketing*, *29*(7), 567–573. doi:10.1080/0965254X.2021.1937284

Alexander, B., & Cano, M. B. (2020a). Store of the Future: Towards a (re) Invention and (re) Imagination of Physical Store Space in an Omni-Channel Context. *Journal of Retailing and Consumer Services*, *55*, 101913. doi:10.1016/j.jretconser.2019.101913

Alexander, B., & Kent, A. (2020b). Change in Technology-Enabled Omnichannel Customer Experiences in-store. *Journal of Retailing and Consumer Services*, *65*, 102338. Advance online publication. doi:10.1016/j.jretconser.2020.102338

Ameen, N., Tarhini, A., Shah, M., & Madichie, N. O. (2021). Going with the flow: Smart shopping malls and omnichannel retailing. *Journal of Services Marketing*, *35*(3), 325–348. doi:10.1108/JSM-02-2020-0066

Arora, S., Parida, R. R., & Sahney, S. (2020). Understanding Consumers' Showrooming Behaviour: A Stimulus–Organism–Response (SOR) Perspective. *International Journal of Retail & Distribution Management*, *48*(11), 1157–1176. doi:10.1108/IJRDM-01-2020-0033

Arora, S., Sahney, S., & Parida, R. R. (2022). Drivers of showrooming behaviour: insights from integrated perspectives. *International Journal of Retail & Distribution Management, 50*(3), 398-413. doi: 10.1108/IJRDM-09-2020-0374/

Aw, E. C. X. (2019). Understanding the webrooming phenomenon: Shopping motivation, channel-related benefits and costs. *International Journal of Retail & Distribution Management, 47*(10), 1074-1092. doi: 10.1108/IJRDM-01-2019-0026/

Banik, S. (2021). Exploring the involvement-patronage link in the phygital retail experiences. *Journal of Retailing and Consumer Services, 63*, 102739. doi:10.1016/j.jretconser.2021.102739

Barann, B., Hermann, A., Heuchert, M., & Becker, J. A. (2022). Can't touch this? Conceptualizing the Customer Touchpoint in the Context of Omni-Channel Retailing. *Journal of Retailing and Consumer Services*, *65*, 102269. Advance online publication. doi:10.1016/j.jretconser.2020.102269

Barwitz, N., & Maas, P. (2018). Understanding the Omnichannel Customer Journey: Determinants of Interaction Choice. *Journal of Interactive Marketing*, *43*, 116–133. doi:10.1016/j.intmar.2018.02.001

Batat, W. (2022). What does phygital really mean? A Conceptual Introduction to the Phygital Customer Experience (Ph-Cx) Framework. *Journal of Strategic Marketing*, **•••**, 1–24. doi:10.1080/09652 54X.2022.2059775

Beck, N., & Rygl, D. (2015). Categorization of Multiple Channel Retailing in Multi-, Cross-, and Omni-Channel Retailing for Retailers and Retailing. *Journal of Retailing and Consumer Services*, *27*, 170–178. doi:10.1016/j.jretconser.2015.08.001

Berman, B., & Thelen, S. (2018). Planning and implementing an effective omnichannel marketing program. *International Journal of Retail & Distribution Management, 46*(7), 598-614. doi: 10.1108/IJRDM-08-2016-0131/

Bolton, R. N., Gustafsson, A., Tarasi, C. O., & Witell, L. (2022). Designing Satisfying Service Encounters: Website Versus Store Touchpoints. *Journal of the Academy of Marketing Science, 50*(1), 85-107. doi: 10.1007/s11747-021-00808-9

Cai, Y. J., & Lo, C. K. (2020). Omni-channel management in the new retailing era: A systematic review and future research agenda. *International Journal of Production Economics, 229*, 107729. doi:10.1016/j.ijpe.2020.107729

Cao, L., & Li, L. (2018). Determinants of Retailers' Cross-Channel Integration: An Innovation Diffusion Perspective on Omni-Channel Retailing. *Journal of Interactive Marketing, 44*, 1–16. doi:10.1016/j.intmar.2018.04.003

Chang, Y. P., & Li, J. (2022). Seamless Experience in the Context of Omnichannel Shopping: Scale Development and Empirical Validation. *Journal of Retailing and Consumer Services, 64*, 102800. doi:10.1016/j.jretconser.2021.102800

Chatterjee, S., Chaudhuri, R., & Vrontis, D. (2021). Examining the global retail apocalypse during the COVID-19 pandemic using strategic omnichannel management: A consumers' data privacy and data security perspective. *Journal of Strategic Marketing, 29*(7), 617–632. doi:10.1080/0965254X.2021.1936132

Chaudhary, P., Singh, A., & Sharma, S. (2021). Understanding the antecedents of omni-channel shopping by customers with reference to fashion category: The Indian millennials' perspective. *Young Consumers, 23*(2), 304–320. doi:10.1108/YC-05-2021-1327

Chen, T. Y., Yeh, T. L., Wu, H. L., & Deng, S. (2022). Effect of Channel Integration Quality on Consumer Responses within Omni-channel Retailing. *Asia Pacific Journal of Marketing and Logistics*. Advance online publication. doi:10.1108/APJML-04-2021-0270

Chimborazo-Azogue, L. E., Mollá-Descals, A., Miquel-Romero, M. J., & Frasquet, M. (2022). Mobile dependency and uncertainty reduction: Influence on showrooming behaviours and user-generated content creation. *International Journal of Retail & Distribution Management, 50*(8/9), 996–1014. Advance online publication. doi:10.1108/IJRDM-10-2021-0487

Chou, S. Y., Shen, G. C., Chiu, H. C., & Chou, Y. T. (2016). Multichannel service providers' strategy: Understanding customers' switching and free-riding behavior. *Journal of Business Research, 69*(6), 2226–2232. doi:10.1016/j.jbusres.2015.12.034

Cropanzano, R., & Mitchell, M. S. (2005). Social exchange theory: An interdisciplinary review. *Journal of Management, 31*(6), 874–900. doi:10.1177/0149206305279602

Cui, T. H., Ghose, A., Halaburda, H., Iyengar, R., Pauwels, K., Sriram, S., Tucker, C., & Venkataraman, S. (2021). Informational Challenges in Omnichannel Marketing: Remedies and Future Research. *Journal of Marketing, 85*(1), 103–120. doi:10.1177/0022242920968810

Dahana, W. D., Shin, H., & Katsumata, S. (2018). Influence of individual characteristics on whether and how much consumers engage in showrooming behavior. *Electronic Commerce Research, 18*(4), 665–692. doi:10.100710660-017-9277-4

Davis, F. D. (1989). Perceived usefulness, perceived ease of use and user acceptance of information technology. *Management Information Systems Quarterly, 13*(3), 319–339. doi:10.2307/249008

De Keyser, A., Lemon, K. N., Klaus, P., & Keiningham, T. L. (2015). A framework for understanding and managing the customer experience. *Marketing Science Institute working paper series, 85*(1), 15-121

De Keyser, A., Verleye, K., Lemon, K. N., Keiningham, T. L., & Klaus, P. (2020). Moving the Customer Experience Field Forward: Introducing the Touchpoints, Context, Qualities (TCQ) Nomenclature. *Journal of Service Research, 23*(4), 433–455. doi:10.1177/1094670520928390

Fang, J., Liu, H., Li, Y., & Cai, Z. (2021). Retaining customers with in-store mobile usage experience in omni-channel retailing: The moderating effects of product information overload and alternative attractiveness. *Electronic Commerce Research and Applications, 46*, 101028. doi:10.1016/j.elerap.2020.101028

Farah, M. F., Ramadan, Z. B., & Harb, D. H. (2019). The examination of virtual reality at the intersection of consumer experience, shopping journey and physical retailing. *Journal of Retailing and Consumer Services, 48*, 136–143. doi:10.1016/j.jretconser.2019.02.016

Ferm, L. E. C., & Thaichon, P. (2021). Customer pre-participatory social media drivers and their influence on attitudinal loyalty within the retail banking industry: A multi-group analysis utilizing social exchange theory. *Journal of Retailing and Consumer Services, 61*, 102584. doi:10.1016/j.jretconser.2021.102584

Fernández, N. V., Pérez, M. J. S., & Vázquez-Casielles, R. (2018). Webroomers versus showroomers: Are they the same? *Journal of Business Research, 92*, 300–320. doi:10.1016/j.jbusres.2018.08.004

Fiestas, J. C., & Tuzovic, S. (2021). Mobile-assisted showroomers: Understanding their purchase journey and personalities. *Journal of Retailing and Consumer Services, 58*, 102280. doi:10.1016/j.jretconser.2020.102280

Flavián, C., Gurrea, R., & Orús, C. (2016). Choice confidence in the webrooming purchase process: The impact of online positive reviews and the motivation to touch. *Journal of Consumer Behaviour, 15*(5), 459–476. doi:10.1002/cb.1585

Flavián, C., Gurrea, R., & Orús, C. (2019a). Feeling Confident and Smart with Webrooming: Understanding the Consumer's Path to Satisfaction. *Journal of Interactive Marketing, 47*, 1–15. doi:10.1016/j.intmar.2019.02.002

Flavián, C., Gurrea, R., & Orus, C. (2020). Combining Channels to Make Smart Purchases: The Role of Webrooming and Showrooming. *Journal of Retailing and Consumer Services, 52*, 101923. doi:10.1016/j.jretconser.2019.101923

Flavian, C., Gurrea, R., & Orús, C. (2021). Mobile word of mouth (m-WOM): Analysing its negative impact on webrooming in omnichannel retailing. *International Journal of Retail & Distribution Management, 49*(3), 394–420. doi:10.1108/IJRDM-05-2020-0169

Flavián, C., Ibáñez-Sánchez, S., & Orús, C. (2019). The Impact of Virtual, Augmented and Mixed Reality Technologies on the Customer Experience. *Journal of Business Research*, *100*, 547–560. doi:10.1016/j.jbusres.2018.10.050

Følstad, A., & Kvale, K. (2017). Customer journeys: A systematic literature review. *Journal of Service Theory and Practice*, *28*(2), 196–227. doi:10.1108/JSTP-11-2014-0261

Frasquet-Deltoro, M., Molla-Descals, A., & Miquel-Romero, M. J. (2021). Omnichannel retailer brand experience: Conceptualisation and proposal of a comprehensive scale. *Journal of Brand Management*, *28*(4), 388–401. doi:10.105741262-021-00233-x

Gao, L., Melero, I., & Sese, F. J. (2020). Multichannel Integration along the Customer Journey: A Systematic Review and Research Agenda. *Service Industries Journal*, *40*(15-16), 1087–1118. doi:10.1080/02642069.2019.1652600

Gao, M., & Huang, L. (2021). Quality of channel integration and customer loyalty in omnichannel retailing: The mediating role of customer engagement and relationship program receptiveness. *Journal of Retailing and Consumer Services*, *63*, 102688. doi:10.1016/j.jretconser.2021.102688

Gao, W., Fan, H., Li, W., & Wang, H. (2021). Crafting The Customer Experience in Omnichannel Contexts: The Role of Channel Integration. *Journal of Business Research*, *126*, 12–22. doi:10.1016/j.jbusres.2020.12.056

Gasparin, I., Panina, E., Becker, L., Yrjölä, M., Jaakkola, E., & Pizzutti, C. (2022). Challenging the "Integration Imperative": A Customer Perspective on Omnichannel Journeys. *Journal of Retailing and Consumer Services*, *64*, 102829. doi:10.1016/j.jretconser.2021.102829

Grewal, D., Noble, S. M., Roggeveen, A. L., & Nordfalt, J. (2020). The future of in-store technology. *Journal of the Academy of Marketing Science*, *48*(1), 96–113. doi:10.100711747-019-00697-z

Grewal, D., & Roggeveen, A. L. (2020). Understanding retail experiences and customer journey management. *Journal of Retailing*, *96*(1), 3–8. doi:10.1016/j.jretai.2020.02.002

Guo, Y., Zhang, M., & Wang, V. L. (2021). Webrooming or showrooming? The moderating effect of product attributes. *Journal of Research in Interactive Marketing*. Advance online publication. doi:10.1108/JRIM-08-2020-0161

Hajdas, M., Radomska, J., & Silva, S. C. (2020). The omni-channel approach: A utopia for companies? *Journal of Retailing and Consumer Services*, *102131*. Advance online publication. doi:10.1016/j.jretconser.2020.102131

Hajdas, M., Radomska, J., & Silva, S. C. (2020). The omni-channel approach: A utopia for companies? *Journal of Retailing and Consumer Services*, *102131*. Advance online publication. doi:10.1016/j.jretconser.2020.102131

Hamouda, M. (2019). Omni-channel banking integration quality and perceived value as drivers of consumers' satisfaction and loyalty. *Journal of Enterprise Information Management*, *32*(4), 608–625. doi:10.1108/JEIM-12-2018-0279

Herrero-Crespo, A., Viejo-Fernández, N., Collado-Agudo, J., & Pérez, M. J. S. (2021). Webrooming or showrooming, that is the question: Explaining omnichannel behavioural intention through the technology acceptance model and exploratory behaviour. *Journal of Fashion Marketing and Management, 26*(3), 401–419. doi:10.1108/JFMM-05-2020-0091

Herrero-Crespo, A., Viejo-Fernández, N., Collado-Agudo, J., & Sanzo Pérez, M. J. (2022). Webrooming or showrooming, that is the question: Explaining omnichannel behavioural intention through the technology acceptance model and exploratory behaviour. *Journal of Fashion Marketing and Management, 26*(3), 401–419. doi:10.1108/JFMM-05-2020-0091

Hickman, E., Kharouf, H., & Sekhon, H. (2020). An omnichannel approach to retailing: Demystifying and identifying the factors influencing an omnichannel experience. *International Review of Retail, Distribution and Consumer Research, 30*(3), 266–288. doi:10.1080/09593969.2019.1694562

Hilken, T., Heller, J., Chylinski, M., Keeling, D. I., Mahr, D., & de Ruyter, K. (2018). Making Omnichannel an Augmented Reality: The Current and Future State of the Art. *Journal of Research in Interactive Marketing, 12*(4), 509–523. doi:10.1108/JRIM-01-2018-0023

Hilken, T., Chylinski, M., Keeling, D. I., Heller, J., de Ruyter, K., & Mahr, D. (2022). How to strategically choose or combine augmented and virtual reality for improved online experiential retailing. *Psychology & Marketing, 39*(3), 495-507. doi: 10.1002/

Hossain, T. M. T., Akter, S., Kattiyapornpong, U., & Dwivedi, Y. (2020). Reconceptualizing integration quality dynamics for omnichannel marketing. *Industrial Marketing Management, 87*, 225-241. doi:10.1016/j.indmarman.2019.12.006

Huré, E., Picot-Coupey, K., & Ackermann, C. L. (2017). Understanding omni-channel shopping value: A mixed-method study. *Journal of Retailing and Consumer Services, 39*, 314–330. doi:10.1016/j.jretconser.2017.08.011

Hüseyinoğlu, I. Ö. Y., Galipoğlu, E., & Kotzab, H. (2017). Social, local and mobile commerce practices in omni-channel retailing: Insights from Germany and Turkey. *International Journal of Retail & Distribution Management, 45*(7-8), 711–729. doi:10.1108/IJRDM-09-2016-0151

Ieva, M., & Ziliani, C. (2018). Mapping touchpoint exposure in retailing: Implications for developing an omnichannel customer experience. *International Journal of Retail & Distribution Management, 46*(3), 304–322. doi:10.1108/IJRDM-04-2017-0097

Inman, J. J., & Nikolova, H. (2017). Shopper-facing retail technology: A retailer adoption decision framework incorporating shopper attitudes and privacy concerns. *Journal of Retailing, 93*(1), 7–28. doi:10.1016/j.jretai.2016.12.006

Jacob, F., Pez, V., & Volle, P. (2021). Shifting to phygital experience management with design science: A six-step method to manage customer journeys. *Journal of Strategic Marketing*, 1–22. doi:10.1080/0965254X.2021.2016894

Jain, S., Schweiss, T., Bender, S., & Werth, D. (2021). Omnichannel Retail Customer Experience with Mixed-Reality Shopping Assistant Systems. In *Advances in Visual Computing. International Symposium on Visual Computing 2021. Lecture Notes in Computer Science* (vol. 13017). Springer. 10.1007/978-3-030-90439-5_40

Jones, A. L., Miller, J. W., Griffis, S. E., Whipple, J. M., & Voorhees, C. M. (2021). An examination of the effects of omni-channel service offerings on retailer performance. *International Journal of Physical Distribution & Logistics Management*, *52*(2), 150–169. doi:10.1108/IJPDLM-06-2020-0175

Joo, H., & Lee, E. J. (2016). A study on the user acceptance model of omni channel service based on Unified Theory of Acceptance and Use of Technology (UTAUT). *Family and Environment Research*, *54*(4), 405–414. doi:10.6115/fer.2016.031

Juaneda-Ayensa, E., Mosquera, A., & Sierra Murillo, Y. (2016). Omnichannel customer behavior: Key drivers of technology acceptance and use and their effects on purchase intention. *Frontiers in Psychology*, *7*, 1117. doi:10.3389/fpsyg.2016.01117 PMID:27516749

Juaneda-Ayensa, E., Mosquera, A., & Sierra Murillo, Y. (2016). Omnichannel customer behavior: Key drivers of technology acceptance and use and their effects on purchase intention. *Frontiers in Psychology*, *7*, 1117. doi:10.3389/fpsyg.2016.01117 PMID:27516749

Kaabachi, S. B., Mrad, S. B., Fiedler, A., & Kpossa, M. R. (2021). The Role of Personal Innovativeness in French Omnichannel Banking. *Journal of Euromarketing*, *30*, 72–94. doi:10.9768/jem.v30i3and4.35

Kang, J. Y. M. (2018). Showrooming, Webrooming, and User-Generated Content Creation in the Omnichannel Era. *Journal of Internet Commerce*, *17*(2), 145–169. doi:10.1080/15332861.2018.1433907

Kang, J. Y. M. (2019). What drives omnichannel shopping behaviors? Fashion lifestyle of social-local-mobile consumers. *Journal of Fashion Marketing and Management*, *23*(2), 224–238. doi:10.1108/JFMM-07-2018-0088

Kaur, K., Bakar, E. A., & Singh, J. (2020). Theoretical framework development on users adoption of omni-channel retailing of fashion apparels based on UTAUT2 and the role of personal innovativeness, brand image and fashion involvement. In *20th Kuala Lumpur International Business, Economics and Law Conference* (pp. 67-81). Academic Press.

Kazancoglu, I., & Aydin, H. (2018). An investigation of consumers' purchase intentions towards omni-channel shopping: A qualitative exploratory study. *International Journal of Retail & Distribution Management*, *46*(10), 959–976. doi:10.1108/IJRDM-04-2018-0074

Kim, B., & Park, M. J. (2019). An examination of the effect of consumer personal traits on show rooming behavior: The effect of product type. *Journal of Internet Commerce*, *18*(1), 24–44. doi:10.1080/15332861.2018.1547569

Kim, J. H., Kim, M., Yoo, J., & Park, M. (2020). Consumer decision-making in a retail store: The role of mental imagery and gender difference. *International Journal of Retail & Distribution Management*, *49*(3), 421–445. doi:10.1108/IJRDM-10-2019-0353

Kleinlercher, K., Linzmajer, M., Verhoef, P. C., & Rudolph, T. (2020). Antecedents of webrooming in omnichannel retailing. *Frontiers in Psychology, 3342*, 606798. Advance online publication. doi:10.3389/fpsyg.2020.606798 PMID:33329282

Larke, R., Kilgour, M., & O'Connor, H. (2018). Build touchpoints and they will come: transitioning to omnichannel retailing. *International Journal of Physical Distribution & Logistics Management, 48*(4), 465-483. doi: 10.1108/IJPDLM-09-2016-0276/

Lawry, C. A., & Bhappu, A. D. (2021). Measuring consumer engagement in omnichannel retailing: the mobile in-store experience (MIX) index. *Frontiers in Psychology, 12*. doi: 10.3389/fpsyg.2021.661503

Lazaris, C., Sarantopoulos, P., Vrechopoulos, A., & Doukidis, G. (2021). Effects of Increased Omnichannel Integration on Customer Satisfaction and Loyalty Intentions. *International Journal of Electronic Commerce*, *25*(4), 440–468. doi:10.1080/10864415.2021.1967005

Lazaris, C., Vrechopoulos, A., Sarantopoulos, P., & Doukidis, G. (2022). Additive omnichannel atmospheric cues: The mediating effects of cognitive and affective responses on purchase intention. *Journal of Retailing and Consumer Services*, *64*, 102731. doi:10.1016/j.jretconser.2021.102731

Le, A. N. H., & Le, X. D. N. (2020). A moderated mediating mechanism of omnichannel customer experiences. *International Journal of Retail & Distribution Management*, *49*(5), 595–615. doi:10.1108/IJRDM-02-2020-0054

Lee, Z. W., Chan, T. K., Chong, A. Y. L., & Thadani, D. R. (2019). Customer engagement through omnichannel retailing: The effects of channel integration quality. *Industrial Marketing Management*, *77*, 90–101. doi:10.1016/j.indmarman.2018.12.004

Lemon, K. N., & Verhoef, P. C. (2016). Understanding Customer Experience Throughout the Customer Journey. *Journal of Marketing*, *80*(6), 69–96. doi:10.1509/jm.15.0420

Liao, Y., Deschamps, F., Loures, E. D. F. R., & Ramos, L. F. P. (2017). Past, present and future of Industry 4.0-a systematic literature review and research agenda proposal. *International Journal of Production Research*, *55*(12), 3609–3629. doi:10.1080/00207543.2017.1308576

Lim, X. J., Cheah, J. H., Dwivedi, Y. K., & Richard, J. E. (2022). Does retail type matter? Consumer responses to channel integration in omni-channel retailing. *Journal of Retailing and Consumer Services*, *67*, 102992. doi:10.1016/j.jretconser.2022.102992

Mehrabian, A., & Russell, J. A. (1974). *An Approach to Environmental Psychology*. The MIT Press.

Mele, C., Russo-Spena, T., Tregua, M., & Amitrano, C. C. (2021). The millennial customer journey: A Phygital mapping of emotional, behavioural, and social experiences. *Journal of Consumer Marketing*, *38*(4), 420–433. doi:10.1108/JCM-03-2020-3701

Melero, I., Sese, F. J., & Verhoef, P. C. (2016). Recasting the customer experience in today's omni-channel environment. *Universia Business Review*, (50), 18–37. doi:10.3232/UBR.2016.V13.N2.01

Mimoun, M. S. B., Miltgen, C. L., & Slama, B. (2022). Is the shopper always the king/queen? Study of omnichannel retail technology use and shopping orientations. *Journal of Retailing and Consumer Services*, *65*, 102844. doi:10.1016/j.jretconser.2021.102844

Mishra, R., Singh, R. K., & Koles, B. (2021). Consumer Decision-Making in Omnichannel Retailing: Literature Review and Future Research Agenda. *International Journal of Consumer Studies*, *45*(2), 147–174. doi:10.1111/ijcs.12617

Mishra, S., Malhotra, G., Arora, V., & Mukhopadhyay, S. (2021b). Omnichannel retailing: Does it empower consumers and influence patronage? *International Journal of Retail & Distribution Management*, *50*(2), 229–250. doi:10.1108/IJRDM-04-2021-0199

Mishra, S., Malhotra, G., Chatterjee, R., & Shukla, Y. (2021a). Consumer retention through phygital experience in omnichannel retailing: Role of consumer empowerment and satisfaction. *Journal of Strategic Marketing*, 1–18. doi:10.1080/0965254X.2021.1985594

Moher, D., Liberati, A., Tetzlaff, J., & Altman, D. G. (2009). Preferred Reporting Items for Systematic Reviews and Meta Analyses: The PRISMA Statement. *PLoS Medicine*, *6*(7), e1000097. doi:10.1371/journal.pmed.1000097 PMID:19621072

Morgan, R. M., & Hunt, S. D. (1994). The commitment-trust theory of relationship marketing. *Journal of Marketing*, *58*(3), 20–38. doi:10.1177/002224299405800302

Mosquera, A., Olarte Pascual, C. & Juaneda Ayensa, E. (2017): Understanding the customer experience in the age of omni-channel shopping, *Icono 14*, *15*(2), 166-185. doi:10.7195/ri14.v15i2.1070

Murfield, M., Boone, C. A., Rutner, P., & Thomas, R. (2017). Investigating logistics service quality in omni-channel retailing. *International Journal of Physical Distribution & Logistics Management*, *47*(4), 263–296. doi:10.1108/IJPDLM-06-2016-0161

Neslin, S. A. (2022). The Omnichannel Continuum: Integrating Online and Offline Channels Along the Customer Journey. *Journal of Retailing*, *98*(1), 111–132. doi:10.1016/j.jretai.2022.02.003

Neslin, S. A., Grewal, D., Leghorn, R., Shankar, V., Teerling, M. L., Thomas, J. S., & Verhoef, P. C. (2006). Challenges and Opportunities in Multichannel Customer Management. *Journal of Service Research*, *9*(2), 95–112. doi:10.1177/1094670506293559

Neslin, S. A., & Shankar, V. (2009). Key Issues in Multichannel Customer Management: Current Knowledge and Future Directions. *Journal of Interactive Marketing*, *23*(1), 70–81. doi:10.1016/j.intmar.2008.10.005

Nguyen, A., McClelland, R., Hoang Thuan, N., & Hoang, T. G. (2022). Omnichannel Marketing: Structured Review, Synthesis, and Future Directions. *International Review of Retail, Distribution and Consumer Research*, *32*(3), 1–45. doi:10.1080/09593969.2022.2034125

Palazón, M., López, M., Sicilia, M., & López, I. (2022). The customer journey: A proposal of indicators to evaluate integration and customer orientation. *Journal of Marketing Communications*, *28*(5), 1–32. doi:10.1080/13527266.2022.2051584

Pangarkar, A., Arora, V., & Shukla, Y. (2022). Exploring phygital omnichannel luxury retailing for immersive customer experience: The role of rapport and social engagement. *Journal of Retailing and Consumer Services*, *68*, 103001. doi:10.1016/j.jretconser.2022.103001

Pantano, E., & Viassone, M. (2015). Engaging consumers on new integrated multichannel retail settings: Challenges for retailers. *Journal of Retailing and Consumer Services*, *25*, 106–114. doi:10.1016/j.jretconser.2015.04.003

Park, J., & Kim, R. B. (2022). Importance of offline service quality in building loyalty of OC service brand. *Journal of Retailing and Consumer Services*, *65*, 102493. doi:10.1016/j.jretconser.2021.102493

Park, J., & Stoel, L. (2005). Effect of brand familiarity, experience and information on online apparel purchase. *International Journal of Retail & Distribution Management*, *33*(2), 148–160. doi:10.1108/09590550510581476

Picot-Coupey, K., Huré, E., & Piveteau, L. (2016). Channel design to enrich customers' shopping experiences: Synchronizing clicks with bricks in an omni-channel perspective - the Direct Optic case. *International Journal of Retail & Distribution Management*, *44*(3), 336–368. doi:10.1108/IJRDM-04-2015-0056

Qin, H., Peak, D. A., & Prybutok, V. (2021). A virtual market in your pocket: How does mobile augmented reality (MAR) influence consumer decision making? *Journal of Retailing and Consumer Services*, *58*, 102337. doi:10.1016/j.jretconser.2020.102337

Quach, S., Barari, M., Moudrý, D. V., & Quach, K. (2022). Service Integration in Omnichannel Retailing and its Impact on Customer Experience. *Journal of Retailing and Consumer Services*, *102267*, 102267. Advance online publication. doi:10.1016/j.jretconser.2020.102267

Rahman, S. M., Carlson, J., Gudergan, S. P., Wetzels, M., & Grewal, D. (2022). Perceived Omnichannel Customer Experience (OCX): Concept, measurement, and impact. *Journal of Retailing*. Advance online publication. doi:10.1016/j.jretai.2022.03.003

Rigby, D. (2011). The future of shopping. *Harvard Business Review*, *89*(12), 65–76. https://hbr.org/2011/12/the-future-of-shopping

Rodríguez-Torrico, P., Cabezudo, R. S. J., San-Martín, S., & Apadula, L. T. (2021). Let it flow: The role of seamlessness and the optimal experience on consumer word of mouth in omnichannel marketing. *Journal of Research in Interactive Marketing*. Advance online publication. doi:10.1108/JRIM-06-2021-0154

Rodríguez-Torrico, P., Trabold Apadula, L., San-Martín, S., & San Jose Cabezudo, R. (2020). Have an omnichannel seamless interaction experience! Dimensions and effect on consumer satisfaction. *Journal of Marketing Management*, *36*(17-18), 1731–1761. doi:10.1080/0267257X.2020.1801798

Rogers, E. M. (1995). *The Diffusion of Innovations*. Free Press.

Saghiri, S., Wilding, R., Mena, C., & Bourlakis, M. (2017). Toward a three-dimensional framework for omni-channel. *Journal of Business Research*, *77*, 53–67. doi:10.1016/j.jbusres.2017.03.025

Sahu, K. C., Naved Khan, M., & Gupta, K. D. (2021). Determinants of webrooming and showrooming behavior: A systematic literature review. *Journal of Internet Commerce*, *20*(2), 137–166. doi:10.1080/15332861.2020.1863041

Salvietti, G., Ziliani, C., Teller, C., Ieva, M., & Ranfagni, S. (2022). Omnichannel retailing and post-pandemic recovery: Building a research agenda. *International Journal of Retail & Distribution Management*, *50*(8/9), 1156–1181. Advance online publication. doi:10.1108/IJRDM-10-2021-0485

Schenk, B., Mateusz, D., Schwabe, C., & Schwabe, G. (2021). What Citizens Experience and How Omni-Channel Could Help-Insights from a Building Permit Case. *Information Technology & People*, *2021*, 277–299. doi:10.1108/ITP-06-2020-0374

Shankar, A., Gupta, M., Tiwari, A. K., & Behl, A. (2021). How does convenience impact showrooming intention? Omnichannel retail strategies to manage global retail apocalypse. *Journal of Strategic Marketing*, 1–22. doi:10.1080/0965254X.2021.1965189

Sheth, J. N. (2021). Future of brick and mortar retailing: How will it survive and thrive? *Journal of Strategic Marketing*, *29*(7), 598–607. doi:10.1080/0965254X.2021.1891128

Shi, S., Wang, Y., Chen, X., & Zhang, Q. (2020). Conceptualization of omnichannel customer experience and its impact on shopping intention: A mixed-method approach. *International Journal of Information Management*, *50*, 325–336. doi:10.1016/j.ijinfomgt.2019.09.001

Silva, S. C. E., Martins, C. C., & Sousa, J. M. D. (2018). Omnichannel approach: Factors affecting consumer acceptance. *Journal of Marketing Channels*, *25*(1-2), 73–84. doi:10.1080/1046669X.2019.1647910

Tueanrat, Y., Papagiannidis, S., & Alamanos, E. (2021). A Conceptual Framework of the Antecedents of Customer Journey Satisfaction in Omnichannel Retailing. *Journal of Retailing and Consumer Services*, *61*, 102550. doi:10.1016/j.jretconser.2021.102550

Tyrväinen, O., Karjaluoto, H., & Saarijärvi, H. (2020). Personalization and hedonic motivation in creating customer experiences and loyalty in omnichannel retail. *Journal of Retailing and Consumer Services*, *57*, 102233. doi:10.1016/j.jretconser.2020.102233

Ürgüplü, Ö., & Yumurtacı Hüseyinoğlu, I. Ö. (2021). The mediating effect of consumer empowerment in omni-channel retailing. *International Journal of Retail & Distribution Management*, *49*(11), 1481–1496. doi:10.1108/IJRDM-10-2020-0403

Van Nguyen, A. T., McClelland, R., & Thuan, N. H. (2022). Exploring Customer Experience During Channel Switching in Omnichannel Retailing Context: A Qualitative Assessment. *Journal of Retailing and Consumer Services, 64*, 102803. doi:10.1016/j.jretconser.2021.102803

Vannucci, V., & Pantano, E. (2019). Digital or human touchpoints? Insights from consumer-facing in-store services. *Information Technology & People, 33*(1), 296-310. doi: 10.1108/ITP-02-2018-0113/

Verhoef, P. C., Kannan, P. K., & Inman, J. J. (2015). From Multi-Channel Retailing to Omni-Channel Retailing: Introduction to the Special Issue on Multi-Channel Retailing. *Journal of Retailing*, *91*(2), 174–181. doi:10.1016/j.jretai.2015.02.005

Verhoef, P. C., Neslin, S. A., & Vroomen, B. (2007). Multichannel Customer Management: Understanding the Research-Shopper Phenomenon. *International Journal of Research in Marketing*, *24*(2), 129–148. doi:10.1016/j.ijresmar.2006.11.002

Verhoef, P. C., Stephen, A. T., Kannan, P. K., Luo, X., Abhishek, V., Andrews, M., Bart, Y., Datta, H., Fong, N., Hoffman, D. L., Hu, M. M., Novak, T., Rand, W., & Zhang, Y. (2017). Consumer Connectivity in A Complex, Technology-Enabled, and Mobile-Oriented World with Smart Products. *Journal of Interactive Marketing*, *40*(1), 1–8. doi:10.1016/j.intmar.2017.06.001

Viejo-Fernandez, N., Sanzo-Pérez, M. J., & Vázquez-Casielles, R. (2020). Is showrooming really so terrible? Start understanding showroomers. *Journal of Retailing and Consumer Services*, *54*, 102048. doi:10.1016/j.jretconser.2020.102048

Von Briel, F. (2018). The future of omnichannel retail: A four-stage Delphi study. *Technological Forecasting and Social Change*, *132*, 217–229. doi:10.1016/j.techfore.2018.02.004

Wang, J., & Wang, S. (2022). Revisiting the showrooming effect on online and offline retailers: The strategic role of in-store service. *Journal of Retailing and Consumer Services*, *66*, 102884. doi:10.1016/j.jretconser.2021.102884

Yrjölä, M., Spence, M. T., & Saarijärvi, H. (2018). Omni-channel retailing: Propositions, examples and solutions. *International Review of Retail, Distribution and Consumer Research*, *28*(3), 259–276. doi:10.1080/09593969.2018.1445657

Zhang, M., Li, Y., Sun, L., & Moustapha, F. A. (2022). Integrated store service quality measurement scale in omni-channel retailing. *International Journal of Retail & Distribution Management*, *50*(7), 839–859. Advance online publication. doi:10.1108/IJRDM-02-2021-0056

Zhang, M., Ren, C., Wang, G. A., & He, Z. (2018). The impact of channel integration on consumer responses in omni-channel retailing: The mediating effect of consumer empowerment. *Electronic Commerce Research and Applications*, *28*, 181–193. doi:10.1016/j.elerap.2018.02.002

KEY TERMS AND DEFINITIONS

Customer Journey: Whole stages of interaction of the customer and the retailer from pre-purchase to post-purchase.

Cross-Channel Shopping: Customer's aptitude to migrate among different channels in their consumption journey.

Omni-Channel Shopping Experience: Combination of all activities before during and after purchase decision in omnichannel customer journey.

Phygital Customer Experience: Unique combination of physical experience and virtual experience in different shopping context.

Showrooming: Customer's practice of searching for information in online channels and then purchasing in offline channels.

Touchpoint: Point of contact between the retailer and customer.

Webrooming: Customer's practice of searching for information online and then purchasing offline.

Chapter 5
Challenges in Adapting to Mobile Banking From the Perspective of the Customers

Carolina Afonso
https://orcid.org/0000-0002-8851-1571
Advance/CSG, ISEG, Universidade de Lisboa, Portugal

João Duarte Teixeira
ISEG, Universidade de Lisboa, Portugal

ABSTRACT

This chapter aims to understand the challenges in transitioning to the digital era of banking from the perspective of customers. This study uses an exploratory and quantitative approach. Quantitative data was collected from 329 bank clients, and then descriptive and correlation analysis was performed. The findings showed that performance expectancy, effort expectancy, and facilitating conditions have a positive relation, whereas social influence was found to have weak relation. Academically, this study presents valuable insights to mobile banking utilization and to the adaptation of the UTAUT model to banking sector. From a managerial perspective, this study provides insights into clients' challenges and suggestions to better integrate online platforms into branch service and to improve customer experience.

INTRODUCTION

Within the scope of digital transformation, self-service technologies (SSTs) have been increased. SSTs enables customers to produce a service without direct employee involvement (Meuter, Ostrom, Roundtree, & Bitner, 2000). They are commonly seen at self-pumping petrol stations, ticket machines at bus stations, and ATMs (Lovelock & Wirtz, 2016). SSTs have been present in the Portuguese banking sector since the introduction of the ATM in the 1980s (SIBS, 2021). Since then, the sector has been an early adopter of technology to differentiate itself from competitors (Devlin, 1995). Other examples of SSTs in the sector include telephone, internet, and mobile banking.

DOI: 10.4018/978-1-6684-5538-8.ch005

In the retail banking sector, the main advantages of adopting SSTs include costs reduction, service improvement efficiency, and effectiveness (Proença & Rodrigues, 2011). For customers, the ease of use, convenience, cost and time savings, and control over services increase the SST adoption rate (Curran & Meuter, 2005). Organizations need to create positive customer experiences, encourage the use of SSTs, and assist first-time users, which improves long-term customer satisfaction and loyalty and reduces waiting times (Djelassi, Diallo, & Zielke, 2018). However, they are important aspects to explore and monitor (Iqbal, Hassan, & Habibah, 2018).

When facing technology adoption, there are many challenges perceived by consumers. Despite several studies reporting on barriers regarding technology adoption regarding mobile banking (Al-Saedi & Al-Emran, 2021; Jadil, Rana & Dwivedi, 2021), there is a gap in thorough understanding of technology adoption applied to mobile banking within the context of Portugal.

Hence, the objective of the present study is first, to better understand the Portuguese context of mobile banking technologies adoption and, secondly to explore the relation between performance expectancy, effort expectancy, social influence and facilitating conditions regarding mobile banking technology adoption in Portugal.

To the best of our knowledge, we are the first to empirically explore the interrelations of the dimensions of mobile banking technology adoption within Portuguese market.

BACKGROUND

Mobile Banking

Many authors define mobile banking as an application of m-commerce that allows customers to access bank accounts through mobile devices, such as smartphones or personal digital assistants, to perform transactions, such as checking the account balance, making transfers, payments, and selling stocks (Alafeef, Singh, & Ahmad, 2012). Opening a bank account using a smartphone and its camera is possible in many banks as the industry begins to serve the generation of mobile-first clients (Zhang, Lu, & Kizildag, 2018). Technological advancements have made bank visits less frequent as the digital banking experience capitalizes on disruptive technologies, such as artificial intelligence, chatbots, and voice-based payments (Singh & Srivastava, 2020).

According to a report on financial and digital inclusion and banking products choice in Portugal (Banco de Portugal, 2020), approximately 46% of people with bank accounts utilized digital channels related to financial services. In the same survey, it is possible to note that 19.6% of inquired people used home banking and the apps provided by the financial institution, 10% used home banking, and 16.5% only used the apps. The main users of digital channels are aged between 25 and 39 years, have higher education and have high-income levels (Banco de Portugal, 2020). The main reasons given in the survey of the previous report for not using digital channels were difficulty using technology, preferring to use ATMs or personal contact in an agency, not trusting security, and not accessing the internet regularly.

The urgency of mobile banking adoption comes from dematerialization and cost reductions in the banking industry, as physical branches and human resources become scarce (Balkan, 2021). From 2017 to 2020, the Portuguese banking sector saw a 15% reduction in physical locations (Associação Portuguesa de Bancos, 2020). In the same period, the Euro area witnessed an 18% decrease in banking branches and, in 2020 alone, an 8% decrease from the previous year (European Central Bank, 2021).

Fintech and Regulation

Fintech, short form for Financial Technology, includes innovative financial solutions facilitated by IT, often developed by start-up companies and entities in the financial services sector (Puschmann, 2017). Vives (2017) argues that fintech has a welfare-enhancing disruptive capability to automate financial activities, provide more cost-effective products, and revolutionize payment systems. The author also states that these technologies reduce the cost of intermediation and increase access to financial services by reducing the information asymmetries of banking and have a larger innovating capacity than traditional businesses.

New financial institutions have emerged under the app-based banking concept, such as Revolut and N26, offering a fully remote service with reduced bureaucracy, few fees, openness for all nationalities, and a wide variety of transactions available (Moştcanu, Faccia, & Bhatia, 2020). Other nonbanks, such as Apple, Google, and PayPal, threaten traditional banking institutions by offering innovative payment solutions, such as virtual wallets, to their extensive number of clients (Thakor, 2020). New entrants can capitalize on millennials' reluctance toward traditional banking (Deloitte, 2015) and the possibility of being protected by the government's deposit insurance as banks (Vives, 2017) compose the strategies of new players.

The new era of digital financial services raises financial stability concerns, and without proper regulation, it can compromise the intended benefits of fintech (Vives, 2019). Regulation faces balancing market integrity challenges and fostering innovation (Wójcik, 2020). In other words, as the traditional banking and financial services sector is being introduced to vast amounts of products, processes, and information, market stability, and consumer protection may be jeopardized (Brummer & Yadav, 2019).

To protect consumers' data privacy and improve the safety of payment systems, the EU has implemented Payment Services Directive 2 (PSD 2), a directive that contributes to the integration of electronic payments within the EU through stricter security, more transparency, and information. The PSD 2 constitutes the legal foundation for the Single Euro Payments Area (SEPA), an initiative that harmonizes transfers of funds within its members using a single currency, the euro (EUR-Lex, 2015).

Blockchain technology, a decentralized network that records transactions without any intermediary, has valuable applications in banking as it improves security, recordkeeping, transaction speed, and transparency while reducing processing costs, risk of errors, fraud, and money laundering (Garg, et al., 2021). Blockchain technology can improve financial inclusion in developing countries with unbanked or underbanked populations by providing access to affordable, transparent financial services (Kshetri, 2017).

Internet habits of the Portuguese population

There were 8.58 million internet users in Portugal in January 2021. Internet penetration in Portugal stood at 84.2% in January 2021. The number of internet users in Portugal increased by 62 thousand (+0.7%) between 2020 and 2021 and is forecasted to continue increasing over the years (Statista, 2022).

According to Marktest (2021), 78.4% of the Portuguese population use the internet in 2021, increasing 75.7% of registered users in 2020. Smartphones, PCs, smart TVs, and tablets continue to be the preferred platforms for accessing the internet. In 2020, there were approximately 7.2 million smartphone users, which corresponds to a smartphone penetration rate of 84.2% in the Portuguese mobile phone market (Marktest, 2020).

In 2018, 39% of all individuals in Portugal used the internet for online banking (Statista, 2021). The upward trend in internet usage can be supported by the increase in domestic households with internet connections in Portugal, increasing from 80.9% in 2019 to 84.5% in 2020 (INE, 2020), below the EU average of 93%. There was also a 55% increase in internet traffic in 2020 (ANACOM, 2021) due to the COVID-19 lockdowns since people were working/studying remotely and entertaining themselves with digital content. The COVID-19 crisis accelerated digitalization and technological opportunities (Amankwah-Amoah, Khan, Wood, & Knight, 2021). With many countries under strict social distancing restrictions, online platforms began to host sectors such as education, health, the press, commerce, and finance (Ishfaq & Mengxing, 2021).

Technology Adoption Theories

Several studies discuss the applicability of technology adoption theories in the context of mobile banking. The most recurrent theories are the technology acceptance model (TAM) and the unified theory of acceptance and usage of technology (UTAUT), with the others being the initial trust model (ITM), theory of planned behavior (TPB), and innovation diffusion theory (IDT).

TAM was first introduced by Davis in 1986 and used in 1989 to explain the determinants of computing technology acceptance (Davis, Bagozzi, & Warshaw, 1989). It uses as determinant factors Perceived Usefulness, which is the likelihood that the potential user believes the use of a given system will benefit performance, and Perceived Ease of Use, which is the degree to which the user believes in the usage of a certain system being effortless (Davis, 1989).

In 2000, Davis and Venkatesh proposed an extension of the initial TAM, whose purpose is to include social influence effects in the two determinants, enabling an understanding of how determinants change with user experience and over time (Perdigoto & Picoto, 2012). Through TAM2, it is possible to conclude that social influence (subjective norms, voluntariness, and image) and cognitive instrumental processes (job relevance, output quality, and result demonstrability) influence technology acceptance more than the two initial determinant factors (Venkatesh & Davis, 2000).

UTAUT is an extension of TAM, proposed in 2003 by Venkatesh et al. (2003), with four determinants of intention and usage: performance and effort expectancies, social influence, and facilitating conditions (Venkatesh, Morris, Davis, & Davis, 2003). Performance expectancy comprises five constructs: perceived usefulness, extrinsic motivation, job fit, relative advantage, and outcome expectations, while effort expectancy is perceived ease of use, ease of use, and complexity. Social influence has proved to influence behavior because of potential social status gains and social pressure in the early stages of adoption. Facilitating conditions refer to the degree to which an individual believes that an organization provides the necessary support for using the system, from removing barriers to use to having support infrastructure. For the purpose of this study four dimensions are explored: performance expectancy, effort expectancy, social influence and facilitating conditions. Based on the literature review, the following proposition is formulated:

P1. There is a positive relation between performance expectancy, effort expectancy, social influence and facilitating conditions regarding mobile banking technology adoption.

Methodology and Data Collection

This study follows an abduction approach to theory development. It intends to explore a phenomenon and explain patterns to modify existing theories that can be tested through further data collection (Saunders, Lewis, & Thornhill, 2016).

The exploratory nature of the research implies a search through a problem to provide a better understanding and identify alternative courses of action. In contrast, the descriptive nature entails describing the characteristics of relevant groups and service characteristics (Malhotra, 2012). This study employs quantitative research method. An online questionnaire is used for data collection (Saunders, Lewis, & Thornhill, 2016). The time horizon of this study is cross-sectional because the research relates to a certain phenomenon at a particular time.

The study population represents all elements that share a common set of characteristics within the universe of the research problem (Malhotra, 2012). In this study, the population comprises all individuals aged over 18 years with bank accounts in Portugal. The sample is a subgroup of the population selected to participate in the study.

A non-probability sampling technique is adopted for practical reasons, which implies that the sampling relies on the researcher's judgment, meaning that the estimated results are not statistically representative of the population (Malhotra, 2012).

The non-probability technique selected is convenience sampling, as it is the least expensive and time-consuming technique and allows for the participation of easily accessible and cooperative elements (Malhotra, 2012).

Primary data is collected using an online and self-administered questionnaire for quantitative data. An online survey platform supports the questionnaire, Google Forms, completed by the respondents via a browser or mobile device. The link to the questionnaire was shared online, and the criteria were people aged up to 18 years living in Portugal who are bank clients. Data is collected between November 14th and November 26th, 2021.

The questionnaire includes 14 mandatory closed-ended questions where respondents indicate their degree of agreement or disagreement using five-point Likert scales (1 = strongly disagree, 5 = strongly agree) to measure the constructs. All constructs are adapted from existing literature on mobile banking adoption and UTAUT, see Table 1. The final segment of the questionnaire concerned socio-demographic questions.

Table 1. Scales and authors

	Performance Expectancy	
P1	Mobile banking is useful.	(Hanafizadeh et al., 2012)
P2	Mobile banking makes banking tasks quicker to accomplish.	(Venkatesh, Thong, & Xu, 2012)
P3	Mobile banking improves the management of my personal finances.	(Venkatesh, Thong, & Xu, 2012)
	Effort Expectancy	
E1	Mobile banking is easy to use.	(Gu, Lee, & Suh, 2009)
E2	My interaction with Mobile Banking is clear and understandable.	(Venkatesh, Thong, & Xu, 2012)
	Social Influence	
S1	I use Mobile Banking because many people use it.	(Gu, Lee, & Suh, 2009)
S2	I use Mobile Banking because my family and friends use it	(Farah, Hasni, & Abbas, 2018)
	Facilitating Conditions	
F1	I have the knowledge necessary to use Mobile Banking.	(Gu, Lee, & Suh, 2009)
F2	I can get help when I have difficulties using Mobile Banking.	(Venkatesh, Thong, & Xu, 2012)
F3	My bank provides sufficient support in case I have difficulties using the app (support line, virtual assistant, at the branch)	(Oliveira et al., 2014)
F4	Mobile Banking is compatible with my banking activities.	(Hanafizadeh, et al., 2012)

Results Analysis

The statistical analysis is performed using the IBM SPSS Statistics 28 software. The age variable is recoded into different age groups, from "<20 years old" to ">75 years old," with the remaining groups divided into 5-year intervals, such as "20-24 years old" and "25-29 years old." The recording is performed to facilitate the analysis in cases where the option "Other" is provided for the other demographic variables. Additionally, the mobile banking usage frequency variable is recoded as a numeric variable for the quantitative analysis, where the most frequent option, "Several times a day," corresponded to 5 and the least frequent "2 times a month or less" to 1.

First, the internal consistency of all the dimensions is measured using Cronbach's alpha coefficient, as shown in Table 2.

Table 2. Internal consistency and reliability

			Reliability Analysis		
				Item-Total Statistics	
Construct	Items	No. of items	Cronbach's Alpha	Cronbach's Alpha if item deleted	Corrected Item-Total Correlation
Performance Expectancy	P1	3	0.740	0.647	0.596
	P2			0.557	0.665
	P3			0.792	0.792
Effort Expectancy	E1	2	0.911	-	-
	E2			-	-
Social Influence	S1	2	0.840	-	-
	S2			-	-
Facilitating Conditions	F1	4	0.534	0.488	0.300
	F2			0.676	0.214
	F3			0.295	0.515
	F4			0.421	0.445

From Table 2, all constructs have values between 0.676 and 0.911, indicating acceptable to very good internal consistency (Sarstedt & Mooi, 2019). F2 was removed from the Facilitating Conditions construct in order to improve the internal consistency of the construct to a Cronbach's alpha coefficient of 0.676. The lower value for internal consistency demonstrated for the construct Facilitating Conditions may be explained by the variety of banks present in the questionnaire's answers; therefore, different conditions, systems, and support.

The Kaiser-Meyer-Olkin (KMO) measure and Bartlett's test of sphericity are performed to measure the data's suitability for PCA. According to Sarstedt and Mooi (2019), the KMO statistic varies between 0 and 1, with values greater than 0.5, which are considered adequate for correlation inferences. Bartlett's test of sphericity is significant ($p<0.05$), indicating that PCA is adequate (Pallant, 2016).
The results of the PCA can be depicted in Table 3.

Table 3. Principal Component Analysis

			Bartlett's Test of Sphericity	
Construct	**KMO**	**Total Variance Explained (%)**	**Approx. Chi-Square**	**Sig.**
Performance Expectancy	0.656	68.767	243.346	< 0.001
Effort Expectancy	0.500	91.849	333.392	< 0.001
Social Influence	0.500	86.230	205.872	< 0.001
Facilitating Conditions	0.659	62.024	146.001	< 0.001

All the analyzed constructs present KMO values between 0.5 and 0.659, while Bartlett's test of sphericity demonstrates significant values ($p<0.001$). Both measures confirm the suitability of the data for ACP. Additionally, the total variance explained is higher than 62% for all constructs.

The sample of this study comprises 329 respondents of the online questionnaire; 52% of the respondents are female, and 48% are male. The ages range from 19 to 84 years, with the largest concentration of respondents ages 45 and 59 years (53.5%); 32.5% are 44 years or younger, and 14% are 60 years or older. Setubal is the highest distribution of respondents (36.5%), followed by Lisbon (25.3%) and Faro (10.6%). Regarding academic qualifications, 52.6% of the respondents hold a bachelor's degree, 28.6% have not reached superior education, and 18.8% hold a master's or doctorate degree. Regarding occupation, 74.5% of the respondents are employed (69.3% are employed by others and 5.2% are self-employed), 12.5% are retired, 6.7% are studying, and 4.6% are working and studying simultaneously. The remaining 1.7% of the sample are either unemployed or domestic. Finally, 49.9% of respondents report an individual net monthly income between €1001 and €1900, 28% earn more than €1900 a month, and 13.7% receive less than €1000. The rest of the sample states that they do not have income or prefer not to respond to this question. Table 4 provides the full details of the previous analysis.

Table 4. Sample Characterization

Indicator	Options	*n*	%	Indicator	Options	*n*	%
Gender	Feminine	171	52	District of Residence	Aveiro	1	0.3
(*N*=329)	Masculine	158	48		Beja	3	0.9
Age	< 20	1	0.3	(*N*=329)	Braga	1	0.3
(*N*=329)	20 - 24	38	11.6		Bragança	0	0
	25 - 29	9	2.8		Castelo Branco	2	0.6
	30 - 34	6	1.8		Coimbra	3	0.9
	35 - 39	19	5.8		Évora	14	4.3
	40 - 44	34	10.3		Faro	35	10.7
	45 - 49	67	20.4		Guarda	2	0.6
	50 - 54	57	17.3		Leiria	3	0.9
	55 - 59	52	15.8		Lisboa	83	25.2
	60 - 64	28	8.5		Portalegre	16	4.9
	65 - 69	10	3		Porto	8	2.4
	70 - 74	3	0.9		Santarém	6	1.8
	> 75	5	1.5		Setúbal	120	36.5
Education	Until 9th grade	3	0,9		Viana do Castelo	2	0.6
(*N*=329)	9th grade	3	0,9		Vila Real	0	0
	Secondary degree	88	26.8		Viseu	7	2.1
	Undergraduate	173	52.6		R. A. Açores	16	4.9
	Masters	57	17.3		R. A. Madeira	5	1.5
	Doctorate	5	1.5		Outside Portugal	2	0.6

Continued on following page

Table 4. Continued

Indicator	Options	*n*	%	Indicator	Options	*n*	%
Occupation	Student	22	6.7	Individual Net Monthly Income (*N*=329)	No Income	14	4.3
(*N*=329)	Student worker	15	4.6		Up to 400€	10	3
	Dependent worker	228	69.3		401€ to 700€	10	3
	Self-employed	17	5.2		701€ to 1000€	25	7.6
	Unemployed	5	1.5		1001€ to 1300€	55	16.7
	Retired	41	12.4		1301€ to 1600€	57	17.3
	Domestic	1	0.3		1601€ to 1900€	52	15.8
					1901€ to 2200€	39	11.8
					2201€ to 2500€	14	4.3
					More than 2500€	39	11.9
					Prefer not to say	14	4.3

To better understand the profile of the respondents, questions regarding the usage of mobile banking and financial service apps are included. All respondents state that they have at least one bank account, 57.1% own only one account, and 32.2% have two different banks. One respondent indicates to have accounts in six banks. Most of the bank's present in the sample are Montepio (188) and Caixa Geral de Depósitos (111), followed by Santander (44), Millennium BCP (42), and BPI (41). Regarding mobile banking, the majority of the sample (84.8%) uses the bank's app; of those, 40.5% resort to the app daily, and 35.5% two or three times a week. The most common uses for the app include basic transactions, such as checking the account balance and movements (92.1%), making payments and transfers (94.6%), checking digital documents (39.4%), increasing and mobilizing savings (36.2%), and managing cards (34.8%). In contrast, more complex transactions, such as buying/selling shares/securities/funds and simulating credit and insurance, are executed by an inferior proportion of the sample (4.7% and 1.8%, respectively).

Regarding payment and financial services apps, such as MB Way, PayPal, Apple Pay, and Google Pay, 72.3% of the respondents admit using at least one. MB Way is the most popular among the respondents (90.8%), followed by PayPal (39.5%), Apple Pay (8.8%), and Google Pay (5.5%). By analyzing the intersection between respondents who use mobile banking and financial services apps, it is possible to understand that 67.5% of the sample uses both services, while 10.3% use neither. Respondents who use the bank's app but not financial services apps make up 17.3% of the sample, whereas respondents who only resort to financial services apps account for 4.9%.

The respondents' characteristics regarding utilization and frequency of mobile banking users are further analyzed by gender, age, education, occupation, and income. Regarding utilization, a larger proportion of mobile banking non-users is recorded in females. This trend can be transposed to the frequency of utilization, which was also lower for this gender. Respondents aged between 25 and 34 register a 100% rate of mobile banking adoption, while respondents aged 45–49 and 55–59 has adoption rates of over 90%. Nonetheless, respondents in the age groups of 20–24, 40–44, 50–54, and 60–64 years have adoption rates of over 80%. Concerning frequency, respondents aged between 25 and 29 have the highest values, followed by those aged 30 to 34, 40 to 44, and 20 to 24. Respondents in the age groups 45 to 49 and 55 to 64 have a frequency of at least 75% usage two or three times a week and more frequently. The proportion of the sample with secondary education or higher academic degrees has a mobile banking adoption rate above 80% and a frequency of at least 75% usage two or three times a week or more frequently. The sample proportion whose occupation fits within the active population (i.e., student, student worker, and the employed) have mobile banking adoption rates above 80%, while the retired respondents only record a 70.7% adoption rate, which can be attributed to the later age of this proportion. Lastly, the income variable presents a trend with regard to mobile banking adoption, with respondents earning up to €1000 a month having an adoption rate between 70-80%, respondents with earnings between €1001 and €2200 a month an adoption rate between 84-88% and the proportion with earnings over €2201 a month with a rate between 92-93%.

Among the 50 respondents who do not use mobile banking, 34% affirm the willingness to use it in the future, 26% refuse to be interested in using it, and 40% are indecisive about using it.

The scales used in this study are adapted from the UTAUT model and the existing mobile banking literature. Performance expectancy is the construct that displayed the highest levels of agreement (M = 4.56; SD = 0.65), indicating that it is the factor that affects mobile banking adoption and usage the most. The second dimension regarding agreement levels is effort expectancy (M = 4.51; SD = 0.69), followed by facilitating conditions (M = 4.35, SD = 0.78). Social influence is the construct with the lowest level of agreement (M = 1.98; SD = 1.15). Table 5 presents descriptive statistics of the constructs and items.

Table 5. Descriptive statistics of constructs and items

Construct	Item	*n*	Mean		Standard Deviation		Median	Mode	Min.	Max.
			Construct	Item	Construct	Item				
Performance Expectancy	P1	329	4.56	4.70	0.65	0.516	5	5	3	5
	P2	329		4.64		0.562	5	5	2	5
	P3	329		4.34		0.790	5	5	1	5
Effort Expectancy	E1	329	4.51	4.51	0.69	0.703	5	5	1	5
	E2	329		4.50		0.687	5	5	2	5
Social Influence	S1	329	1.98	1.98	1.15	1.170	2	1	1	5
	S2	329		1.98		1.133	2	1	1	5
Facilitating Conditions	F1	329	4.35	4.53	0.78	0.712	5	5	1	5
	F3	329		4.04		0.897	4	5	1	5
	F4	329		4.48		0.610	5	5	2	5

Pearson's correlation coefficient is calculated to assess the correlation between the constructs and other variables. This coefficient comprises values between -1 and 1, indicating full negative or positive correlation, respectively, and 0 corresponds to no correlation (Schober, Boer, & Schwarte, 2018).

According to Table 6, it can be assumed that there is a moderate positive correlation between performance expectancy and effort expectancy, performance expectancy and facilitating conditions as well as effort expectancy and facilitating conditions. It can also be assumed that there is a weak positive correlation between social influence and facilitating conditions. No significant correlation was found between social influence and performance expectancy and effort expectancy.

Table 6. Pearson Correlation Coefficient

		Performance Expectancy	Effort Expectancy	Social Influence	Facilitating Conditions
Performance Expectancy	Pearson Cor.	1	.668**	.114	.593**
	Sig. (2-tailed)		<.001	.057	<.001
	N	329	329	329	329
Effort Expectancy	Pearson Cor.		1	.150*	.651**
	Sig. (2-tailed)			.012	<.001
	N		329	329	329
Social Influence	Pearson Cor.			1	.141*
	Sig. (2-tailed)				.018
	N			329	329
Facilitating Conditions	Pearson Cor.				1
	Sig. (2-tailed)				
	N				329

**. Correlation is significant at the 0.01 level (2-tailed)
*. Correlation is significant at the 0.05 level (2-tailed)

Based on this, P1 was partly confirmed, meaning that all variables are correlated, except social influence and performance expectancy and social influence and effort expectancy.

SOLUTIONS AND RECOMMENDATIONS

The banking sector continues to be hampered by the Portuguese population's slow adoption of internet technology, which results in lower adoption of online solutions when compared to their European counterparts. The problem is exacerbated by the fact that the older generation is accustomed to personal interaction or has someone helping them satisfy their banking needs and is often wary of ATMs, cards, and mobile banking. Lack of technological skills leads to a lack of resources to develop them, worsened by having no support to safely navigate technology and widespread internet and banking data skepticism.

Mobile banking and peer-to-peer payment solutions are beginning to be integrated into mainstream society for younger generations, as bank visits and physical forms of currency become obsolete. The number of banking branches is decreasing, and with that comes the challenge of serving the aging population and elevating the burden on its human resources. Banking agencies are shifting their efforts to more commercial functions and redirecting other affairs to their digital channels. Nevertheless, personal interaction continues to be valued for matters related to the treasury, online difficulties, and credit, occupying a significant portion of a branch's daily activities.

To stay current and retain customers from a young age, Portuguese banks have begun to invest in innovative mobile solutions as a new era of fintech, decentralized, and non-banking institutions threaten the stability of traditional banks. These new players bring a fresh perspective to the market on managing personal finances, with less bureaucracy and costs, widespread access to their services, and an online-only focus. Many of these, not directly related to financial services, have the advantage of being platforms for millions of users across the globe.

Other concerns for consumer safety are raised in the digital age, as data have become the most valuable commodity of the 21st century, especially when linked to financial detains, further pressing the necessity for large investments in cybersecurity. Blockchain and other technologies may improve EU payment systems regarding safety, transparency, and overall functionality; however, its implementation in traditional banking poses difficulties.

Financial institutions are now faced with the challenge of maintaining an older clientele whose inheritance is of higher value while adapting to a future that operates according to the technological demands of younger generations.

The present study showed that regarding mobile banking technology adoption there is an associating between performance expectancy and effort expectancy, performance expectancy and facilitating conditions as well as effort expectancy and facilitating conditions. Social influence and facilitating conditions are also correlated.

Based on these conclusions, some recommendations to foster mobile banking adoption can be ascertained.

For instance, clients can be granted more assistance to improve their technical skills in the form of more time attributed to agencies for customer support, potentially an employee exclusively dedicated to this, aided by tablets or PCs for the client to safely access the platform. Online platforms can complement agency services as they require less bureaucracy and time to execute certain operations. In the future, with fewer agencies, it will be an even more important tool for the client. Additionally, it is noted that the tendency to employ the internet for entertainment purposes instead of more practical ones, such as banking, can be solved by improving the digital and financial literacy of the Portuguese population. Banks should support initiatives to provide digital literacy to customers, as a significant part of their clients could benefit from these programs.

Clients tend to value intuitiveness and simple utilization of the platforms offered by banks, a wide range of available operations, and features such as a virtual assistant and several security mechanisms. Some suggestions include access personalization to activities, where the customer would have access to a suggested level of features based on their technological proficiency yet choosing to alter its level. Each level progresses various features, starting with a consulting model, followed by an increase/mobilize savings feature, payments and transfers, and access to funds and credit simulations. The further development of tutorials on bank websites are also recommended to help customers how to use certain features and become educated in financial matters.

Employee training is the final step in improving service, as employees in the agency network are on the frontline for all products/services of the bank and therefore must understand its characteristics to be in contact with the client. Banks regularly use e-learning courses to provide training to their employees. However, these can be useful if there is a desire to learn and use real-world applications. Despite the rise of digitalization and remote learning, employees continue to value in-person training, which could be more dynamic and practical, increasing their interest in the materials and allowing them to clarify potential doubts on the spot.

Solutions for habilitating the client to employ online platforms to satisfy its banking needs go hand in hand with dematerialization and personnel and branch cost reductions. Investments in agency services and resources to support clients' technological difficulties will contribute to the long-term efficiency of banking institutions, which will maintain physical branches for commercial purposes. In this respect, operations that do not result in business, such as increasing/mobilizing savings and requesting checks, should be carried out by the client via available online platforms, which can also be adapted to complement other operations. Furthermore, innovative infrastructure, such as intuitive platforms with adequate customer support, is crucial for the sector's future, ensuring greater customer satisfaction.

FUTURE RESEARCH DIRECTIONS

Although the UTAUT model considers experience as a moderator, future research might focus on the importance of experience in the intention and usage of technology and how different generations and age groups have adapted to employ these resources in their everyday lives and work. Future studies could focus on one bank, its products/services, and its clients to obtain more specific results. Collecting field testimonies from clients in agencies or conducting focus groups to discuss difficulties and find solutions would greatly enrich our findings. The difficulties observed by clients, such as lack of experience, support, and resources and fear of the internet and online fraud, may be investigated further to understand their origins and possible strategies to address them. Other relevant topics to research in this area include the future implementation of technology in banking, the threats of fintech and P2P payment systems to traditional banking, the applications of blockchain in financial services, and the regulatory challenges of new technologies.

CONCLUSION

This study founded that performance expectancy, effort expectancy, social influence and facilitating conditions important constructs that help to explain mobile banking challenges today. More specifically, from consumers perspective, there is an association between performance expectancy and effort expectancy, performance expectancy and facilitating conditions as well as effort expectancy and facilitating conditions. It can also be assumed that there an association between social influence and facilitating conditions.

This means that the usefulness and how mobile banking improves the task at hand, the ease and intuitiveness of using this technology, and the perceived support that customers receive from organizations to use said service. Social influence is found not to have relation with performance and effort expectancy, implying that the experience of others around us, such as friends and family members who use this service, does not relate to clients' expectations regarding performance and effort regarding mobile banking technology.

This study provides meaningful insights into the existing literature on the challenges of mobile banking adoption, as the data support the findings that performance and effort expectancies and facilitating conditions contribute the most to the phenomenon, while social influence does not. The study's quantitative approach allows for measurable analysis of the constructs. It strengthens the knowledge of the profile of mobile banking users and those who lack the means and skills to adopt it. However, the limited timeframe hinders the progress of the report, and the non-probabilistic sampling method compromises the reliability of the data.

The solutions to mitigate the challenges of mobile banking usage and improve consumer experience can be divided into three categories: client support, platform simplification, and employee training. These solutions derive from the conclusion that clients lack the skills and support to handle online platforms properly and that the conditions provided by organizations contribute to higher adoption and utilization of those platforms.

The present study had the limitation of using of non-probabilistic convenience sample. Thus, in future research, a study with a quantitative approach, with a broader and diversified sample, should be developed. Finally, other variables such as behavioral intent regarding mobile banking adopting, should also be explored.

ACKNOWLEDGMENT

We gratefully acknowledge financial support from FCT- Fundação para a Ciencia e Tecnologia (Portugal), national funding through research grant UIDB/04521/2020.

REFERENCES

Al-Saedi, K., & Al-Emran, M. (2021). A systematic review of mobile payment studies from the lens of the UTAUT model. *Recent Advances in Technology Acceptance Models and Theories*, 79-106.

Alafeef, M., Singh, D., & Ahmad, K. (2012). The Influence of Demographic Factors and User Interface on Mobile Banking Adoption: A Review. *Journal of Applied Sciences*, *12*(20), 2082–2095. doi:10.3923/jas.2012.2082.2095

Amankwah-Amoah, J., Khan, Z., Wood, G., & Knight, G. (2021). COVID-19 and digitalization: The great acceleration. *Journal of Business Research*, *136*, 602–611. doi:10.1016/j.jbusres.2021.08.011 PMID:34538980

ANACOM. (2021). *Pandemic COVID-19 - Impacto na utilização dos serviços de comunicações.* ANACOM.

Associação Portuguesa de Bancos. (2020). *Boletim Informativo APB*. Associação Portuguesa de Bancos.

Balkan, B. (2021). Impacts of Digitalization on Banks and Banking. In The Impact of Artificial Intelligence on Governance, Economics and Finance, Volume 1 (pp. 33-50). Springer. doi:10.1007/978-981-33-6811-8_3

Banco de Portugal. (2018). *Aviso n.º 2/2018*. Retrieved from October 9, 2021, from https://www.bportugal.pt/aviso/22018

Banco de Portugal. (2020). *Inclusão Financeira e Digital e Escolha de Produtos Bancários em Portugal*. Author.

Brummer, C., & Yadav, Y. (2019). Fintech and the Innovation Trilemma. *The Georgetown Law Journal*, *107*(235), 235–307.

Caixa Geral de Depósitos. (2019). *Missão e Estratégia do Grupo CGD*. Retrieved October 9, 2021, from https://www.cgd.pt/Institucional/Governo-Sociedade-CGD/Pages/Missao-Estrategia.aspx

Caixa Geral de Depósitos. (2020). *Grupo CGD*. Retrieved October 10, 2021, from https://www.cgd.pt/Institucional/Grupo-CGD/Documents/Organograma-do-Grupo-CGD.pdf

Caixa Geral de Depósitos. (2021). *CGD*. Retrieved October 10, 2021, from https://www.cgd.pt/Institucional/Patrimonio-Historico-CGD/Historia/Pages/Historia-CGD.aspx

Caixa Geral de Depósitos. (2021). *EUSOUDIGITAL - Programa de Capacitação Digital de Adultos*. Retrieved September 29, 2021, from https://www.cgd.pt/Institucional/eusoudigital/Pages/eusoudigital.aspx

Caixa Geral de Depósitos. (2021). *Institucional: 9M2021 - Resultados*. Retrieved November 5, 2021 from https://www.cgd.pt/institucional/sala-de-imprensa/2021/pages/atividade-consolidada-30-setembro-2021.aspx

Curran, J. M., & Meuter, M. L. (2005). Self-service technology adoption: Comparing three technologies. *Journal of Services Marketing*, *19*(2), 103–113. doi:10.1108/08876040510591411

Davis, F. D. (1989). Perceived Usefulness, Perceived Ease of Use, and User Acceptance of Information Technology. *Management Information Systems Research Center, 13*(3), 319-340.

Davis, F. D., Bagozzi, R. P., & Warshaw, P. R. (1989). User Acceptance of Computer Technology: A Comparison of Two Theoretical Models. *Management Science*, *35*(8), 982–1003. doi:10.1287/mnsc.35.8.982

Deloitte. (2015). *Millenials and wealth management - Trends and challenges of the new clientele*. Retrieved October 16, 2021 from https://www2.deloitte.com/content/dam/Deloitte/lu/Documents/about-deloitte/Inside/lu-inside9-full.pdf

Devlin, J. F. (1995). Technology and innovation in retail banking distribution. *International Journal of Bank Marketing*, *13*(4), 19–25. doi:10.1108/02652329510082915

Djelassi, S., Diallo, M. F., & Zielke, S. (2018). How self-service technology experience evaluation affects waiting time and customer satisfaction? A moderated mediation model. *Decision Support Systems*, *111*, 38–47. doi:10.1016/j.dss.2018.04.004

EUR-Lex. (2015). *Payment services (PSD 2) - Directive (EU) 2015/2366*. Retrieved October 27, 2021 from https://eur-lex.europa.eu/legal-content/EN/TXT/HTML/?uri=LEGISSUM:2404020302_1

European Central Bank. (2021). *EU structural financial indicators: end of 2020*. Retrieved November 20, 2021 from https://www.ecb.europa.eu/press/pr/date/2021/html/ecb.pr210526~7469dedaaf.en.html

Farah, M. F., Hasni, M. J., & Abbas, A. K. (2018). Mobile-banking adoption: Empirical evidence from the banking sector in Pakistan. *International Journal of Bank Marketing*, *36*(7), 1386–1413. doi:10.1108/IJBM-10-2017-0215

Garg, P., Gupta, B., Chauhan, A. K., Sivarajah, U., Gupta, S., & Modgil, S. (2021). Measuring the perceived benefits of implementing blockchain technology in the banking sector. *Technological Forecasting and Social Change*, *163*, 163. doi:10.1016/j.techfore.2020.120407

Gu, J.-C., Lee, S.-C., & Suh, Y.-H. (2009). Determinants of behavioral intention to mobile banking. *Expert Systems with Applications*, *36*(9), 11605–11616. doi:10.1016/j.eswa.2009.03.024

Hanafizadeh, P., Behboudi, M., & Koshksaray, A. A. (2012). Mobile-banking adoption by Iranian clients. *Telematics and Informatics*, *31*(1), 62–78. doi:10.1016/j.tele.2012.11.001

INE. (2020). *Inquérito à Utilização de Tecnologias da Informação e da Comunicação pelas Famílias 2020*. INE.

Iqbal, M. S., Hassan, M. U., & Habibah, U. (2018). *Impact of self-service technology (SST) service quality on customer loyalty and behavioral intention: The mediating role of customer satisfaction*. Cogent Business & Management.

Ishfaq, N., & Mengxing, H. (2021). Consumer usage behavior of internet-based services (IBS) in Pakistan during COVID-19 crisis from the perspective of technology acceptance model. *Environmental Science and Pollution Research International*, 1–16. doi:10.100711356-021-15868-1 PMID:34550524

Jadil, Y., Rana, N. P., & Dwivedi, Y. K. (2021). A meta-analysis of the UTAUT model in the mobile banking literature: The moderating role of sample size and culture. *Journal of Business Research*, *132*, 354–372. doi:10.1016/j.jbusres.2021.04.052

Kshetri, N. (2017). Potential roles of blockchain in fighting poverty and reducing financial exclusion in the global south. *Journal of Global Information Technology Management*, *20*(4), 201–204. doi:10.1080/1097198X.2017.1391370

Lee, K. C., & Chung, N. (2009). Understanding factors affecting trust in and satisfaction with mobile banking in Korea: A modified DeLone and McLean's model perspective. *Interacting with Computers*, *21*(5-6), 385–392. doi:10.1016/j.intcom.2009.06.004

Lovelock, C. H., & Wirtz, J. (2016). Services Marketing: people, technology, strategy. In C. H. Lovelock & J. Wirtz (Eds.), *Services Marketing: people, technology, strategy* (pp. 470–480). World Scientific Publishing Co. Inc.

Malhotra, N. K. (2012). *Marketing Research: an applied orientation* (Vol. 6). Prentice Hall.

Marktest. (2020). *7,2 milhões de portugueses usam smartphone*. Retrieved September 26, 2021, from https://www.marktest.com/wap/a/n/id~2700.aspx

Marktest. (2021). *Bareme Internet 2021*. Grupo Marktest.

Meuter, M. L., Ostrom, A. L., Roundtree, R. I., & Bitner, M. J. (2000). Self-Service Technologies: Understanding Customer Satisfaction with Technology-Based Service Encounters. *Journal of Marketing*, *64*(3), 50–64. doi:10.1509/jmkg.64.3.50.18024

Moody's. (2021). *Rating Action: Moody's takes rating actions on six Portuguese banks*. Retrieved September 24, 2021, from https://www.moodys.com/research/Moodys-takes-rating-actions-on-six-Portuguese-banks--PR_454388

Moody's Investors Service. (2021). *Rating Symbols and Definitions*. Author.

Moşteanu, N., Faccia, A. C., & Bhatia, S. (2020). Digital Technologies' Implementation within Financial and Banking System During Socio Distancing Restrictions - Back to the Future. *International Journal of Advanced Research in Engineering and Technology*, *11*(6), 307–315.

Oliveira, T., Faria, M., Thomas, M. A., & Popovic, A. (2014). Extending the understanding of mobile banking adoption: When UTAUT meets TTF and ITM. *International Journal of Information Management, 34*(5), 689–703. doi:10.1016/j.ijinfomgt.2014.06.004

Pallant, J. (2016). *SPSS Survival Manual: A Step By Step Guide to Data Analysis Using SPSS Program* (6th ed.). McGraw-Hill Education.

Parry, E., & Urwin, P. (2011). Generational Differences in Work Values: A Review of Theory and Evidence. *International Journal of Management Reviews, 13*(1), 79–96. doi:10.1111/j.1468-2370.2010.00285.x

Perdigoto, M., & Picoto, W. (2012). Analysing Intention and Action in Mobile Banking Services. *Portuguese Journal of Management Studies, 17*(2), 133–152.

Pordata. (2021). *Assinantes / equipamentos de utilizadores do serviço móvel*. Retrieved October 9, 2021, from https://www.pordata.pt/Portugal/Assinantes+++equipamentos+de+utilizadores+do+servi%c3%a7o+m%c3%b3vel-1180

Pordata. (2021). *Indicadores de envelhecimento*. Retrieved October 9, 2021, from https://www.pordata.pt/Portugal/Indicadores+de+envelhecimento-526

Pordata. (2021). *População residente, média anual: total e por grupo etário*. Retrieved October 10, 2021, from https://www.pordata.pt/Portugal/Popula%c3%a7%c3%a3o+residente++m%c3%a9dia+anual+total+e+por+grupo+et%c3%a1rio-10

Proença, J. F., & Rodrigues, M. A. (2021). A comparison of users and non-users of banking self-service technology in Portugal. *Journal of Service Theory and Practice*, 192–210.

Proença, J. F., & Silva, M. M. (2008). Os efeitos da desmaterialização das relações bancárias devido ao uso do Internet Banking e do Marketing Relacional. *Revista Portuguesa e Brasileira de Gestão*, 88–102.

Público. (2021). *Portugal entre os países da UE com mais dificuldades no acesso à Net*. Retrieved October 20, 2021, from https://www.publico.pt/2021/02/20/infografia/transicao-digital-uniao-europeia-585

Puschmann, T. (2017). Fintech. *Business & Information Systems Engineering, 59*(1), 69–76. doi:10.100712599-017-0464-6

Roos, J. M., & Kazemi, A. (2021). Personality traits and Internet usage across generation cohorts: Insights from a nationally representative study. *Current Psychology (New Brunswick, N.J.), 40*(3), 1287–1297. doi:10.100712144-018-0033-2

Sarstedt, M., & Mooi, E. (2019). *A Concise Guide to Market Research: The Process, Data, and Methods* (3rd ed.). Springer. doi:10.1007/978-3-662-56707-4

Saunders, M., Lewis, P., & Thornhill, A. (2016). *Research Methods for Business Students* (Vol. 7). Pearson.

Schober, P., Boer, C., & Schwarte, L. A. (2018). Correlation Coefficients: Appropriate Use and Interpretation. *Anesthesia and Analgesia, 126*(5), 1763–1768. doi:10.1213/ANE.0000000000002864 PMID:29481436

Sibs. (2021). *Multibanco*. Retrieved October 9, 2021, from https://www.sibs.com/marcas/multibanco/

Singh, S., & Srivastava, R. K. (2020). Understanding the intention to use mobile banking by existing online banking customers: An empirical study. *Journal of Financial Services Marketing*, *25*(3-4), 86–96. doi:10.105741264-020-00074-w

Statista. (2021). *Online banking penetration in Portugal from 2005 to 2018*. Retrieved May 29, 2022 from https://www.statista.com/statistics/380873/online-banking-penetration-in-portugal/

Statista. (2022). *Forecast of internet user numbers in Portugal from 2017 to 2026*. Retrieved May 29, 2022, from https://www.statista.com/map/europe/portugal/internet

Tavares, E. (2021). Lucros da banca crescem 49 milhões para quase 540 milhões no primeiro semestre. *Dinheiro Vivo*. Retrieved September 29, 2021, from https://www.dinheirovivo.pt/empresas/bancos-lucram-mais-49-milhoes-ate-junho-13993544.html

Thakor, A. V. (2020). Fintech and banking: What do we know? *Journal of Financial Intermediation*, *41*, 1–13. doi:10.1016/j.jfi.2019.100833

Venkatesh, V., & Davis, F. D. (1996). A Model of the Antecedents of Perceived Ease of Use: Development and Test. *Decision Sciences*, *27*(3), 451–481. doi:10.1111/j.1540-5915.1996.tb01822.x

Venkatesh, V., & Davis, F. D. (2000). A Theoretical Extension of the Technology Acceptance Model: Four Longitudinal Field Studies. *Management Science*, *46*(2), 186–204. doi:10.1287/mnsc.46.2.186.11926

Venkatesh, V., Morris, M. G., Davis, G. B., & Davis, F. D. (2003). User Acceptance of Information Technology: Toward a Unified View. *Management Information Systems Research Center, University of Minnesota*, *27*(3), 425–478. doi:10.2307/30036540

Venkatesh, V., Thong, J. Y., & Xu, X. (2012). Consumer Acceptance and Use of Information Technology: Extending the Unified Theory of Acceptance and Use of Technology. *Management Information Systems Quarterly*, *36*(1), 157–178. doi:10.2307/41410412

Vives, X. (2017). The Impact of Fintech on Banking. *European Economy – Banks, Regulation, and the Real Sector*, 97-105.

Vives, X. (2019). Digital Disruption in Banking. *Annual Review of Financial Economics*, *11*(1), 243–272. doi:10.1146/annurev-financial-100719-120854

Wei, M.-F., Luh, Y.-H., Huang, Y.-H., & Chang, Y.-C. (2021). Young Generation's Mobile Payment Adoption Behavior: Analysis Based on an Extended UTAUT Model. *Theoretical and Applied Electronic Commerce Research*, 618-637.

Wójcik, D. (2020). Financial geography II: The impacts of FinTech - Financial sector and centres, regulation and stability, inclusion and governance. *Progress in Human Geography*, *20*(10), 1–12.

Zhang, T., Lu, C., & Kizildag, M. (2018). Banking "on-the-go": Examining consumer's adoption of mobile banking services. *International Journal of Quality and Service Sciences*, *10*(3), 279–295. doi:10.1108/IJQSS-07-2017-0067

KEY TERMS AND DEFINITIONS

Effort Expectancy: It relates to the consumer's perception of easiness to perform a task.

Facilitating Conditions: The perception of the existence of a system or resources that enable a service to be performed.

FinTech: Is a combination of "financial technology" and is used to describe new tech companies that disrupted finance by improving and automating the delivery and use of financial services.

Mobile Banking: Is the usage of a mobile device such as a smartphone or a tablet to make financial transactions such as bank transfers or payments.

Performance Expectancy: It refers to the extent to which a customer perceives that using the system will help him/her to achieve benefits in a certain task.

Social Influence: The influence of other people (family, colleagues or friends) to form a perception.

Technology Adoption: Is the incorporation of new technology into your business and its usage by the consumer. Adoption means that the consumer uses the technology at its fullest potential and sees the benefits of using the new systems.

Chapter 6
Metaverse Applications in Business:
A Systematic Literature Review and Integrative Framework

Julien Diogo
https://orcid.org/0000-0002-5326-2093
Polytechnic Institute of Viseu, Portugal

Pedro Mota Veiga
NECE Research Unit in Business Sciences - University of Beira Interior, Portugal & Universidade Católica Portuguesa, Portugal, & Instituto de Gestão e das Organizações de Saúde, Portugal &, School of Education, Polytechnic Institute of Viseu, Portugal

ABSTRACT

This study aims to identify the main applications of metaverse in business based on scientific publications. More specifically, the chapter aims to identify conceptual frameworks used in the literature. The authors conducted a systematic review based on the Web of Science and Scopus databases to achieve the objectives. This study followed a systematic literature review methodology consisting of protocol development, identified inclusion, and exclusion criteria for relevant publications; data extraction; and synthesis. The research resulted in 26 articles, which were identified and classified according to applications to the business environment, concluding that the main implications and applications of the metaverse in the business environment focus on an integrative framework that includes ethical issues, innovation management, marketing management, and knowledge management. This research has put down the foundation for establishing metaverse as a research topic in business research. With this study, the contribution lies both at the theoretical and practical levels.

DOI: 10.4018/978-1-6684-5538-8.ch006

INTRODUCTION

The growth of virtual worlds - underpinned by the constant diffusion of virtual reality technology - offers opportunities for companies to innovate and for new digital entrepreneurs to germinate from the redefinition of the concept of reality in a metaverse (Zhou et al., 2018). When in October 2021, Mark Zuckerberg announced that Facebook would change its name to Meta and underpin its business model on the Metaverse, many people saw the Metaverse as a new technology or innovation (Damar, 2021). Despite its current popularity, the Metaverse is not a new term. Stephenson (2003) popularised the term Metaverse in the novel, in which the concept of the avatar was used to describe the simulation of the human form in a virtual environment that the author calls the 'Metaverse', a virtual reality version of the Internet. The term Metaverse first appeared in Neal Stephenson's book Snow Crash, released in 1992 in its original version. In recent years, the Metaverse has broadened this vision to include aspects of the physical world, objects, actors, interfaces and networks that develop and interact with virtual environments (Park & Kim, 2022b; Rauschnabel et al., 2022).

Since this first appearance, the metaverse as a computer-generated universe has been defined

Metaverse research has been implemented mainly in the context of gaming and learning, emphasizing social utility (Park et al., 2021; Park & Kim, 2022), and little attention has been paid to the impact of the metaverse on business (Park & Kim, 2022). Early research on the topic directed its attention to Second Life (Bardzell & Shankar, 2007; Papagiannidis et al., 2008), but the current Metaverse is based on the social values of Generation Z, where online and offline are intertwined. (Harrisson-Boudreau & Bellemare, 2022). Thus, a systematic approach is needed for which metaverse applications in a business context are needed.

The main objective of the present study is to conduct a systematic literature review to map the scientific publications, intellectual structure and research trends related to Metaverse in a business context. The secondary objectives are: (1) to delineate the intellectual structure of research on metaverse represented in the scientific literature and (2) to identify the main applications of a metaverse in a business context.

To meet the study's objectives, the articles were collected from two major academic databases, Scopus and Web of Science (WoS). The search was conducted in April 2022. We performed a topic-based search (including the titles, abstracts and keywords) with the following terms: "metaverse*" or "metaverse*. During the search process, the following inclusion criteria were used: journal articles, full text available, papers written in English and included in research domains such as computer science, information systems, social sciences, business, economics and management. The search resulted in 99 articles published in Scopus and/or WoS. After removing the articles not related to the theme, the sample resulted in 26 articles.

The chapter will be organised as follows. After this introduction to the contextualisation and objectives of the chapter, the literature review is presented in the second section. The third section describes the methodological procedures for the systematic literature review. In the fourth section, the research literature is analysed according to relevant approaches and business applications. The results are discussed and a conceptual model is outlined in the fifth section, including contributions of the study, limitations and future lines of research.

LITERATURE REVIEW

Since this first appearance, the metaverse as a computer-generated universe has been defined through vast, diverse concepts such as lifelogging, collective space in virtuality, embedded internet/space internet, a mirror world, an omniverse: a place of simulation and collaboration - Figure 1 (Lee et al., 2021). The Metaverse proposes a new way of relating to and understanding life through technology.

Smart et al. (2007) have posed an axis of understanding the metaverse and scenarios is based on two axes: the spectrum of technologies and applications ranging from augmentation to simulation; and the spectrum ranging from intimate (identity-focused) to external (world-focused)

Figure 1. Metaverse Understanding Scenarios (Smart et al., 2007)

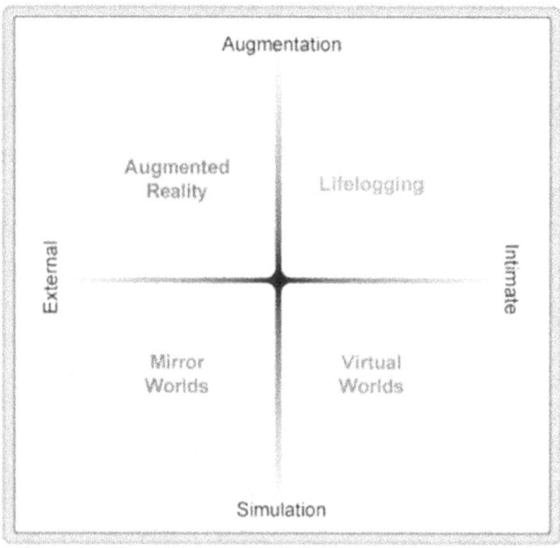

Currently, and based on Figure 2 - Evolution Spectrum for the Metaverse of Gartner (2022), we live in the era of the emergence of the Metaverse, inspired by Web 3.0, apps and a technological market, in which Early Adopters are the range of consumers most involved in the concept. This dimension is guided by principles such as Networking, Interoperable, Open and Hardware Agnostic (Mystakidis, 2022).

Figure 2. Gartner's Evolution Spectrum for the Metaverse (2022)

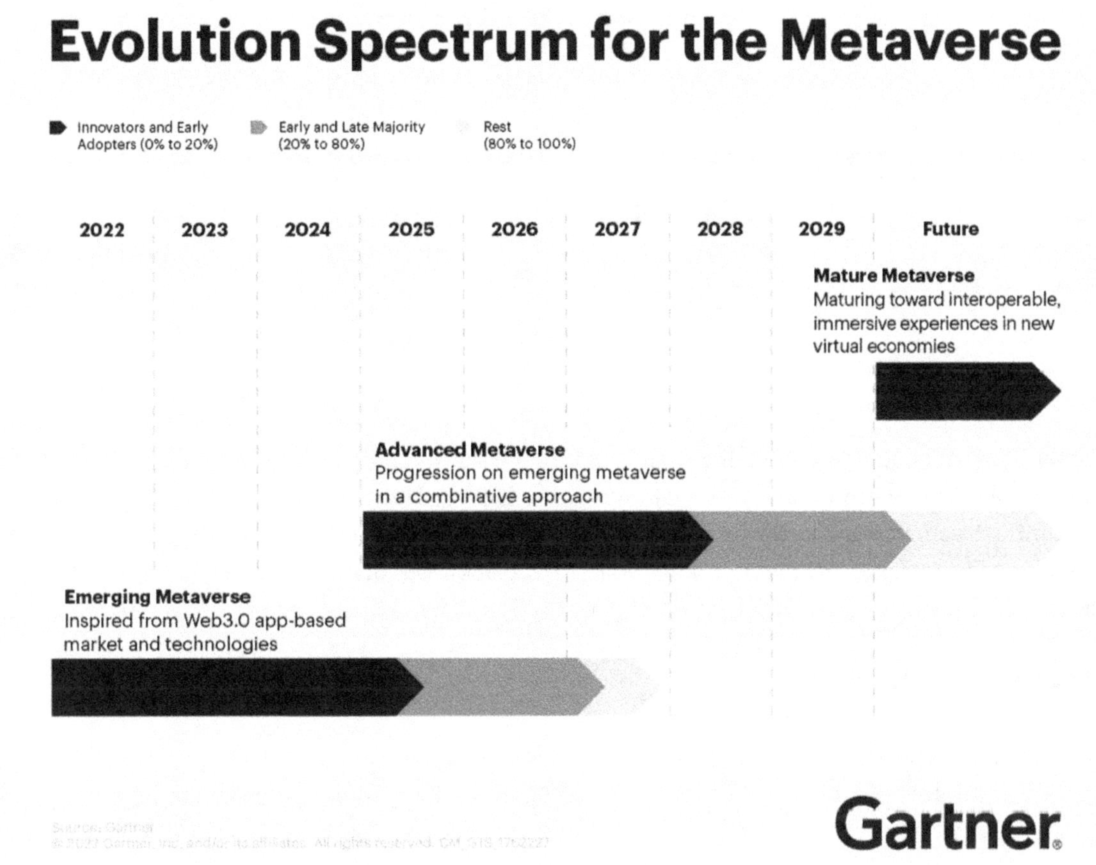

For a broader visualisation of the concept of the Metaverse, it is necessary to perceive the concept not as a virtual space but as the junction or nexus of the physical and virtual worlds ((Smart et al., 2007). However, several concepts dictate the innovation agendas in recent times, such as: virtual reality (VR), augmented reality (AR) and mixed reality (MR). VR presents itself as a human-machine interface that simulates a real environment and allows participants to interact with it (Latta & Oberg, 1994). The AR allows the user to perceive the real context augmented with virtual objects, with a prior computational development, aiming to improve our perception of the real world (Malbos et al., 2014). To understand the concept of RM, it is essential to identify the Virtuality Continuum (Milgram & Kishino, 1994), a continuous scale that varies between completely real reality and completely virtual reality. The scale also frames the augmented reality and augmented virtuality, and the interval between both is called mixed reality - MR - Figure 3. In this mixed reality, we can frame the concept of the Metaverse (Mystakidis, 2022).

Figure 3. Virtual Continuum representation. Adapted from Milgram (1994)

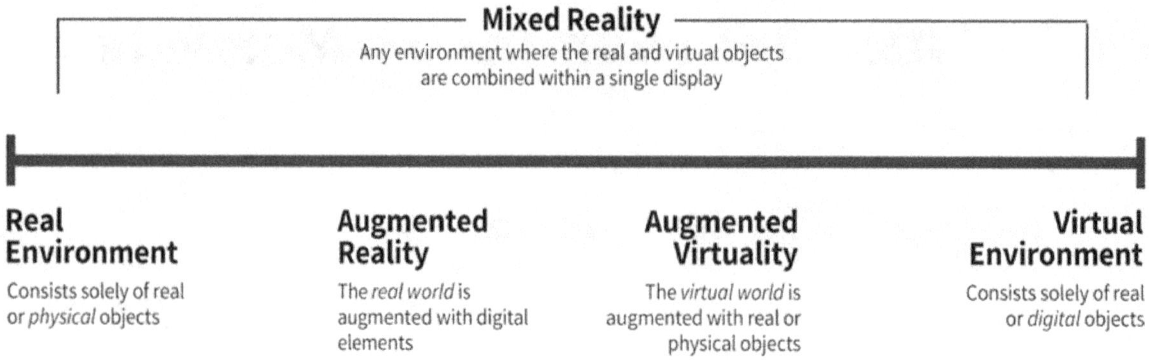

Several constituent elements of the Metaverse can be considered, such as Digital Currency, Online Shopping, Workplace Automation, Social Networks, Digital Humans, Natural Language Processing (NLP), Infrastructure or even Device Independence. The technologies capable of supporting and developing this ecosystem are AR, Flexible work styles, Head-mounted displays (HMDs), An AR cloud, The Internet of Things (IoT), 5G, Artificial Intelligence (AI) and even Space technologies (Gartner, 2022).

Metaverse and Society

Consumer centricity is the reality of the moment, characterised by real-time access to information, plus the use of disruptive platforms - such as Uber and Netflix - have forever transformed the way people consume (Fernandes et al., 2022). On the other hand, centricity is supported by strong leadership commitment, organisational realignment, systems and process support and financial metrics realigned with consumer needs (Shah et al., 2006). Customer focus offers the long-term possibility of differentiation, emerges as more difficult to copy and represents a significant barrier for those wishing to enter the market (Uhl & MacGillavry, 2016).

To talk about centricity is to talk about innovation and the Metaverse. The Metaverse allows a strong personalization of experiences: the game experience, the user experience, as well as the customer experience. Users can personalise their avatars and live totally unique experiences. Personalisation contributes to the creation of memorable experiences and strengthens customer loyalty. For brands, this is an opportunity to improve customer knowledge and a "customer-centric" approach due to data collection (Gallemard, 2022). An excellent example of application falls on an example of April 2020, Fortnite players spent over 3.2 billion hours in the game (Prasad, 2022), allowing for data collection and consequent adaptation.

The Metaverse is an interconnected web of networked social immersive environments on persistent multi-user platforms (Mystakidis, 2022), of which the consumer wants to be a part. Scientist Jeremy Bailenson, in Experience on Demand (Bailenson, 2018) that time spent with virtual reality glasses is psychologically far more powerful than any medium ever produced and that it will transform our lives. For several people, the illusion of reality is so strong that the limbic system (region of the brain involved with emotions and memory) goes into a state of intense activity (Bailenson, 2018). It is curious the data advanced by a study by consultancy GWI (Morris, 2022), which states that only 37% of consumers have heard of the Metaverse and know what it is about. But 51% of respondents show interest in participating in the Metaverse (Morris, 2022). From a social perspective, developing more engaging virtual experiences could help people build communities based on shared values and express themselves more authentically (JP Morgan Chase, 2022). The Metaverse may also negatively impact the ability to engage in non-virtual life, whether in the dimension of self-confidence, belonging, or social anxiety (The New York Times, 2022).

One of the first environments created was the game Second Life, launched in 2003 by the US-based company Liden Lab (Republic Realm, 2021). Over the years, there have been key milestones such as Microsoft's acquisition of the Minecraft platform (games and education) for $2.5 billion, as well as Amazon's acquisition of Twitch (games) in 2014, and Microsoft's acquisition of Activision Blizzard (games) for $68.7 billion. (JP Morgan Chase, 2022). The diversity of metaverse applications is enormous, rapper Travis Scott performed on Epic Games' Fortnite platform in a demonstration of what the future of music events could look like in the metaverse, it had 12.3 million real-time players (BBC, 2021). Sectors and industries such as film (Disney), television (Discovery Channel), sports (Fox Sports), music (Universal Music Group, Live Nation), streaming platforms (Netflix) and newspapers (The New York Times) begin their experiences of immersion in the Metaverse (BBC, 2021). The first experiments of the metaverse did not achieve the results because of technological limitations, requiring processing capacity at a time when internet connections were slow and reduced, and the expansion of social networks (Radfahrer, 2022).

The education and training sector can also benefit from the Metaverse, despite the growth in the use of platforms such as Microsoft Teams, Google Classroom, Adobe Connect, and resources such as Zoom during Pandemic Covid19 (Boltz et al., 2021), the process is 2D and has its limitations for interaction and immersion, which can lead to fatigue, saturation and social isolation (Anderson & Rivera-Vargas, 2020). However, the Metaverse allows a convergence of technologies that enable multisensory interactions and engagement with virtual environments, digital objects and people, such as virtual reality (VR) and augmented reality (AR) (Mystakidis, 2022). In a 2020 PWC study, "How virtual reality is redefining soft skills training", the impact of the metaverse on the learning process is perceived, indicating that the learning process is four times faster, those involved are 275% more confident in their skills, appear 3.75 times more emotionally connected to the content than learners in the physical classroom or even four times more focused than traditional eLearning peers (PWC, 2020) (Figure 4).

Figure 4. VR Soft Skills Training Efficacy (PWC, 2020)

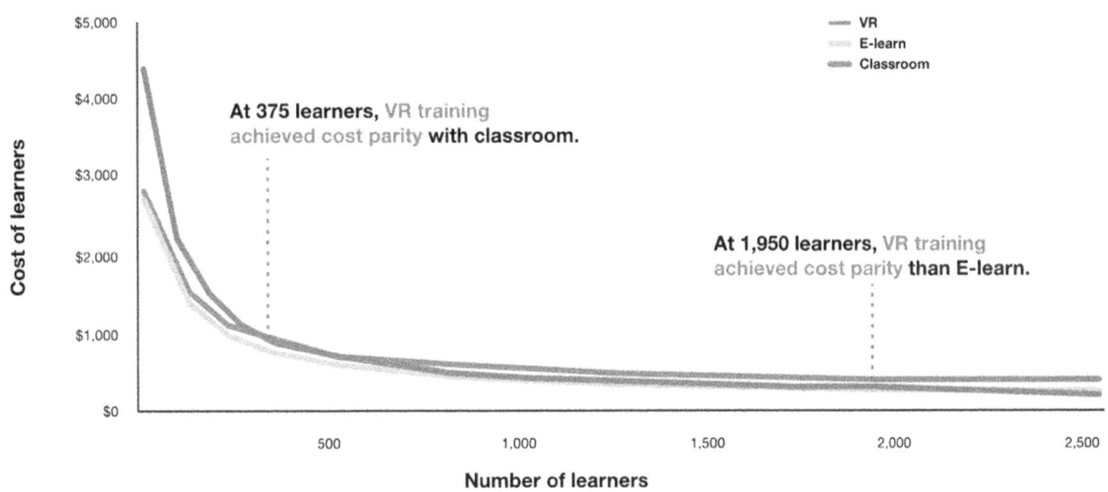

Remote working will be largely impacted by the Metaverse, where the keyboard and mouse will give way to holograms and phones (Upadhyay & Khandelwal, 2022). Co-workers in the Metaverse will not be limited to avatars of real-world colleagues, and we will be joined by a host of digital colleagues - highly realistic, AI-powered, human-like bots (Purdy, 2022). The Meta group has emerged as a pioneer in producing haptic (touch) gloves that allow users to interact with 3-D virtual objects and experience sensations such as movement, texture and pressure. Teamwork and collaboration in the Metaverse are the focus of companies like India-based Nextmeet, which emerges as an avatar-based immersive reality platform focused on interactive work, collaboration and learning solutions. The mission is to reduce the isolation and disconnection of human resources that may result from remote and hybrid working (Purdy, 2022). Another example of the exponential growth of the Metaverse and remote working is PixelMax. This UK start-up supports organisations in designing immersive workplaces designed to improve team cohesion, employee well-being and collaboration (Lee et al., 2021).

The virtual economy is a natural offshoot of content creation and sharing activities in the Metaverse. Shopping spaces in the Metaverse are being built, allowing shopping in shops in the virtual world (Lee et al., 2021). In real life, Alo's black and white GOAT jacket costs US$268.

In the realm of consumption, the Metaverse has brought about a change in behavioural relations, according to the agency Gartner a quarter of consumers will use the Metaverse daily by 2026 (Gartner, 2022a). In Metaverse, individuals can create virtual avatars (age, sexual gender), collect data about other users, and shop in Metaverse through, for example, with multiple copies of users' avatars that can be placed in different locations (Lee et al., 2021). In this new consumer behaviour, cryptocurrency will be the simplest, most convenient and least expensive method to embark on a global Metaverse shopping journey (Abel & Wong, 2010).

In addition to collaboration and large-scale technological advances, the Metaverse will require intense user experience, sophisticated telecommunications infrastructure and human-machine interfaces for a new economic relationship across borders (Mystakidis, 2022). Virtual worlds (VWs) - Metaverse - have emerged as a promising platform for doing business on a global scale (Hassouneh & Brengman, 2011). Companies are in the process of creating consumer spaces, offices, meetings for remote workers or even job interviews that will be conducted in the Metaverse world (Garon, 2022). The IT industry sector will be a strong economic asset in this dynamic, forecasting a market value of US$ 800 billion by 2024, and will have a community of one billion by 2030 (Hassouneh & Brengman, 2011). It is estimated that 10% of the global market will migrate to Metaverse by 2030 (Garon, 2022). The Metaverse has been used by brands like IBM, Toyota, Starwood Hotels and even L'Oréal Paris in internal and international marketing actions (Hassouneh & Brengman, 2011). Metaverse supports rapid internationalisation processes, in which there may be a general market, process, and political risks that result in a significant disruption to the internationalisation process (Hollensen et., 2022).

In terms of governance and framework, the issues of jurisdictions and restrictions are a concern, as well as the tools to control members or still the blockchain and democratic techniques to control illegal behaviour (L.-H. Lee et al., 2021). In terms of ethics, questions are raised about respecting the privacy of the model, however, is the selling agent declared/identified, is the output viable and trustworthy, or still liability issues ultimately dictate the concerns of those involved (Ammanath, 2022). For some authors, the most important aspect that defines true humanity is ethics itself (Vartak, 2021), however, the Metaverse can assert itself as being a potential threat to humanity and ethics (Soepeno, 2021). It is considered ultimately unethical to experience a digital, unethical, over-the-top life in a second world inauthentic to our own (L.-H. Lee et al., 2021). The ethical issues related to the Metaverse go through privacy, neural rights, hidden virtual objects, bodily autonomy, data governance structures or even pro-social/anti-social behaviour (Cotton, 2022). For example, biometric data is available through virtual reality headsets, which track a user's environment, physical movements and dimensions when using a device (Hackl, 2020). The challenge for companies is ultimately to orient processes, employees and the technology ecosystem around the dimensions of trust (Ammanath, 2022). Concerns may also run through issues such as Deepfakes, big data and cyber-attacks, which can damage a brand, customers and privacy (Hackl, 2020).

Metaverse and Economy

The Metaverse concept brings the idea of Decentralised Finance, characterised by an open system and non-dependence on a single company or entity, the need for a common economic system and also smart contracts and DeFi applications, such as decentralised exchanges, enabling liquidity, lending platforms, stable currencies or tokens (L.-H. Lee et al., 2021).

The economy in the Metaverse allows users to play and earn MANA (Decentraland's native token, with which users can buy NFTs, including LAND or collectibles, and vote on governance policies) or create NFTs, giving them real-world and game-world interoperability. In terms of products or services transacted in the Metaverse, the spectrum ranges from Infinite Office, a workplace devised by Facebook, to Nike's NIKELAND, via Roblox technology, where the user fantasises and has fun with various avatars. Currently, besides Facebook, technological companies such as Nvidia, Samsung, Hyundai, Microsoft or consumer companies such as Coca-Cola, Nike and Gucci are trying to occupy a privileged place in this market (Damar, 2021; Kim, 2021) and obtain competitive advantages. There are also several platforms like Axie Infinity, Voxels (formerly Cryptovoxels), or Decentraland developing and enabling access to the Metaverse. The complexity of the Metaverse ecosystem is greater and can be defined in seven categories: Experience, Discovery, Creator Economy, Spatial Computing, Decentralize, Human Interface and Infrastructure. (Wang et al., 2022).

In the various sectors of economic activities, various products are mobilised and transacted through cryptocurrencies in several cases. Cryptocurrency economies are the next investment frontier in emerging markets, and the Metaverse is at the forefront of this Web 3.0 Internet evolution (Grider & Maximo, 2021).

Cryptocurrency emerges as part of an anonymous, distributed electronic money system (Chaum, 1983). Cryptocurrencies rely on two different types of data structures: transactions and blocks. Transactions are grouped into blocks. Blocks are connected through hashes of their predecessors, giving rise to an authenticated data structure, the blockchain (Narayanan et al., 2016), that enables transactions in the Metaverse. Blockchain technology optimises the reconcilability between various platforms, examples such as Polkadot (DOT) and Avalanche (AVAX) enable the creation of singularized blockchains that can interact with each other. A singular metaverse will connect multiple projects, and blockchain will enable such (Binance Academy, 2022).

Cryptocurrencies make it possible to create a digital economy with different types of utility tokens and virtual collectibles (NFTs). The NFT (non-fungible token) appears as an asset that cannot be exchanged for something similar, is unique and exclusive and has its own value. NFTs are not mutually interchangeable. A fungible item, such as money, can be exchanged for another (Mazur, 2021). One of the most publicised examples is the Adidas brand with Bored Ape Yacht Club, these NFTs can be traded by their holders on marketplaces like OpenSea and BakerySwap (Binance Academy, 2022).

In economic terms, business opportunities are based on virtual events, immersive learning experiences, immersive buying experiences, social networks or even involvement/sense of belonging among human resources (Weston, 2022).

The average price of a virtual land parcel doubled in six months in 2021. It went from $6,000 in June to $12,000 by December on the four main Web 3.0 metaverses (Every Realm, 2022). The restaurant Bored and Hungry in the US accepts payments with cryptocurrencies (UOL, 2022). Cryptocurrency has already been used to buy real estate, as was the case with a property in Braga, a city in Portugal, for 3 Bitcoins. (Jornal de Negócios, 2022). The transactions can also pass through objects within games, as was the case of a boat in the game The Sandbox, sold for $ 650 thousand (News Times, 2021).

The metaverse presents itself as a driving force of a new economy, which brings elements for an immersive experience. One of the great possibilities of the Metaverse is the possibility of a massive expansion of consumer access to emerging markets without economic borders. According to consulting firm Bloomberg Intelligence, the market opportunity for the metaverse could reach $800 billion by 2024. Bank of America has included the metaverse in its 14 technologies that will revolutionise consumer lives. The big breakthrough for the Metaverse is expected to happen when the next tier of brands, the regional and local brands, begin to penetrate the Metaverse (Hollensen et al., 2022).

According to Eric Schmidt, chairman of Alphabet, the concept of the Metaverse refers to the citizen being more satisfying, richer, more elegant, more aesthetic and faster worlds. The world will become more digital than physical. And it may not necessarily be the best thing for human society (Insider, 2022). IBM's Thomas Frey points to the following as the main challenges to extending the use of the Metaverse: the internet infrastructure, the possibility of having many participants interacting in real time, language barriers and latency problems. Data protection, economic policies or even fiscal regulation and accounting will still be strong challenges (JP Morgan Chase, 2022). Currencies, health services or even training/education will be a reality and business opportunities (Lee et al., 2021). But it will not be only private companies that will be interested in the metaverse, states need to implement national policies on this topic, South Korea being the only country that has started to implement policies and create its metaverse (Damar, 2021).

METHODOLOGY

In this study, the systematic literature review methodology established by Kitchenham & Charters (2007). First, the review protocol is developed. Second, inclusion and exclusion criteria for relevant publications are identified. Third, an in-depth search of studies is conducted, followed by a critical appraisal, data extraction and synthesis.

Review Protocol

The first step of the systematic literature review was to develop a protocol (fFgure 5) for the next steps.

Figure 5. Review protocol

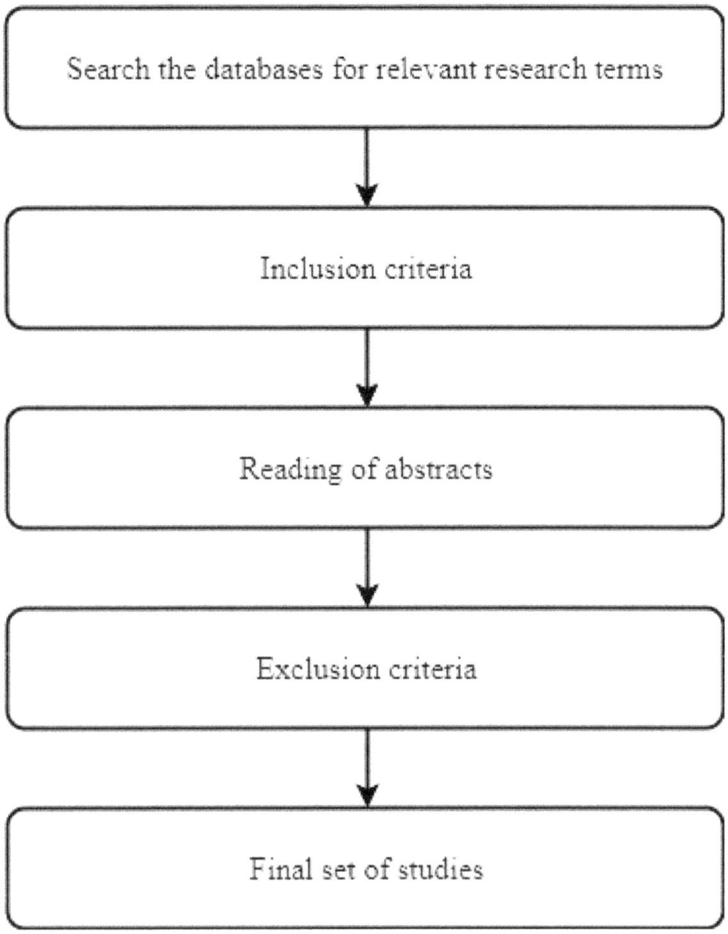

The Review Protocol begins by searching the literature, defining the terms that will be used to determine chosen datasets to retrieve the maximum number of studies that can answer the research questions formulated. The relevant articles for this study were collected and compiled in the online databases Web of Science™ Core Collection (WoS) and Scopus® (Scopus), owned by the Clarivate Analytics and Elsevier companies, respectively, which include thousands of active journals and provide a comprehensive overview of research production worldwide (Ferreira et al., 2022). The WoS and Scopus databases provide detailed metadata for indexed publications, including publication title, English abstract, keywords, authors, affiliations, document type, source information and citation count, and replication searching is possible, which is an advantage compared to searching publisher databases (Mohamed Shaffril et al., 2021). In the next phase, the inclusion and exclusion criteria were defined (Table 1).

Table 1. Inclusion and exclusion criteria

Inclusion criteria	Exclusion criteria
Related to the metaverse	Not related to the business environment
English language writings	The complete document is not available
Published in scientific journals	

To identify relevant articles, searches were conducted using the following expressions included in the title, abstract or keywords "metaverse*" or "meta-verse*", refined to articles published in English and included in the areas of business, management, economics, social sciences and humanities and computer and information sciences. The search of the key terms in the databases selected in the review protocol resulted in 118 articles (64 publications in the Scopus database and 54 publications in the WoS database). After exclusion of duplicate articles (19 articles), 99 articles resulted. After reading the articles, the exclusion criteria were applied, obtaining a total of 26 studies relevant to our research. The selection of studies, based on our inclusion and exclusion criteria, is shown in Figure 2.

Figure 6. Study selection process

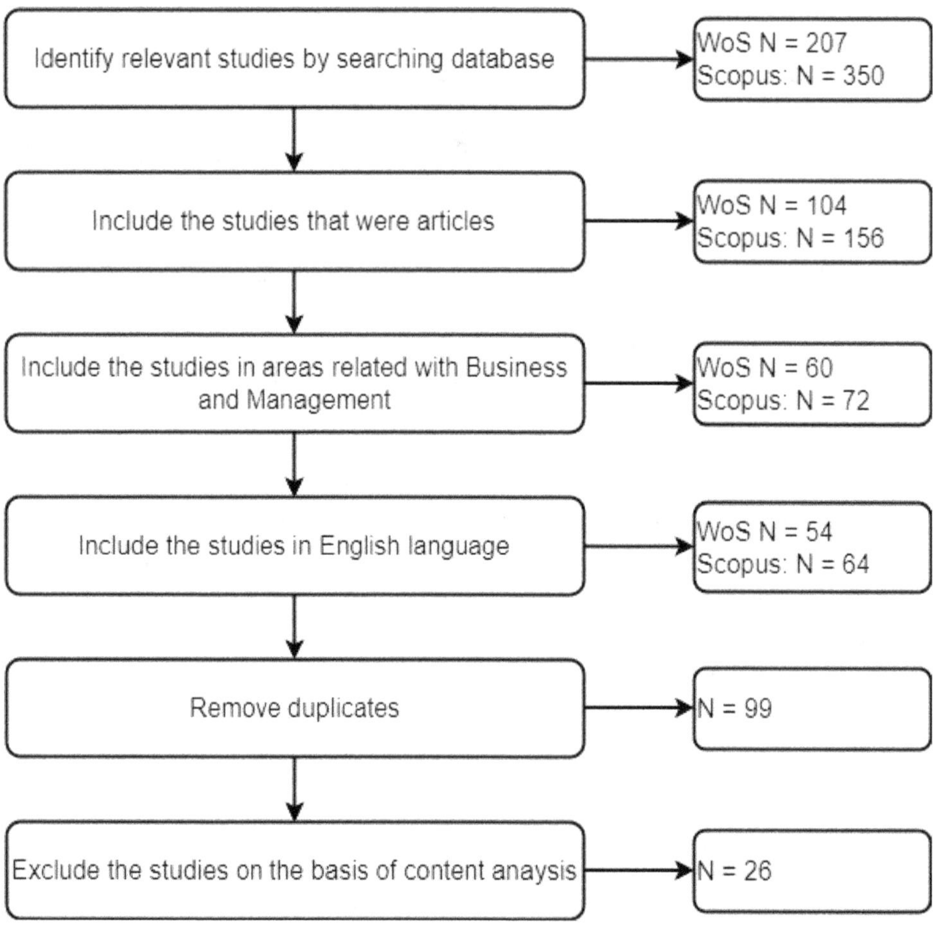

Data Extraction Analysis

The Data Extraction Analysis presents the analysis of various parameters of the selected studies, such as their chronological distribution, the sources and references, the research methodology, as well as the authors' countries of origin.

Table 2 presents a descriptive summary of the articles included in the present study. The articles involved were produced by 96 authors (107 author participations), 6 articles were published by a single author, and there were 1541 references included in our set composed of 26.

Table 2. Summary of articles included in the study

DESCRIPTION	RESULTS
Main Information About Data	
Timespan	2006:2022
Sources (Journals, Books, etc)	23
Documents	26
Average years from publication	7,7
Average citations per document	10,0
Average citations per year per doc	1,1
References	1541
Document Contents	
Author's Keywords (DE)	88
Authors	
Authors	96
Author Appearances	107
Authors of single-authored documents	6
Authors of multi-authored documents	90
Authors Collaboration	
Single-authored documents	6
Documents by Author	0,3
Authors per Document	3,3
Co-Authors per Documents	3,7

Figure 7 shows the evolution of the selected articles over the years and shows that there has been an increasing number of publications in this area. The year 2022 has the highest number of publications (7 articles), followed by 2011 (5 articles). The average year of publication of the articles was 2014.5, 27% were published in 2022, showing that this area is in its nascent phase.

Figure 7. Evolution in the number of selected studies

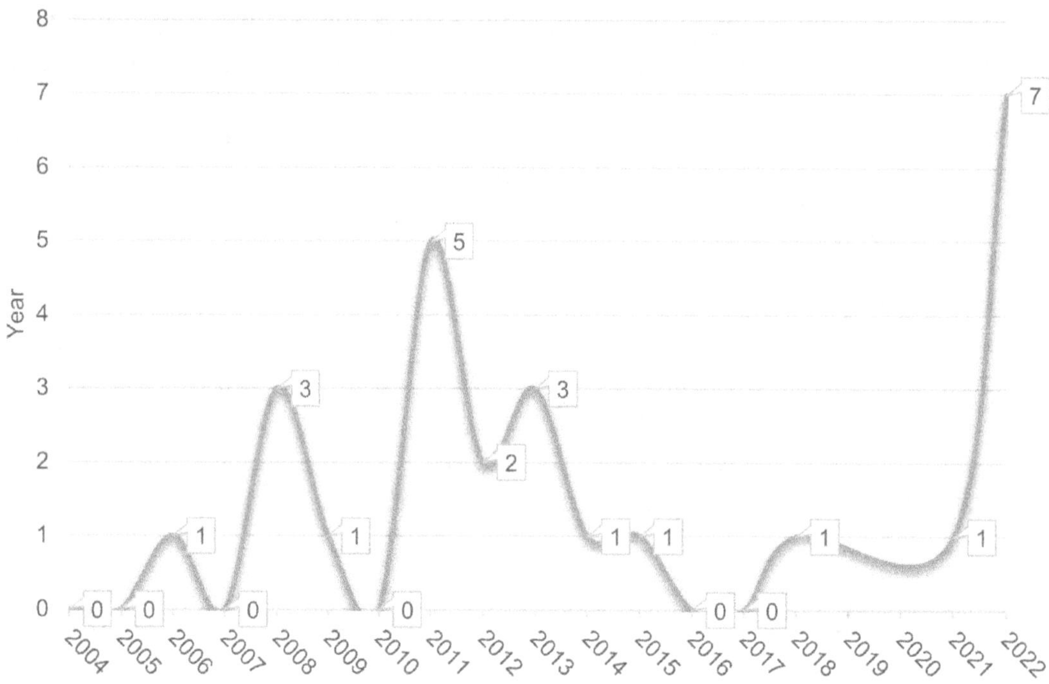

Regarding the sources where the articles were published, the journals Electronic Commerce Research, Harvard Business Review, Journal of Brand Management, Journal of Marketing Management, Technological Forecasting and Social Change stand out with two articles published on the subject. Table 3 shows the most referenced sources in the 26 articles. The Journal of Business Research was referenced in 19 articles, the Journal of Consumer Research and the Journal of Retailing and Consumer Services were cited in 15 articles, the International Journal of Information Management had references in 13, and the journals Computers in Human Behavior and Journal of Marketing had citations in 12 articles.

Table 3. Sources cited in the articles

SOURCES	ARTICLES
Journal of Business Research	19
Journal of Consumer Research	15
Journal of Retailing and Consumer Services	15
International Journal of Information Management	13
Computers in Human Behavior	12
Journal of Marketing	12
Harvard Business Review	11
Psychology & Marketing	11
Technological Forecasting and Social Change	10
Journal of the Academy of Marketing Science	8
Journal of Service Management	8

Fig. 4 shows the number of articles published according to the research methodology and the authors' country of origin. There is an equidistribution of articles published with quantitative and qualitative methodology (9 articles for both methodologies), four articles were theoretical conceptualisations, two corresponded to systematic literature reviews, and two articles used a mixed methodology (both qualitative and quantitative). Regarding the countries (Figure 4), the UK (8 articles), USA (6 articles), and Finland (5 articles) stand out with the most significant number of publications with authors from these countries.

Figure 8. Articles published according to the methodology and the authors' country of origin

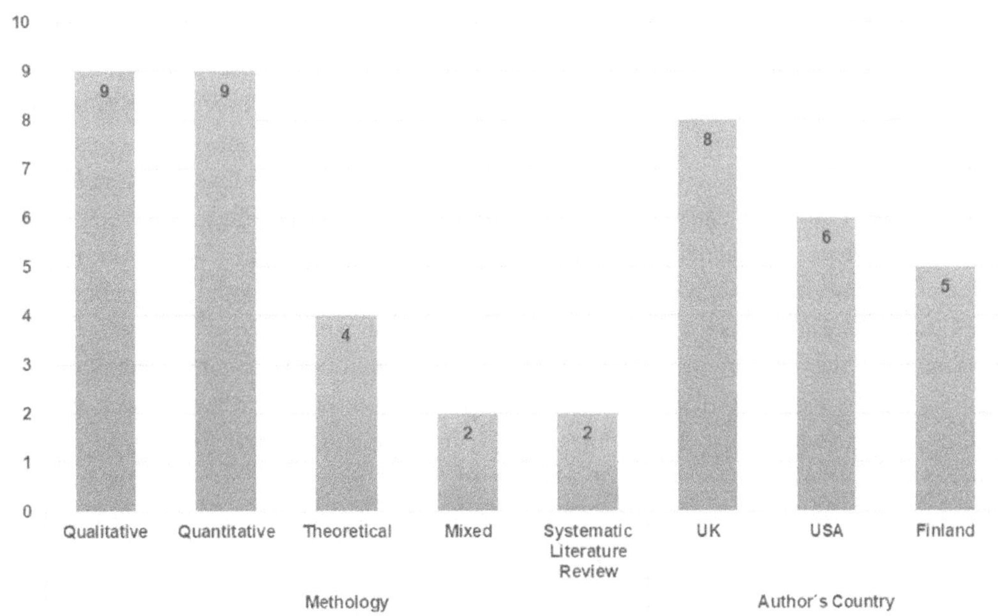

FINDINGS

To synthesize and categorize the studies based on their objective, an analysis of the different research streams was conducted, identifying the main approaches of the metaverse analysis in a business context. As indicated in Table 4, the metaverse is mainly examined for Innovation (11 articles), Marketing (11 articles), Knowledge Management, Business Models and Business Ethics (5 articles each).

Table 4. Business context

Background	Articles
Innovation (n = 11, 42.3%)	(Jeon, 2021; Kozinets, 2022; Kraus et al., 2022; S.-G. Lee et al., 2011; Masferrer et al., 2014; Riar et al., 2022; Tan & Salo, 2021; Vidal-Tomás, 2022; Wallace, 2006)
Marketing (n = 10, 38.5%)	(Bourlakis et al., 2009; Chen & Yao, 2022; Gadalla et al., 2013; Hassouneh & Brengman, 2011; Hollensen et al., 2022; Jeon, 2021; Kozinets, 2022; Scarle et al., 2012; Tan & Salo, 2021; Vidal-Tomás, 2022)
Ethics (n = 5, 19.2%)	(Kozinets & Kozinets, 2022; Papagiannidis et al., 2013; Scarle et al., 2012; Tan et al., 2022; Tan & Salo, 2021)
Learning (n = 5, 19.1%)	(Gonzalez Crespo et al., 2013; Hassouneh & Brengman, 2011, 2015; Kanematsu et al., 2012; Masferrer et al., 2014)
Business Models (n = 5, 19.2%)	(Hassouneh & Brengman, 2011; Kraus et al., 2022; Scarle et al., 2012; Tan & Salo, 2021; Zhou et al., 2018)

Innovation

Innovation is the most frequent business theme in the articles resulting from the research conducted. Jeon (2021) uses User Experience-Based Design Innovativeness (UXBDI) - characterised by novelty in product and service design that satisfies user experience - in Metaverse user-platform relationships. Jeoon (2021) concludes that UXBDI of Metaverse platforms are a key driver of ongoing user-Metaverse relationships and that it impacts relationships differently based on the type of platform Metaversec Kraus et al. (2022) analyse Facebook's entrepreneurial journey in changing its name to Meta and whether communicated strategic reorientation corresponds to a pattern of radical innovation. This research concludes that the rationale behind direct communicative efforts primarily serves to improve the external perception of the company and disseminate a signal of change within the organisation. Lee et al. (2011) examine the effects of innovation and imitation on Metaverse service adoption based on Bass' model. They conclude for several services that imitation effects for all Metaverse services are larger than innovation effects. Masferrer et al. (2014) analysed the teaching process resulting from the use of innovative learning platforms based on metaverse sessions. Riar et al. (2022), based on a systematic literature review, investigated various characteristics of augmented reality technology, namely its novelty and innovation, as well as cognitive innovation. Based on a systematic review of the literature, Tan & Salo (2021) identify the main elements of blockchain features, blockchain attributes and the economic theories underlying blockchain, such as the diffusion of innovations theory and stakeholder theory. Wallace (2006) mentions that the metaverse represents an opportunity to innovate through a technology that will change the world since it has an interface with immense expressive power.

Marketing

Another of the most prominent themes in the analyzed articles alluded to research in the marketing field, namely associated with consumer behaviour and brand management. Bourlakis et al. (2009) examine the evolution of retailing, from traditional retailing to electronic and metaverse, and how metaverses influence that evolution. One of the conclusions is that retailers need to use a holistic and comprehensive approach when designing their promotional strategies, especially if they also intend to operate in the metaverse stage. Chen & Yao (2022) study the roles of telepresence and transportation, two processes that underpin an immersive media viewing experience in the context of virtual marketing. Chen & Yao (2022) separate the impacts of virtual reality-induced telepresence on cognitive processing from the effects of narrative-induced transportation on persuasive outcomes concluding that a highly immersive media experience, combined with transportation deep in the narrative, can strengthen viewers' positive attitude toward the overall experience, pointing to a message-matching strategy to achieve marketing goals. Hassouneh & Brengman (2011) examine the value of virtual worlds in the internationalisation process of SMEs based on literature from the field of international marketing and information systems, concluding that virtual worlds offer SMEs a unique opportunity to internationalise faster and wiser. Hassouneh & Brengman (2015) identify and categorise atmospheric elements in shops designed virtual environments to propose a typology of atmospheres for these types of shops and describe how atmospheres are being used by retail companies Hollensen et al. (2022) explore the "building blocks" of Metaverse and what the benefits are for the customer, inferring that the: Metaverse will be the new marketing platform of the future to present and bring all kinds of brands to life in 3D interactive digital space. Tan et al. (2022) investigate the consumer values that lead to using a sharing economy platform, particularly the intention-behaviour of consumers for their sustainable behaviour. Tan & Salo (2021) conclude that companies need to change their centralised to decentralised marketing management logic in the blockchain-based sharing economy. In this regard, companies should practice the principles of stakeholder engagement, as decentralised marketing management requires the strong engagement of multiple stakeholders for value creation. Vidal-Tomás (2022) analyse the innovative combination of blockchain technologies and the gaming industry at the origin of metaverses, which incorporate their own economy, trade and currencies (NFTs) and the income earned through investment in these markets.

Ethic

The dark side of online communities is a widely researched topic, and subsequently, ethical issues are also present in the included articles. Kozinets (2022) reflects on the dramatically different and innovative context of the metaverse that the legal and ethical standards for conducting ethnographic research in these contexts will evolve and be clarified as these technologies and applications evolve. Papagiannidis et al. (2008) discuss the virtual world's business opportunities and challenges, focusing on ethical and regulatory issues. These authors conclude examining which "ethics" code actually applies is vitally important to ethical issues in virtual worlds and that some thematic metaverses may encourage "unethical" behaviour in line with real-world behaviour for the user to achieve the desired goal. Papagiannidis et al. (2008) suggested addressing the ethical issues of users and businesses with a three-level framework: micro, meso and macro. Tan & Salo (2021) proposed a shift in the logic of ethical marketing in the blockchain-based shared economy that aimed to guide marketing scholars on how blockchain technology enables an institutionally embedded vision of ethical marketing activities and practices that enhance collaborative marketing. Tan et al. (2022) analyse sustainable consumption behaviour. In the research, they conclude that there is a spillover effect among ethical consumers engaged in resale behaviour, and three sustainable resale behaviours were identified: planned resale behaviour, strategic resale behaviour towards unused items and reseller behaviour towards unused items. Tan et al. (2022) further infer that ethical consumer may sometimes manifest environmentally hostile consumption behaviour for temporary and self-centred benefits that are inconsistent with their values.

Learning

Collaborative and learning contexts also correspond to one of the most researched topics. Gonzalez Crespo et al. (2013) analyse the implementation of a metaverse aimed at virtual educational environments and the dissemination of scientific knowledge. Hassouneh & Brengman (2011) evaluate virtual worlds as a generator of market knowledge, a strategic tool for the implementation of marketing and brand building programmes and as a business platform for real and in-world products. Kanematsu et al. (2012) analyse a nuclear energy safety learning project carried out through e-learning and a metaverse as a classroom environment. Kanematsu et al. (2012) concluded that this type of training is effective as a clear and precise understanding by the trainees was observed, as well as their high discussion activity and interest. As a way to reinforce face-to-face classes, especially for continuous assessment, Masferrer et al. (2014) analysed the use of remote conferencing and metaverse-based tutoring, obtaining better learning results on platforms not specially designed for computer-assisted learning.

Business Model

Kraus et al. (2022) address the rebranding that Facebook adopted, mentioning that it corresponds to an option to gradually shift from a business model based on advertising revenues to a model based on transaction revenues and that Meta is strongly focused on hardware and software development to make the metaverse accessible to the mass market and encourage user loyalty by increasing switching costs. However, they conclude that this change did not represent a radical but rather an incremental change in the business model. Tan & Salo (2021) refer that blockchain companies may need a sustainable collaborative business model that serves multiple stakeholders, essential to generate substantial revenue through alliances in the blockchain-based sharing economy, lead to innovations in value chains and create sustainable business models in the sharing economy as well as in the metaverse. Hassouneh & Brengman (2011) assess the change that virtual worlds have on the business model, namely in the internationalisation process.

DISCUSSION

The present systematic literature review contributes to current research on metaverse in a business environment by providing an overview and synthesis of the empirical literature (n = 24). In addition to reporting in which business context the metaverse is used, this review provides an important contribution to current research by synthesising and structuring these contexts.

Figure 4 sets out a conceptualisation of the metaverse according to the present study's findings, consolidating a framework for the business context. Specifically, the conceptual framework is mapped according to the findings presented in the previous section.

Overall, this framework indicates that in the metaverse, as in the real world, high ethical-legal standards (Kozinets, 2022; Papagiannidis et al., 2008), codes of ethics (Papagiannidis et al., 2008), incorporating ethical marketing activities and practices into business since even ethical consumers may manifest environmentally hostile consumption behaviour for temporary and self-centred benefits that are inconsistent with their values (Tan et al., 2022; Tan & Salo, 2021). Given these facts, one component to consider in the metaverse in a business context alludes to Ethical Issues. Metaverses have challenged ethical issues about the security and privacy of users' information. Such a situation makes its users concerned about their privacy and the control of the information they share, as the perceived control in information sharing by the users is crucial to increase user trust. With the democratisation of the metaverse, more users will be connected, and more data will circulate across metaverse platforms. These features may align to allow tracking and surveillance within the metaverse, which can compromise user privacy and personal information. In many cases, users will not even understand such logging, and subsequent analysis, occurs and therefore privacy may be at risk in a number of ways. It is important that privacy is safeguarded for any information that reveals anything about physical, health, economic and socio-cultural status. At another level information relating to behaviour, such as information about lifestyles, activities, values, should also be subject to strict data protection, as should any data and metadata relating to personal communications, how we physically move and potentially the patterns of our neural activities.

Although imitation effects are more significant than service innovation effects associated with the metaverse (Lee et al., 2011), the metaverse represents an opportunity to innovate through a technology that is changing the world (Wallace, 2006). Figure 5 incorporates Innovation Management, which is oriented towards designing products and services that satisfy the user experience (Jeoon; 2021) and diffusing innovations (Tan & Salo; 2021) and fostering co-creation. Innovation management within the metaverse has the potential to bring about radical changes in business models (Kraus et al., 2022). The metaverse can lead to radical changes in business models (Tan & Salo; 2021) and to process innovation, for example, in the process of internationalisation (Hassouneh & Brengman; 2011). Thus, innovation management in metaverse environments must be customer centric. As businesses seek a greater level of knowledge of their customers in this environment, they should be able to use this knowledge to provide them with the goods and services that best meet their needs. Therefore, a better understanding of customers serves as an incentive to develop innovation to keep up with customer needs. Customer orientation will consistently lead to higher levels of innovation in the metaverse. In addition to companies identifying innovation needs for customer satisfaction through internal knowledge creation, customers can also be active in encouraging companies to focus more on innovation in co-creation processes using metaverse platforms. Companies' relationships with customers have innovation and recognition functions that help them develop process and product innovations by getting ideas from customers and accessing information about how market trends are developing. The use of metaverses also increases a company's resources in collecting, processing and using market and customer data and provides new business models based on personalised offerings implemented based on machine learning and artificial intelligence, which can match a product or service. The metaverse can also enable better evaluation and planning during the design and testing phases, as product developers and users interact with different components more effectively using virtual prototypes.

Metaverse will be the new marketing platform of the future to present and bring all types of brands to life in the digital space (Hollensen et al., 2022), having a huge impact on how companies will communicate in the future and perform Marketing Management (Figure 5). The Metaverse context presents opportunities for retailers to enhance the consumer experience, improve service quality and increase purchase intent in 3D shops (Gadalla et al., 2013; Hassouneh & Brengman, 2011; Hollensen et al., 2022; Tan et al., 2022; Tan & Salo, 2021), namely through the atmosphere (Hassouneh & Brengman, 2011; 2015). Metaverse is also gaining popularity in global brand management (Hollensen et al., 2022; Scarle et al., 2012; Tan & Salo, 2021). It is expected that the major breakthrough for the Metaverse to happen when regional and local brands begin to penetrate the Metaverse (Hollensen et al., 2022) and should use the principles of engagement to positively influence user satisfaction when choosing apparel products and, in turn, user satisfaction positively influenced the intention to purchase these products (Papagiannidis et al., 2013; Riar et al., 2022; Scarle et al., 2012). The qualitative and quantitative methodological issues associated with marketing research within the metaverse are essential to consider within the marketing functions (Kozinets, 2022). The metaverse is a tool for the internationalisation process of SMEs and is a key concept, especially in a post-pandemic context. For companies operating in turbulent international environments, sustaining customer-centric relationships is particularly important, as effective product/service development in the metaverse can complement international customer relationships, increasing mutual understanding and benefits in customer relationships. The metaverse may also serve as an enabler in creating international cooperation networks and joint ventures, as well as being a technology that will give rise to more born global businesses. The metaverse also offers SMEs a unique opportunity to internationalise faster and smarter. By having a successful presence in the metaverse, SMEs can acquire market knowledge that will minimise their perceived risk, increasing their capabilities to also have a successful presence in the real market. Furthermore, having a successful presence in the metaverse will mean that the company has built a customer base that is loyal to the brand or at least have positive attitudes toward the brand. Similar to entering any new market, in the case of the metaverse, companies need to understand their culture and locate their marketing activities in the right place to succeed. It is found that it is particularly relevant for SMEs embarking on internationalisation processes based on the metaverse that the business also considers the foreign markets' culture. That way, it will be well received by its customers in those markets and consequently be successful. As metaverse platforms have been used mainly as entertainment venues where people want to have fun and escape from real life, they must be able to offer an experience that customers feel as real regardless of where they are. Digital technologies such as the metaverse also facilitate new business models based on new pricing strategies. Brands will tend to increase digital product lines in the metaverse coupled with the use of blockchain and NFT technology, also responding to the highly differentiating experiences desired by customers of luxury brands. Metaverse applications will propel appreciable investments in automation and improve the interactivity between physical systems and human experiences. All this leads to possibilities of significantly positive outcomes in terms of value creation for customers as well as value creation for businesses.

The last component, transversal to the whole organisation, is related to learning and knowledge, one of the most explored themes in the general context of metaverses. Thus, Knowledge Management practices - creation (Hassouneh & Brengman, 2011; Kanematsu et al., 2012), dissemination (Gonzalez Crespo et al., 2013; Masferrer et al., 2014) and knowledge use (Hassouneh & Brengman, 2011; Masferrer et al., 2014) - should ensure that knowledge is used effectively and efficiently for the benefit of the organisation, namely knowledge used for innovation and marketing. Organisations see the metaverse as a way to respond to the hybrid working environments that have grown exponentially with the COVID-19 pandemic.

Explore the possibilities that avatars and holograms offered by metaverses to increase employee engagement, collaboration and inclusion and for greater productivity in remote working. Metaversos provides richer and more effective collaboration platforms compared to traditional collaboration platforms. One of the aspects that distinguishes metaverse from other collaboration technologies is its ability to create a sense of community among group members as a result of a combination of their skills. The metaverse has the ability to increase task performance as it helps workers learn faster and with fewer mistakes compared to training with physical equipment. The metaverse as a technology for training has been used for many years in various scenarios successfully, such as retail, travel and tourism and sports chaos. Most employees are not comfortable with monitoring and surveillance while working online, and this reluctance to record and monitor while working is a significant challenge. Metaverse platforms offer equivalent or better learning effectiveness compared to traditional learning. The adoption of metaverse also has high potential in the post-pandemic era, as it provides the execution of different activities as in the real world, but without any physical interaction of participants, and there is no risk of disease transmission.

Figure 9. Integrative framework

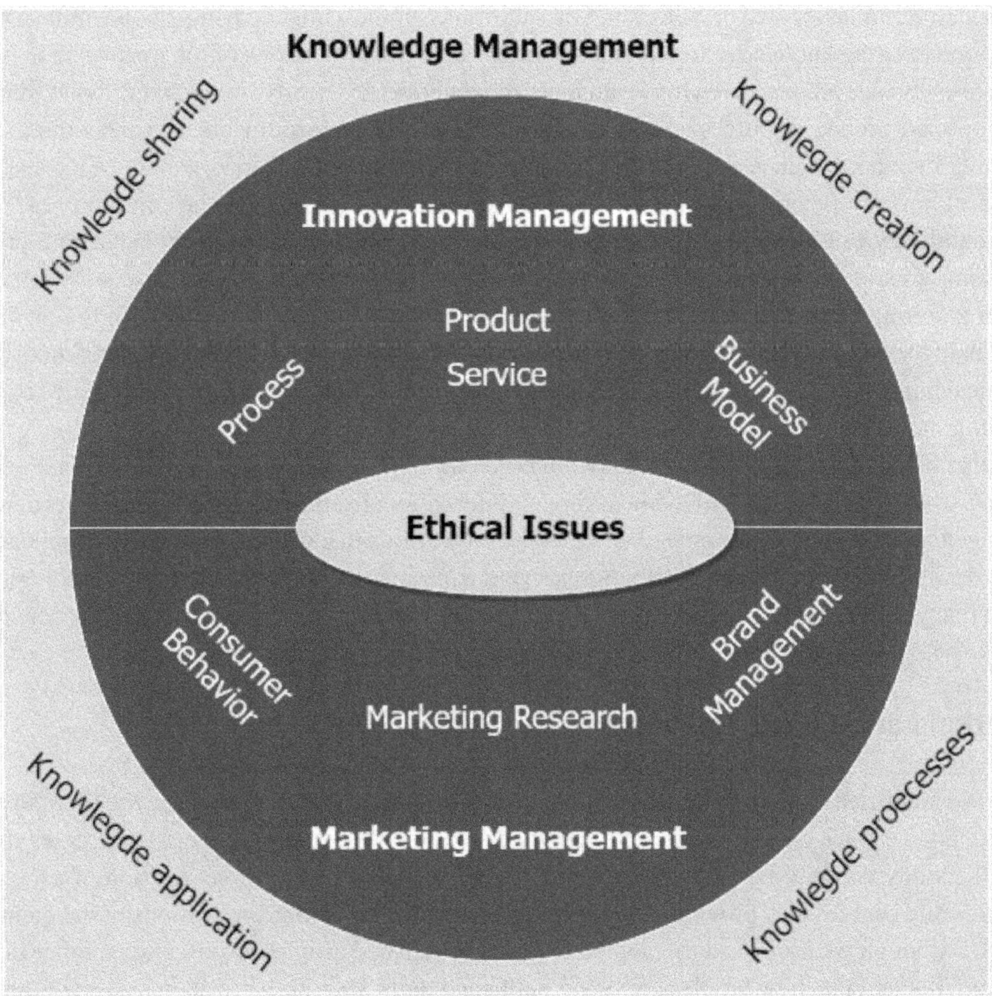

Based on Figure 9, it is concluded that the implications and applications of the metaverse in the business sphere focus on: Ethical Issues; Innovation Management; Marketing Management and Knowledge Management.

Implications

This research has put down the foundation for establishing metaverse as a research topic in business research. More importantly, this review revealed that only a small part of the metaverse applications in business research have been investigated so far, so there are many unexplored domains. With this study, our contribution lies both at the theoretical and practical levels.

At the theoretical level, this study may constitute the mechanism that enables companies to change and use capabilities to take full advantage of the metaverse. First, the review raised the issue that new digital spheres such as the metaverse may require new theoretical approaches and conceptualisations in the scientific business research domain. Second, the review highlighted that existing research has mainly addressed the metaverse in business contexts relating to innovation, marketing, learning, new business models and ethics. Third, a conceptual contribution was made by identifying the main antecedent and consequence variables studied in the existing literature and synthesising them into a conceptual model that goes beyond existing knowledge to integrate a wider range of applications of the metaverse in business.

At the practical level, entrepreneurs, managers and practitioners need to understand "what" and "how" to develop their unique resources and capabilities in metaverse environments. Ensuring the privacy of community members, even as it meets the government's legal responsibilities, should be a consideration in the development of metaverses. Privacy mechanisms that do not preserve the continuity of the metaverse are unlikely to be accepted by the customers they are intended to protect. Thus, data protection mechanisms should be created, such as allowing a user to be "invisible" to other users for some period or to use a private copy of some part of the metaverse or for the user to interact with the private copy without being seen by others. Metaverse environments are conducive to customer-centric innovation, so companies should have relationships with customers in a bidirectional way. Thus, in addition to companies identifying innovation needs for customer satisfaction through internal knowledge creation, customers should also be encouraged to be more active in co-creation processes. The use of metaverse platforms could be extremely useful for SMEs embarking on internationalisation processes. Managers can use the big data generated to investigate consumer responses to various marketing stimuli, with implications for product development, branding, distribution/service experiences, adoption of new technologies, tourism and many other areas. The metaverse offers managers new ways to realistically test concepts and processes without the costs and risks associated with more traditional methods.

Limitations and Future Research

Despite the identified contributions, the study presents limitations, mainly because the literature review is not exhaustive since only the WoS and Scopus databases were used to collect the articles that were part of the study. Despite being the two international reference bases, the collection of articles from other bases and sources may provide new perspectives of analysis on the applications of the metaverse in business. Although we have tried to integrate as many published articles as possible, another limitation stems from the fact that new articles are being published daily with high prevalence. To fill the significant theoretical and empirical gaps, we suggest that future research, which in the first instance, seeks theoretical, conceptual and qualitative contributions. Thus, more empirical evidence such as participants (business managers and users), case studies of businesses of different sizes from various countries should be included to assess any significant differences in business transformation by the metaverse. Future research should investigate the factors affecting the adoption of the metaverse in the marketing context, including the existence of potential mediating and mediating effects. Empirical research should also be conducted on businesses that have already transformed and engaged with metaverse using cross-sectional and longitudinal studies to better understand the current pattern of implementations of this technology.

ACKNOWLEDGMENT

This study received national funding through FCT - Fundação para a Ciência e Tecnologia, under the project UID/GES/04630/2020.

REFERENCES

Abel, S. M., & Wong, A. (2010). Is There A Second Life For Trademarks. In *Second Life*. Fenwick & West LLP.

Ammanath, B. (2022, March 21). An Ethical Metaverse Takes Trustworthy AI. *The Wall Street Journal*. https://deloitte.wsj.com/articles/an-ethical-metaverse-takes-trustworthy-ai-01647884164

Anderson, T., & Rivera-Vargas, P. (2020). A Critical Look at Educational Technology from a Distance Education Perspective. *Digital Education Review*, *37*(37), 208–229. doi:10.1344/der.2020.37.208-229

Bailenson, J. (2018). *Experience on demand: What virtual reality is, how it works, and what it can do*. WW Norton & Company.

Bardzell, S., & Shankar, K. (2007). Video game technologies and virtual design: A study of virtual design teams in a metaverse. Lecture Notes in Computer Science, 4563, 607–616. doi:10.1007/978-3-540-73335-5_65

BBC. (2021). *Metaverso: o que é a economia do mundo paralelo e como ela pode ser explorada nos próximos anos* [Metaverse:What is the shadow world economy and how it can be exploited in the coming years]. https://www.bbc.com/portuguese/internacional-59431866

Binance Academy. (2022). *Retrospectiva 2021*. https://academy.binance.com/pt/articles/binance-academy-2021-year-in-review

Boltz, L. O., Yadav, A., Dillman, B., & Robertson, C. (2021). Transitioning to remote learning: Lessons from supporting K-12 teachers through a MOOC. *British Journal of Educational Technology*, *52*(4), 1377–1393. doi:10.1111/bjet.13075

Bourlakis, M., Papagiannidis, S., & Li, F. (2009). Retail spatial evolution: Paving the way from traditional to metaverse retailing. *Electronic Commerce Research*, *9*(1–2), 135–148. doi:10.100710660-009-9030-8

Chen, C., & Yao, M. Z. (2022). Strategic use of immersive media and narrative message in virtual marketing: Understanding the roles of telepresence and transportation. *Psychology and Marketing*, *39*(3), 524–542. doi:10.1002/mar.21630

Damar, M. (2021). Metaverse Shape of Your Life for Future : A bibliometric snapshot. *Journal of Metaverse*, *1*(1), 1–8.

Every Realm. (2022). *Top Metaverses o Follow in 2022*. https://everyrealm.com/blog/education/top-metaverses-to-follow-in-2022

Fernandes, C., Ferreira, J. J., Veiga, P. M., Kraus, S., & Dabić, M. (2022). Digital entrepreneurship platforms: Mapping the field and looking towards a holistic approach. *Technology in Society*, *70*, 101979. Advance online publication. doi:10.1016/j.techsoc.2022.101979

Ferreira, J. J., Fernandes, C. I., Veiga, P. M., & Hughes, M. (2022). Prevailing theoretical approaches predicting sustainable business models: A systematic review. *International Journal of Productivity and Performance Management*, *71*(3), 790–813. doi:10.1108/IJPPM-12-2020-0653

Gadalla, E., Keeling, K., & Abosag, I. (2013). Metaverse-retail service quality: A future framework for retail service quality in the 3D internet. *Journal of Marketing Management*, *29*(13–14), 1493–1517. doi:10.1080/0267257X.2013.835742

Gallemard, J. (2022). *Understanding the Metaverse and its Potential for Customer Experience*. https://blog.smart-tribune.com/en/metaverse-customer-experience

GaronJ. (2022). Legal Implications of a Ubiquitous Metaverse and a Web3 Future. *Available at* SSRN, 4002551. doi:10.2139/ssrn.4002551

Gartner. (2022a). *Metaverse Hype to Transition into New Business Models that Extend Digital Business*. https://www.gartner.com/en/newsroom/press-releases/2022-02-07-gartner-predicts-25-percent-of-people-will-spend-at-least-one-hour-per-day-in-the-metaverse-by-2026

Gartner. (2022b). *What Is a Metaverse?* https://www.gartner.com/en/articles/what-is-a-metaverse

Gonzalez Crespo, R., Ferro Escobar, R., Joyanes Aguilar, L., Velazco, S., & Castillo Sanz, A. G. (2013). Use of ARIMA mathematical analysis to model the implementation of expert system courses by means of free software OpenSim and Sloodle platforms in virtual university campuses. *Expert Systems with Applications*, *40*(18), 7381–7390. doi:10.1016/j.eswa.2013.06.054

Grider, D., & Maximo, M. (2021). *Metaverse Report: Web 3.0 and Virtual Cloud Economy*. https://en.0xzx.com/metaverse-report-web-3-0-and-virtual-cloud-economy/

Hackl, C. (2020, August 2). Now Is The Time To Talk About Ethics And Privacy In The Metaverse. *Forbes*. https://www.forbes.com/sites/cathyhackl/2020/08/02/now-is-the-time-to-talk-about-ethics--privacy-in-the-metaverse/?sh=36b6ca3fae6c

Harrisson-Boudreau, J.-P., & Bellemare, J. (2022). Going Above and Beyond eCommerce in the Future Highly Virtualized World and Increasingly Digital Ecosystem. *Lecture Notes in Mechanical Engineering*, 789–797. doi:10.1007/978-3-030-90700-6_90

Hassouneh, D., & Brengman, M. (2011). Virtual worlds: A gateway for SMEs toward internationalization. *Journal of Brand Management, 19*(1), 72–90. doi:10.1057/bm.2011.24

Hassouneh, D., & Brengman, M. (2015). Retailing in social virtual worlds: Developing a typology of virtual store atmospherics. *Journal of Electronic Commerce Research, 16*(3), 218–241. https://www.scopus.com/inward/record.uri?eid=2-s2.0-84940050446&partnerID=40&md5=0c56284bbc6c664d43606bf0dcfd3d79

Hollensen, S., Kotler, P., & Opresnik, M. O. (2022). Metaverse – the new marketing universe. *The Journal of Business Strategy*. Advance online publication. doi:10.1108/JBS-01-2022-0014

Insider. (2022). *Former Google CEO says Facebook's metaverse is "not necessarily the best thing for human society" and expresses concerns about safety of artificial intelligence*. https://www.businessinsider.com/former-google-ceo-expresses-concerns-about-facebook-metaverse-

Jeon, J.-E. (2021). The Effects of User Experience-Based Design Innovativeness on User– Metaverse Platform Channel Relationships in South Korea. *Journal of Distribution Science, 19*(11), 81–90. doi:10.15722/jds.19.11.202111.81

Jornal de Negócios. (2022). *Está fechada a primeira aquisição de casa com criptomoedas em Portugal* [The first home Purchase with Cryptocrrencies in Portugal is closed]. https://www.jornaldenegocios.pt/mercados/criptoativos/detalhe/esta-fechada-a-primeira-aquisicao-de-casa-com-criptomoedas-em-portugal

Morgan Chase, J. P. (2022). *Opportunities in the metaverse. How businesses can explore the metaverse and navigate the hype vs. reality*. https://www.jpmorgan.com/content/dam/jpm/treasury-services/documents/opportunities-in-the-metaverse.pdf

Kanematsu, H., Kobayashi, T., Ogawa, N., Fukumura, Y., Barry, D. M., & Nagai, H. (2012). Nuclear Energy Safety Project in Metaverse. *Smart Innovation. Systems and Technologies, 14*, 411–418. doi:10.1007/978-3-642-29934-6_39

Kim, J. (2021). Advertising in the Metaverse: Research Agenda. *Journal of Interactive Advertising, 21*(3), 141–144. doi:10.1080/15252019.2021.2001273

Kitchenham, B., & Charters, S. (2007). *Guidelines for performing systematic literature reviews in software engineering*. doi:10.1145/1134285.1134500

Kozinets, R. V. (2022). Immersive netnography: A novel method for service experience research in virtual reality, augmented reality and metaverse contexts. *Journal of Service Management*. Advance online publication. doi:10.1108/JOSM-12-2021-0481

Kraus, S., Kanbach, D. K., Krysta, P. M., Steinhoff, M. M., & Tomini, N. (2022). Facebook and the creation of the metaverse: Radical business model innovation or incremental transformation? *International Journal of Entrepreneurial Behaviour & Research, 28*(9), 52–77. doi:10.1108/IJEBR-12-2021-0984

Latta, J. N., & Oberg, D. J. (1994). A conceptual virtual reality model. *IEEE Computer Graphics and Applications, 14*(1), 23–29. doi:10.1109/38.250915

Lee, L.-H., Braud, T., Zhou, P., Wang, L., Xu, D., Lin, Z., Kumar, A., Bermejo, C., & Hui, P. (2021). *All one needs to know about metaverse: A complete survey on technological singularity, virtual ecosystem, and research agenda.* ArXiv Preprint ArXiv:2110.05352.

Lee, S.-G., Trimi, S., Byun, W. K., & Kang, M. (2011). Innovation and imitation effects in Metaverse service adoption. *Service Business*, *5*(2), 155–172. doi:10.100711628-011-0108-8

Malbos, A. N. A., Rochadel, W., De Lima, J. P. C., & Da Silva, J. B. (2014). Aplicação da Realidade Aumentada para simulação de experimentos físicos em dispositivos móveis [Application of Augmented Reality for simulation of physical experiments on Mobile Devices]. *2014 11th International Conference on Remote Engineering and Virtual Instrumentation (REV)*, 231–235.

Masferrer, J. Á. R., Sánchez, F. E., & Hernández, D. F.-O. (2014). Experiences complementing classroom teaching with distance seminars in metaverses and videos. *Journal of Cases on Information Technology*, *16*(4), 1–12. doi:10.4018/jcit.2014100101

MazurM. (2021). Non-Fungible Tokens (NFT). The Analysis of Risk and Return. Available at SSRN 3953535. doi:10.2139/ssrn.3953535

Milgram, P., & Kishino, F. (1994). A taxonomy of mixed reality visual displays. *IEICE Transactions on Information and Systems*, *77*(12), 1321–1329.

Mohamed Shaffril, H. A., Samsuddin, S. F., & Abu Samah, A. (2021). The ABC of systematic literature review: The basic methodological guidance for beginners. *Quality & Quantity*, *55*(4), 1319–1346. doi:10.100711135-020-01059-6

Morris, T. (2022). *Unsure about the metaverse? Here's what you (might) need to know*. https://blog.gwi.com/chart-of-the-week/understanding-the-metaverse/

Mystakidis, S. (2022). Metaverse. *Encyclopedia*, *2*(1), 486–497. doi:10.3390/encyclopedia2010031

Narayanan, A., Bonneau, J., Felten, E., Miller, A., & Goldfeder, S. (2016). *Bitcoin and cryptocurrency technologies: a comprehensive introduction.* Princeton University Press.

News Times. (2021). *A virtual yacht sells in the Sandbox metaverse for $ 650,000.* https://www.news-times.com/business/article/A-virtual-yacht-sells-in-the-Sandbox-metaverse-16666199.php

Papagiannidis, S., Bourlakis, M., & Li, F. (2008). Making real money in virtual worlds: MMORPGs and emerging business opportunities, challenges and ethical implications in metaverses. *Technological Forecasting and Social Change*, *75*(5), 610–622. doi:10.1016/j.techfore.2007.04.007

Papagiannidis, S., Pantano, E., See-To, E. W. K., & Bourlakis, M. (2013). Modelling the determinants of a simulated experience in a virtual retail store and users' product purchasing intentions. *Journal of Marketing Management*, *29*(13–14), 1462–1492. doi:10.1080/0267257X.2013.821150

Park, S., & Kim, S. (2022a). Identifying World Types to Deliver Gameful Experiences for Sustainable Learning in the Metaverse. *Sustainability*, *14*(3), 1361. Advance online publication. doi:10.3390u14031361

Park, S., & Kim, Y. G. (2022b). A Metaverse: Taxonomy, Components, Applications, and Open Challenges. *IEEE Access: Practical Innovations, Open Solutions*, *10*, 4209–4251. doi:10.1109/ACCESS.2021.3140175

Park, S., Min, K., & Kim, S. (2021). Differences in learning motivation among bartle's player types and measures for the delivery of sustainable gameful experiences. *Sustainability*, *13*(16), 9121. Advance online publication. doi:10.3390u13169121

Prasad, R. (2022). *The transformation of customer experience within the Metaverse*. https://customerthink.com/the-transformation-of-customer-experience-within-the-metaverse/

Purdy, M. (2022). How the Metaverse Could Change Work. *Harvard Business Review - Digital Articles*. https://hbr.org/2022/04/how-the-metaverse-could-change-work

PwC. (2020). *How Virtual Reality Is Redefining Soft Skills Training*. PwC.

Radfahrer, L. (2022). *Entenda o que é o metaverso e por que ele pode não estar tão distante de você* [Understand what the Metaverse is and Why it might nit be so far away from you]. https://www.cnnbrasil.com.br/business/entenda-o-que-e-o-metaverso-e-por-que-ele-pode-nao-estar-tao-distante-de-voce/

Rauschnabel, P. A., Babin, B. J., tom Dieck, M. C., Krey, N., & Jung, T. (2022). What is augmented reality marketing? Its definition, complexity, and future. *Journal of Business Research*, *142*, 1140–1150. doi:10.1016/j.jbusres.2021.12.084

Republic Realm. (2021). *The 2021 Metaverse Real Estate Report*. https://www.republicrealm.com/post/the-2021-metaverse-real-estate-report

Riar, M., Xi, N., Korbel, J. J., Zarnekow, R., & Hamari, J. (2022). Using augmented reality for shopping: A framework for AR induced consumer behavior, literature review and future agenda. *Internet Research*. Advance online publication. doi:10.1108/INTR-08-2021-0611

Scarle, S., Arnab, S., Dunwell, I., Petridis, P., Protopsaltis, A., & de Freitas, S. (2012). E-commerce transactions in a virtual environment: Virtual transactions. *Electronic Commerce Research*, *12*(3), 379–407. doi:10.100710660-012-9098-4

Shah, D., Rust, R. T., Parasuraman, A., Staelin, R., & Day, G. S. (2006). The Path to Customer Centricity. *Journal of Service Research*, *9*(2), 113–124. doi:10.1177/1094670506294666

Smart, J., Cascio, J., & Paffendorf, J. (2007). Metaverse roadmap: pathways to the 3D web. *Metaverse: A Cross-Industry Public Foresight Project*.

Stephenson, N. (2003). *Snow crash: A novel*. Spectra.

Tan, T. M., Makkonen, H., Kaur, P., & Salo, J. (2022). How do ethical consumers utilize sharing economy platforms as part of their sustainable resale behavior? The role of consumers' green consumption values. *Technological Forecasting and Social Change*, *176*, 121432. Advance online publication. doi:10.1016/j.techfore.2021.121432

Tan, T. M., & Salo, J. (2021). Ethical Marketing in the Blockchain-Based Sharing Economy: Theoretical Integration and Guiding Insights. *Journal of Business Ethics*. Advance online publication. doi:10.100710551-021-05015-8

The New York Times. (2022). *Crypto Is Cool. Now Get on the Yacht*. https://www.nytimes.com/2021/11/05/technology/nft-nyc-metaverse.html

Uhl, A., & MacGillavry, K. (2016). Customer Centricity. In A. Uhl & L. A. Gollenia (Eds.), *Digital Enterprise Transformation* (pp. 149–171). Routledge. doi:10.4324/9781315577166-6

UOL. (2022). *Com tema de arte NFT, hamburgueria recebe pagamento em criptomoedas* [NFT-Themed Art, Hamburgueria receives payment in Cryptocureencies]. UOL.

Upadhyay, A. K., & Khandelwal, K. (2022). Metaverse: The future of immersive training. *Strategic HR Review*, *21*(3), 83–86. doi:10.1108/SHR-02-2022-0009

Vidal-Tomás, D. (2022). The new crypto niche: NFTs, play-to-earn, and metaverse tokens. *Finance Research Letters*, *47*, 102742. Advance online publication. doi:10.1016/j.frl.2022.102742

Wallace, M. (2006). Virtual worlds, virtual lives. *PC World*, *24*(11), 133–136. https://www.scopus.com/inward/record.uri?eid=2-s2.0-33750551484&partnerID=40&md5=fbc08e8b2d59ea37bbbfd49f23c16f57

Wang, J., Wang, T., Shi, Y., Xu, D., Chen, Y., & Wu, J. (2022). Metaverse, SED Model, and New Theory of Value. *Complexity*, *2022*, 4771516. doi:10.1155/2022/4771516

Weston, M. J. (2022). Strategic Planning for a Very Different Nursing Workforce. *Nurse Leader*, *20*(2), 152–160. doi:10.1016/j.mnl.2021.12.021 PMID:35075352

Zhou, M., Leenders, M., & Cong, L. M. (2018). Ownership in the virtual world and the implications for long-term user innovation success. *Technovation*, *78*(June), 56–65. doi:10.1016/j.technovation.2018.06.002

KEY TERMS AND DEFINITIONS

Ethics: Standards of conduct regarding what is acceptable behavior and unacceptable behavior in business.

Innovation Management: Process related to decisions, activities and practices that transform ideas into new opportunities to generate business value.

Knowledge Management: Process related to creation, discovery, acquisition, transfer, sharing, and diffusion of knowledge in an organization.

Marketing Management: Process related to formulating, pricing, promoting and distributing goods and services to achieve business goals.

Metaverse: Digital technology that allows for a shared immersive experience where participants can have diverse experiences that take place in the real world such as socializing, shopping, teaching and training, or even having cultural experiences.

Chapter 7
Gamification Strategies for Social Media

Jorge Esparteiro Garcia
https://orcid.org/0000-0001-5710-5557
Polytechnic Institute of Viana do Castelo, Portugal

Pedro Rodrigues
Polytechnic Institute of Viana do Castelo, Portugal

Jorge Simões
https://orcid.org/0000-0002-7822-0068
Instituto Superior Politécnico Gaya, Portugal

Manuel José Serra da Fonseca
https://orcid.org/0000-0002-3290-8449
Polytechnic Institute of Viana do Castelo, Portugal

ABSTRACT

With the evolution of technology and all the associated paradigms, the business reality had the need to adapt and incorporate all this evolution. Gamification strategies in social networks are used to optimize the use and involvement of users. Thus, in a business context, these strategies are used to increase user engagement in the company/brand's social networks with the aim of increasing and strengthening trust. This chapter's main objectives are to understand if internet users and social media users know the term gamification and if the associated strategies are perceptible in the social media context. Therefore, to better analyze the users' behavior regarding these strategies and their knowledge about gamification, an online survey was developed and applied. After the survey, it was possible to conclude that 16.3% of the respondents already knew the term gamification. Through the survey it was also possible to conclude that most respondents had already been in contact with gamification strategies, even if 90.6% did not adhere to them frequently.

DOI: 10.4018/978-1-6684-5538-8.ch007

INTRODUCTION

Before the creation of computers as we know them today, in the middle of the 20th century, there was a huge evolution caused by technological advances and the growing industrial demand. Due to the need for information storage and automation of processes that until then would have been manual, the creation of the first computers began, quite different from the ones we know today. A few years later, and thanks to the continuous and rapid technological advance, computers became more intelligent and capable of storing more information. At the height of the Cold War between the United States of America and the USSR (from 1947 to 1991), the Americans felt the need to decentralize the information they contained in buildings such as the Pentagon and so began the ideology of backup and data security and consequently the creation of the Internet (Sandroni, 2015).

After the first email was sent, there was an exponential growth of computer systems and the Internet itself. A few years later, after years of research and investment in this type of technology, the World Wide Web, the World Wide Web of Computers, was created in the 1990s - the Internet thus became a global phenomenon, spreading rapidly through all countries on planet Earth. In a fraction of time, it became possible to communicate with any part of the world, which until then would have been impossible (Sandroni, 2015).

Gamification has already been used successfully to achieve improvements and optimizations in a multitude of challenges. Gamification can be seen as the use of game elements for user reward.

Examples include, from creating bronze or marble statues of the winners from the early days of the Olympic games to coming to the present day with improving learning through success badges (Hakulinen, Auvinen & Korhonen, 2013), improving employee engagement (Neeli, 2012) and encouraging users to act more safely (Kroeze & Oliver, 2012). In each of these studies, game-like elements were used to make improvements, rewarding users, or encouraging certain behaviors.

Throughout this chapter, the main concepts, and the evolution that gamification has undergone over the years in the different sectors where it is used will be addressed.

Of the multiple applications that the Internet has brought to the human being, there was the creation of web applications called social networks. It was in 1997 that one of the first social networks was created - it is difficult to classify one as the first - Six Degrees, where in many aspects it fits into the context that we know today as a social network (creating a profile, uploading photos, sharing personal information). It is, however, in 2004 that the social network MySpace reaches one million monthly users, and it is argued that this is the beginning of social networks (Ortiz-Ospina, 2019).

Right now, Facebook, the world's largest social media platform, has 2.4 billion users. Other social media platforms, including Youtube and Whatsapp, also have over a billion users each.

Given that there are almost 8 billion people worldwide, with at least 3.5 billion nodes online, this means that social media platforms are used by more than one in three people in the world, and more than two-thirds of all Internet users (Ortiz-Ospina, 2019).

Social media has changed and will continue to change the world. The rapid and wide adoption of these technologies is changing how we meet partners, how we access information from the news, and how we organize to demand policy change.

In the case of this chapter, the object of study of this research is the knowledge of the term gamification and its strategies in social networks, directly or indirectly, by the user.

The objectives are divided into general and specific, having fundamental roles in the research process.

The general objective of the research is as follows:

- To verify the knowledge of gamification strategies in social networks.

To achieve the general objective, the following research questions (RQ) were determined:

RQ1 - Does the audience of this study recognize the term gamification?

RQ2 - Do the respondents recognize the existence of strategies in social media with specific objectives and are they aware of the strategies applied?

RQ3 - On what device are social media most accessed?

RQ4 - Are individuals willing to participate in gamification strategies?

This chapter is divided into 7 sections. The first section introduces this chapter by contextualizing the theme, defining the problem and objectives, and presenting the structure of the document. Sections 2, 3 and 4 are related to the literature review, with a study of the definition of social networks and the evolution of the World Wide Web in section 2, the definition of gamification and its constituents in section 3, and the contextualization of gamification in networks, the platforms under study and the mobile era in section 4. Section 5 presents the research methodology of this work. Section 6 describes the analysis of the results of the questionnaire survey developed in this work. Finally, the conclusions, the final considerations of this chapter are presented in Section 7, as well as research ideas related to the results of this chapter are proposed.

LITERATURE REVIEW

There is a huge variety of proposals for defining social media. For Ahlqvist, Bäck, Halonen and Heinonen (2008) the definition is based on three key elements: content, communities, and web 2.0. The argument of these authors is built on the development of these elements in social networks where there is creation of content and communities with the same interests of the user himself and this whole process is done through the development of services based on web 2.0 technologies.

Kietzmann et al. (2011) gives us another and different perspective on how to define social networks. Their proposal is called "social network hive". The hive is a metaphor for the construction of social networks. According to the same authors and according to their study, the way social networks are built is divided into seven parts: identity, conversations, sharing, presence, relationships, reputations, and groups. Identity is the amount of information revealed by each user about themselves. Conversations depict the extent of communication between users (Rodrigues, MIM., Fonseca, MJSd., & Garcia, JE., 2022). Sharing depicts the amount of content distributed by users (Kotler, P., Kartajaya, H., Setiawan, I., 2021). Presence refers to the ability of users to know about the accessibility of others. Reputation is about users knowing the standing of others and finally, groups which represent the communities formed by users.

According to Kaplan & Haenlein (2010) social networks refer us to a set of web-based services and applications where users create and exchange content of their own creation (Costa C.R.., Garcia J.E.., Da Fonseca M.J.S.., & Teixeira A., 2021). These authors propose that social media can be divided into six categories. Thus, the categories are based on the level of social presence and self-presentation and are as follows: blogs, social networking websites, virtual social worlds, virtual game worlds, content communities and collaborative projects.

The same authors also admit, and in addition to the concrete definition of social networks, it is necessary to mention two concepts that are related to each other and that underlie the definition of social networks: web 2.0 and user-created content (Kaplan & Haenlein, 2010).

First, user-created content. For the Organization for Economic Cooperation and Development (2007), there are requirements for this content. These are as follows: publication on a public website so that a group of users can access this content and furthermore it must demonstrate a creative effort and must be created outside of professional practices (OECD, 2007). Kaplan & Haenlein (2010) also reinforce these requirements by saying that emails, replicas, and content that is made for a professional purpose are excluded from this category.

In 2004 the term web 2.0 appeared to describe new ways of using web-based technologies in which users create and change content collaboratively. These authors, Kaplan & Haenlein (2010), exemplify web 2.0 with Adobe Flash which allows animation and interactivity, AJAX for asynchronous data transfer without the need to refresh web pages or interfere with their viewing, and RSS, which is a family of web feed formats.

In short, there is no concrete and precise definition for social media, but although they vary somewhat, they complement each other in such a way that many elements are generally accepted. In this way, social media help in the creation and sharing of content by users (Kaplan & Haenlein, 2010; Ahlqvist, Bäck, Halonen & Heinonen, 2008; Kietzmann, Hermkens, McCarthy & Silvestre, 2011).

World Wide Web

Today society still thinks that the web is synonymous with the Internet, but this is a common misconception. The World Wide Web, or better known as the web, is the most prominent part of the Internet. It can be defined, based on technological systems, as a techno-social system for the purpose of human-machine interaction. A techno-social system is defined as a system for the purpose of improvement. Improvement at the cooperative, cognitive and communicational level.
To better understand this process, it is necessary to understand two aspects:

- To communicate it first requires recognition, and in turn the neuron needs to communicate to cooperate.
- Cooperation is a recipe for communication and, in turn, a recipe for cognition.

Web 1.0

It is in 1989 that Tim Burners-Lee suggested the creation of a worldwide hypertext space that, at any time of accessibility, would be driven by a single Universal Document Identifier (CUDI). This suggestion comes with a single purpose and goal: the creation of a pervasive information space in which people communicate by sharing information. Web 1.0 emerged as a read-only web, meaning that it was one-way and static. It was a big step for the business world, as companies could present their catalogs, which were advertised in newspapers and magazines. Thus, a new way for companies to present their range of products and/or services to consumers emerged. The HTML pages of the websites were only static and, in turn, not very regularly updated. The main purpose of the websites was solely and exclusively to publish information to mark the online presence, and the websites were not interactive, resembling brochures or posters. In this era of the web, the following main protocols were used: HTTP, HTML and URI, which made connections too weak, making it impossible for many users to visit.

Web 2.0

In 2004, Dale Doughertv defines the term web 2.0 as the business revolution of the computer industry. Caused by the shift to the internet as a platform and an attempt to understand the rules to achieve their success. "Web 2.0 is the business revolution in the computer industry caused by the move to the internet as platform, and an attempt to understand the rules for success on that new platform. Chief among those rules is this: Build applications that harness network effects to get better the more people use them." (Tim O'Reilly, 2004). People-centric, web 2.0 identically known as a web of wisdom, participatory, reading and writing, thus becoming bidirectional. Web 2.0 does not appear as a new version of web 1.0. It allows flexible design, updates, creative reuse, and the modification of collaborative content, i.e., it is a web as a platform where users can let go of many of the controls, they were used to in web 1.0. Web 2.0 marks this evolution with a feature with great weight to the user which is to support collaboration to gather collective intelligence instead of web 1.0 (Murugesan San, 2007).

Web 3.0

In 2006, John Markoff, suggests web 3.0 as the third generation (Nova, Spivack, 2011). The basis of web 3.0 is to define structural data and link it to new more effective discovery, integration, automation, and reuse across the various applications (Ossi, Nykänen, 2003). Web 3.0 attempts to link, integrate, and analyze data from various data sets to obtain new information flows, support mobile Internet accessibility, improve data management, encourage the factor of globalization phenomena, increase customer satisfaction, help organize collaboration in social networks, and simulate creativity and innovation. The semantic web, the way web 3.0 is known, was conceived by Tim Berners-Lee, inventor of the World Wide Web. Being worked on by a team to extend, improve and standardize the system, publications, languages, and tools that had previously been developed (Sean B, Palmer, 2001). Semantic web is a web that can demonstrate things in the approach that computer can understand, with its main goal which is to make the web readable by machines and not just humans. Currently the web is a web of documents, functioning, in a way, as a global information system that includes the most important problems on it. It was generated for human consumption, where the primary objects are documents and the links between them as well (or part of them). The semantics of content and links are implicit and the degree of structure between objects is quite low (Christian, Bizer & Tom, Heath & Tim, Berners-Lee, 2009; Tim, Berners-Lee & Christian, Bizer & Tom, Heath & Kingsley, Idehen, 2008; Oktie, Hassanzadeh, 2008). The development of the semantic web comes to solve and overcome the problems of the current web. It can be defined as a web of data and in a sense a global database that is mostly represented by the following characteristics: The design goal of the data web is machines first and only then humans. The primary goals are things hence the links between things. Thus, the semantics of content and links is explicit and the degree of structure between objects is high and is based on the RDF model (Christian, Bizer & Tom, Heath & Tim, Berners-Lee, 2009; Tim, Berners-Lee & Christian, Bizer & Tom, Heath & Kingsley, Idehen, 2008; Oktie, Hassanzadeh, 2008). The main difference between web 2.0 and web

3.0 is that web 2.0 targets the content creativity of users and producers while web 3.0 targets the connected data sets.

Web 4.0

As the phase of the web we are witnessing, it can be said that it is an ongoing idea and there is no exact definition like the previous ones. Known as the symbiotic web, it aims at the interaction between humans and machines in symbiosis. In a brief explanation, machines would be intelligent in reading web content, and would react in the way of executing and deciding what to execute first to load websites quickly with higher quality and performance and build interfaces with greater command (Hemnath, 2010). Web 4.0 is predicted to be the read-write-execute-agree web. It will reach a critical mass of participation in online networks that provide global transparency, governance, distribution, participation, collaboration in key communities such as industry, political, social, and other communities (Marcus, Cake, 2008). Web 4.0 will parallel the human brain, and this implies a massive network of highly intelligent interactions (Dan, Farber, 2007). Although web 4.0 is not defined in an exact way nor as all its technologies, this reality is moving towards using artificial intelligence and making web 4.0 an intelligent web. To best describe this statement, the best example will be the Siri and Alexa technology.

GAMIFICATION

Over the past few years, gamification has been a phenomenon that has gained greater importance in academic research contexts. Recently, gamification has been used in many contexts to achieve various goals. It has been used in risk management (Bajdor & Dragolea, 2011), usability problem solving (Saha, Manna & Geetha, 2012), education (Huang & Soman, 2013; Hakulinen, Auvinen & Korhonen, 2013; Muntean, 2011), and marketing (Tillström, 2012), to name a few.

In this chapter the different aspects of gamification are presented, it is important to start with the definition of gamification and closely related concepts, and then focusing further on games, game-like elements, and the effects of gamification - which can be found in real-life use.

Definition

The definition of gamification according to Deterding, Dixon, Khaled and Nacke (2011) refers to the "use of game design elements in non-game contexts", that is, in gamification, game design elements are used in other types of systems, such as social media.

Terms that have been used in the same context are, among others, "applied game" and "play design." (Deterding et al., 2011) It is important to note that gamification involves the elements of game design while a ludic design often lacks structure, goals, or both (Groh, 2012). In terms of what the "non-game contexts" are, (Deterding et al., 2011) suggest not limiting the definition further as it offers no advantage in making such restrictions.

From a service marketing perspective, there is another definition, given by Huotari and Hamari (2011), where according to their definition, gamification is seen as a packaging for a service. The core service is enhanced by a rule-based system incorporating feedback and interaction mechanisms. This definition is more restrictive than the previous one but has many similarities with the various definitions of game and game elements, which will be discussed throughout this chapter. The main goal of gamification is to "support overall user value creation by providing game-like experiences" (Hamari & Koivisto, 2013). From a broader perspective, gamification has some similarities to more traditional marketing tools, such as customer loyalty cards, leading companies to increasingly view their marketing as games (Hamari & Eranti, 2011).

Gamification can sometimes be confused with serious games, which are games made for non-recreational purposes (Deterding et al., 2011; Susi, Johannesson & Backlund, 2007).

A "serious game" is a game based on computer games and simulation approaches and/or technologies for non-entertainment purposes. The term "serious" refers that the game is intended for educational purposes rather than entertainment. They have been widely used in the fields of defense, education, scientific exploration, health services, emergency management, defense, business, tourism, urban planning, engineering, religion, and politics in an immersive or interactive way that can be enjoyed in the best possible way.

Gamified applications use only certain appropriate elements of games (Groh, 2012), while "serious games" are games. Overlapping with gamification in terms of mechanisms used, another closely related topic is that of persuasive technologies, although these aim to change users' attitudes and behavior (Hamari & Koivisto, 2013).

A term closely related to gamification is "games with a purpose," or GWAP (Games with a Porpuse). In GWAPs, users play a game to entertain themselves while simultaneously completing tasks that are difficult or impossible for computers (Von Ahn & Dabbish, 2008). For example, GWAPs can be used to tag subjects in photographs (Von Ahn & Dabbish, 2008) or to collect geospatial data using a location-based mobile game (Matyas et al, 2008).

To fully understand gamification and what it consists of it is first necessary to understand the definition of games and game-like elements.

Badges

A commonly used game element in gamification and one of the most discussed in previous literature is a badge, also known as an achievement or a trophy. Badges are defined by Antin and Churchill as "digital artifacts that have some visual representation that are awarded to users who complete specific activities" (Antin & Churchill, 2011), like medals awarded to military heroes or merit badges used by the Boy Scouts of America.

Achievements, or in English "achievements," can be defined as optional subgoals that are rewarded in a secondary reward system, i.e., the achievement system, separate from the primary reward system, according to Montola et al. (2009).

However, Hamari and Eranti (2011) argue that viewing achievements as optional or secondary can be problematic. They raise the issue of a player explicitly attempting to complete all available achievements, thus making achievements their primary goal. Therefore, these authors propose that achievements should be viewed as games of their own, and not just as secondary reward systems to the main functionality. This view that achievements are a game in themselves contradicts the definition of a game, casting achievements as puzzles rather than games, as mentioned earlier.

Badges have five main functions:

Goal setting - badges challenge users to achieve the goals set by their requirements.

Instruction - badges can instruct users about the activities that are available in the system.
Reputation - badges can be used to assess the reputation of users, for example, expertise
Affirmation - badges function as status symbols and provide personal affirmation by showing users' past accomplishments.
Group identification - badges "communicate a set of shared activities that bind a group of users together" (Antin and Churchill, 2011, p. 3).

Broadly speaking, achievements can fall into several categories (Montola et al, 2009), including tutorial achievements attributed for experiencing the features of the game, completion achievements attributed for completing tasks, collection achievements attributed for obtaining items, virtuosity achievements attributed for excelling exceptionally (e.g., not dying at all), hard mode achievements attributed for succeeding on high difficulty, veteran achievements attributed for playing the game for an extended period of time loyalty achievements awarded to players for being loyal to the game and the community (e.g. subscription for an extended period of time), paragon achievements which are achievements awarded to some top players for accomplishing something extremely difficult or rare (e.g. being the first player in the world to complete a task), and "random" achievements awarded to players who perform fan activities (e.g. attending game conventions or buying collector's editions).

Based on the findings of Montola et al. (2009), achievement systems require certain properties to work well with players. For example, when the player completes an achievement, it must be notified immediately and explicitly to reward the player, to remind the player of the existence of the achievements, and to "pique their curiosity for the achievements".

Game Design Mechanics and Patterns

According to Sicart, the definition of game mechanism has been problematic to achieve (Sicart, 2008). As such, there have been numerous attempts over the years to define game mechanics. In some studies, distinctions have been made between the rules of games and the actions that players have available (Sicart, 2008). An example of this distinction is made by Avedon (1971) who suggests that game mechanics are formed by "action procedures," i.e., "specific operations, necessary courses of action, method of play" (p. 422).

In 2003, Lundgren and Björk present a more informal definition of game mechanics that is given by defining them as "any part of a game's rule system that covers one, and only one, type of possible interaction that takes place during the game" (Lundgren and Björk, 2003). Examples of game mechanics are trading, bidding, negotiation, storytelling, and role-playing.

A few years later, in 2008 and building on the previous definitions, Sicart defines a game mechanic as "methods invoked by agents, designed for interaction with the game state" (Sicart, 2008, Introduction, par. 6). Examples of game design patterns and mechanics include the game of Monopoly: "Roll dice, move counterclockwise around the board, the number of spaces indicated on the dice" (Avedon, 1971, p. 422). Other common mechanics include time constraints, limited resources, and turn-based play (Groh, 2012).

Hunicke et al. (2004) define game mechanics as "the particular components of the game, at the level of data representation and algorithms." According to this definition, mechanics are all the actions, behaviors, and control mechanisms in the context of the game that are available to the player.

In his article, Muntean (2011) lists some of the various game mechanics used in gamification. These mechanisms include points, levels, challenges, virtual goods, leaderboards, donations, and charity. A larger list of examples of game mechanics is listed by Gamification-Research.org (2014). In this list, mechanisms such as achievements, commitments, behavioral dynamics, happy productivity, bonuses, cascading information theory, combos, community collaboration, countdown, discovery, infinite gameplay, levels, points, progression, quests, rewards, status, among others are mentioned. While some of these mechanics are not mentioned by other sources, many of them, such as points, achievements, levels, and progression are (Muntean, 2011; Avedon, 1971; Groh, 2012).

Effects of Gamification

Over the years and to support the research papers claiming the effects of gamification in various contexts, there have been reported results of real-world applications of gamification. Microsoft has applied gamification to increase both internal productivity and end-user productivity (Jacobs, 2011). One such project was "Communicate Hope," which was used to collect feedback from Lync users on product usability and design, as well as to submit bug reports. The scoreboard was linked to a variety of charities, and as users submitted feedback and bug reports, Microsoft donated more money to the charities based on the results of the scoreboard. As a result, users who chose to play the game submitted sixteen times more feedback than users who did not play. Tens of thousands of dollars were also donated to charities ("Facts & Figures," 2013). Microsoft also created two games, Ribbon Hero 1 and 2, to improve end-user productivity. These games aim to improve the skills and practices of Microsoft Office users (Jacobs, 2011).

Autodesk's research team applied gamification to teach beginner users how to use AutoCAD (Li, Grossman & Fitzmaurice, 2012). Called "GamiCAD," the project was integrated into AutoCAD software as an interactive tutorial system. The researchers behind GamiCAD argue that in the case of software applications containing hundreds or thousands of features, new users may gain little or no help from software documentation. To overcome this problem, they incorporated GamiCAD into AutoCAD software, with fantasy game elements, clear goals, feedback and guidance, progressive disclosure, time pressure, rewards, and encouragement. In GamiCAD, the user is given "missions," or tasks, to complete. These missions are related to designing components of a spacecraft for NASA, giving a game context for the actions. Each mission consists of multiple levels, and after completing a level, the system displays a performance feedback screen.

The University of Hawaii has committed to sustainability and energy conservation, even changing the behavior of dorm residents through gamification (Brewer et al, 2011). In their study, Brewer et al created a contest for energy conservation. They used game design elements such as rounds, levels, and prizes to encourage participation. The goals were to increase knowledge about sustainability issues, gain insight into one's behavior and how to change it, build community, and create commitment. Evidence of the effectiveness of gamification includes increased energy knowledge, changes in energy behaviors, short-term energy reductions, and educational contributions (Kukui Cup, 2014).

Gamification in Social Networks

After all the previous definitions and the literature review of some concepts it is now possible to insert and analyze the main topic of this chapter: gamification in social networks. The effects of gamification in the context of social networks have been increasingly the subject of research and analysis. Bista, Nepal, Colineau, and Paris (2012) defined a model of gamification of online communities and proceeded to implement a gamified online service for welfare recipients transitioning back to work. Although the study does not look at the gamification of an actual social media service, but rather an online community, it provides relevant information about how gamification is or could be conducted in this context.

Despite the differences between commercial portals and social communities, "visitor engagement remains at the heart of their success" (Bista et al 2012, p. 611). As such, the benefits of gamification could be leveraged to improve user engagement in social media as well. According to the same authors, the problems in establishing a new online community have three challenges that can be addressed with gamification:

Bootstrapping - process of gathering initial members to the community and keeping them involved during the initial phase;
Monitoring - process of observing the use of the service;
Sustainability - process of maintaining user engagement after the initial phase.

According to Antin and Churchill (2011), badges are commonly used in social media to engage and motivate users. For example, Wikipedia has incorporated "Barnstars," which are badges used by users to reward others for doing valuable work. Other examples given by Antin and Churchill are StackOverflow and Foursquare. A community that shares similar goals and interests is essential in the context of gamification, according to Hamari and Koivisto, the network of users "creates opportunities for meaningful interaction," "enables reciprocal activity," and "increases the perceived benefits of the service". To create engaging gamified services, social elements are essential, so social networking services and gamification support each other simultaneously. However, the authors also noted that a user's attitude toward a gamified service strongly determines whether the user will continue to use the service, as well as intentions to recommend the service to others. Some social media services make use of the real-life contexts surrounding the services. For example, when a Foursquare user checks in at five different Starbucks cafes, they receive a free drink (Hamari & Eranti, 2011). In this case, the feature is called the "out-game" component, referring to the fact that the reward is external to the game or the system itself.

We can admit that, gamification is at its core, entirely related to statistics (Bunchball, 2010), the collection of statistics and their derivatives creates a continuous and extended motivation for participation. Closely related, Bista et al. (2012) gives examples of number of YouTube views, Facebook likes, and Instagram likes as statistics related to status, achievement, competition and reward, all game dynamics listed by Bunchball (2010) as seen earlier. As such, many features of social media such as the counting of views and likes can be considered as game dynamics.

Mobile Era

As it has been possible to do throughout this chapter, the concept of gamification can be discussed as being as old as the idea of the Olympic Games, in which the status of the winner of the Olympic Games was symbolized by a crown of olive leaves and in which the winners of 3 times in a row, received as a reward their own bronze or marble statues made to be displayed in Olympia ("JetBlue Badges Deliver Gartner's Gamification," Comarch, 2014). This view fails to consider a significant gamification landscape, which consists of three elements:

- The rapid adoption of the smartphone;
- The enormous growth of the mobile web;
- Finally, the increased use of social media.

7.2 million people use a smartphone in Portugal, which represents 84.2% of the population and a significant increase compared to 2012, when this percentage was less than 30%. Marktest data for the month of July 2020 also show that among the groups that most use a smartphone are men and the higher social classes.

The values now disclosed leave no doubt: the Portuguese are increasingly using the smartphone. The July figures more than tripled compared to 2012, with a growth of 57.2% over the past eight years. The data and analysis are part of the TGI study. The research collects information for 17 market sectors, 280 product and service categories and over 3,000 brands providing data on consumption, brands, hobbies, lifestyle, and media consumption. More recently, in 2017, 6.5 million Portuguese had a smartphone. All in all, in three years there has been a growth of almost one million users. Smartphones have brought a world of possibilities, allowed Internet access and use on the go. All the information stored in the world's largest library - i.e., the Internet - is now available on a smartphone, which can be easily carried wherever you go. The rapid adoption of the smartphone has led to a growth in mobile data traffic, and this has had a direct and consequent effect on the mobile web - the part of the World Wide Web that mobile device access - as the desktop web competitor. By far the most common smartphone use is access to social networks such as Facebook and Twitter. In 2013, smartphone users spent nine hours and six minutes per month on social networks, compared to one hour and fifteen minutes per month streaming video on the device and about one hour and eleven minutes engaged in sports-related news and videos (Dawn C. Chmielewski, 2103). The combination of these three elements - the smartphone, the mobile web and social networking - has made the popularity of gamification applications possible. The mobile web itself has made real-time information sharing about such gamification applications a reality. People who were used to social networks not only quickly adopted these gamification applications as another means of tracking and sharing their own and their friends' daily activities, but they also began to share and display the points and badges they earned on social networks, thus increasing the popularity of gamification itself.

What is new about gamification is not necessarily the idea of applying game elements to a real-world activity, but rather how these game elements are now applied seamlessly, ubiquitously, and socially. It is now possible to access our contacts, emails, notes, calendar, map, social media accounts, and even history and spending patterns (if using a personal finance app) on a small smartphone. Being placed on the same device where all this information resides and which we carry wherever we go, games can easily infiltrate real-world actuality activities. This is a real difference between today's mobile gamification applications and all past attempts to use games and game elements for a real-world purpose. Commercial video games have been popular for years, and many of them had impressive graphics and sophisticated narratives. But the smartphone, mobile and social networks have completely changed where and how games are played and game dynamics can be applied. It is this new mobile and social era that has allowed games to become ubiquitous in everyday activities, even can be used beyond an imaginary game world confined to a computer or video game console. While there have been "serious games" that have attempted to use game elements and dynamics for education beyond mere entertainment, these serious games have never reached the same level of popularity as gamification. The significance of the smartphone, mobile network, and social media in the wide adoption and popularity of gamification does not mean that gamification should take the form of a mobile application. As we will see in the next chapter, everyday activities like recycling or watching the speed limit while driving can be gamified without the use of a smartphone or the Internet. What makes something a game is never pure technology. However, understanding the significance of these three elements in the recent trend of gamification provides a vantage point from which we can determine where the strength of gamification lies. A simple way to understand the difference between gamification and games is that while games tend to create an imaginary world separate from reality, gamification creates a game layer on top of the real world (Seth Priebatsch, 2010). We do not enter a fictional game world when, for example, we open the Foursquare, Waze, or Nike+ apps. Instead, these gamification apps create a game layer on top of real-life activities, using game elements such as points, badges, and leaderboards. Gamification rewards our behavior on the web, often on the mobile web, with social connections and statuses. It even occasionally offers discounts or freebies that can be used in the real world. As we spend more and more time online, the line between our online and real lives will only become increasingly blurred, and more things will start to cross between these two domains.

RESEARCH METHODOLOGY

Throughout any research, it is necessary to adopt a specific methodology. Carlos Morais says that "it is not easy to find a definition for research, but it is quite consensual to admit that research is being carried out, among other reasons, to solve problems, deepen concepts and build knowledge. The search for consistency and answers to problems is one of the challenges that we can associate with the concept of research". Thus, research "is a procedure of systematic, controlled and critical reflection that allows the discovery of new facts or data, relationships or laws in the field of knowledge" (Ander-Egg, 1978, cited in Marconi & Lakatos, 2003).

Object of study

In this case the object of study of this research is the user's knowledge of the term gamification and its strategies in social networks, directly or indirectly. The objectives are divided into general and specific, having key roles in the research process.
The general objective of the research is as follows:

- To verify the knowledge of gamification strategies in social media.
  To achieve the general objective, the following research questions were determined:

RQ1 - Does the audience of this study recognize the term gamification?

RQ2 - Do the respondents recognize the existence of strategies in social media with specific objectives and are they aware of the strategies applied?

RQ3 - On what device are social media most accessed?

RQ4 - Are individuals willing to participate in gamification strategies?

Methodology

It is essential to use a methodology that allows, as reliable as possible, to obtain an inspection for the evaluation of the main problem.

With the establishment of objectives, it is foreseen to be fulfilled, for this it is necessary to divide it into parts for the accomplishment and, in the end, the success of the same.

First, a survey was constructed with the necessary questions to be able to make a study and draw objective and determined conclusions about this very area of research.

Next, the survey was applied to a specific target audience to determine their knowledge (directly or indirectly) about gamification strategies on social media.

The final phase of this chapter involves data collection and data analysis through a qualitative and quantitative research method, using analytical reports provided by the survey responses in question.

Questionnaire

The various scales and questionnaire model that was used were based on verifying and researching, in the most practical way possible, i.e., by online questionnaire, regarding gamification in social media.

The questionnaire is a research instrument that will include several organized questions aimed at collecting data and information for research (Fonseca, 2022).

There are some advantages and disadvantages in using questionnaires to obtain data. The main advantages are associated with the low costs of operationalization of this type of data collection, especially online questionnaires that still offer a larger scope without the need for printing or the need for researchers. Their practicality and quick results allow for collection and subsequent analysis almost in real time, and it is also possible to readjust response times according to the needs of the project in question. These questionnaires also have the advantage of guaranteeing total anonymity to the respondent, since there is no direct contact with any of the stakeholders.

However, in using these questionnaires there are also some disadvantages, the main one being the likelihood of dishonesty in the survey responses. There is a probability that some results may not correspond faithfully to reality due to dishonesty on the part of the respondent, which may be motivated by social desirability or an attempt to protect individual privacy. Answers may not be given conscientiously and may have been given too quickly, without consideration, or even by not reading them carefully enough. In the case of weighted questions, differences may arise in the understanding and interpretation by the respondents, and predetermined answers may limit the real feeling or meaning of the questionnaire. These questionnaires may also present accessibility problems, such as for visually impaired users.

For these disadvantages to be minimized, it is advisable that the questions asked are simple and easy to interpret, and that answers are provided that do not cause the respondent to feel doubtful. Taking into consideration the points discussed above, it is possible to obtain questionnaires with more reliable results. The questionnaire was structurally divided into four groups, and was conducted on the Google Forms platform, free of charge. Before the respondent enters the groups of questions, an introductory text is written to briefly explain the area, the institute, and what the research study is about. The initial title, "Gamification strategies in social networks", describes the topic that is intended to be investigated. This choice is related to the general object of study and the research questions presented earlier in this chapter.

In group I, the user, a decisive question was asked to continue answering this questionnaire, which is "Are you a user of social networks". In case they answered no, the questionnaire would end here. The single answer selection method was used.

In group II, social network advertisements, we asked directly whether the respondent knows the term gamification and his or her opinion on strategies in this area. Questions 2, 3, 7, 8 and 9 are answerable to all respondents, with questions 4, 5 and 6 being only for respondents who answered "yes" to question 3 which is "Do you know the term Gamification, or do you know what it means". In this group, the single answer selection method was used for questions 2, 3, 4, 7, 8 and 9 and for questions 5 and 6 the Likert scale of values 1 to 5, where values 1 and 2 are negative, i.e., disagree or strongly disagree and values 4 and 5 are positive response, agree or strongly agree. The value 3 is neutral.

In group III, social networks are used to indirectly find out if the respondent knows and interacts with gamification strategies on social networks. In this way 11 questions were asked. The first question in this group, question 10, uses the multiple answer method to find out which social networks the respondent uses. In question 12, 13 and 14 examples of social networks are shown with an illustration for each question to check whether the user recognizes that content. For this, questions 11, 12, 13 and 14 the single answer selection method is used. In the answers to questions 15, 19, 20 and 21 the Likert scale of values 1 to 5 is used, where values 1 and 2 are negative, i.e., disagree or strongly disagree, and values 4 and 5 are positive answers, agree or strongly agree. The value 3 is neutral. And finally, in questions 16, 17, and 18 the frequency is verified through a Likert scale of values 1 to 5, where the values 1 and 2 are negative, i.e., never, and rarely, and the values 4 and 5 are positive, frequent, and very frequent. The value 3 is neutral.

In Group IV, it is intended to collect information regarding socio-demographic elements, which will analyze and characterize the sample, such as gender, age, academic qualifications, income, current employment status, and area of residence.

Ending, the questionnaire, in a new window with a thank you message.

Sampling

To conduct the study for a large audience, the study will be published on various social media platforms, which will further broaden the possibility to respond, and there will be no criteria of any kind. For convenience, the locations of the respondents will be in the north of Portugal as well as in the age groups being in higher education. It will be shared in university, public and private education groups. The choice of sharing in these groups arises in the context that Millennials are the public with more connection, interactivity, intensity, and habit of using social networks. Without underestimating the rest of the individuals who use social networks, Millennials are the audience that receives the most information about the social networks where gamification strategies are applied. In summary, the target audience for this study will be as follows: Female and male individuals, aged 17 to 60, belonging to social classes A, B, and C. Students or workers with a lifestyle that leads them to frequently use social networks and or other applications that come from social networks in their daily lives.

RESULTS

In this section, the results obtained from the questionnaire applied to the respondents will be analyzed. This questionnaire was developed through the Google Forms platform and was available from January 2022 to February of the same year. The main results obtained are presented below:

In general, do you pay attention to the ads and publications of brands, institutions, associations, or companies on social networks?

Of all the participants in this survey, 68% (138 answers) of the respondents answered that they pay attention to the advertisements and publications of brands, institutions, associations, or companies on social networks. And 32% (65 answers) do not. With this, we can conclude that the attention that users have while surfing and using social networks differs from their habits, interests, and lifestyles, such as how they use social networks, for professional issues, direct search for some product or service or even for leisure. These habits, interests, and lifestyles, previously mentioned, are influential factors with great relevance in the answers to this survey, as well as all the conclusions that may arise from it.

Are you familiar with the term Gamification, or do you know what it means?

This question comes as a key to the various conclusions that can be made about this study. Thus, only 16.3% (33 answers) of the total respondents know the term Gamification and, consequently, 83.7% (170 answers) do not. The term Gamification is an uncommon term both in society and in the vast world of marketing, more precisely in digital marketing, because it is still in a phase of introduction.

Do you know any kind of Gamification strategy?

In a first instance the respondents who know the term Gamification, 30.3% (10 answers) do not know any Gamification strategy in social networks, which makes 69.7% (23 answers) know concrete strategies of this area. Concluding even if knowing the term and the area of Gamification, respondents may not know applications of it in social networks. So, they may know strategies by the entities but are not recognizing or relating with the term Gamification.

Do you agree that Gamification strategies are a good practice by the brand/company?

Despite their opinion about these strategies on the reputation of brands, institutions, associations, or companies and how much it affects them, of the respondents 18.2% (6 responses) Totally Agree that these types of strategies are a good practice by entities, and further, 63.6% (21 responses) Agree and 18.2% (6 responses) Neither Agree nor Disagree. It is important to emphasize that no respondents Disagree or Totally Disagree with this statement. Of these 6 answers that Neither Agree nor Disagree are from the same respondents that in the previous question 66.67% (4 answers) answered Neither Agree nor Disagree, and 33.33% (2 answers) Agree that these strategies damage the reputation of brands, institutions, associations, or companies.

Do the brands/companies that you follow on social networks usually have this type of strategies?

Of all the respondents, the brands, institutions, associations, or companies of the personal interests of the respondents of this questionnaire, answering the question whether they adopt these strategies, 3% (6 answers) entities never adopt this type of strategies, and 20.2% (41 answers) rarely, 37.9% (77 answers) occasionally, 33.5% (68 answers) often and 5.4% (11 answers) very often.

Do I usually interact with brands/companies that use this type of strategies in their publications?

Of all respondents 24.6% (50 answers) never interact with these strategies on social networks, so that, 31% (63 answers) rarely, 35% (71 answers) occasionally, 8.4% (17 answers) often and 1% (2 answers) very often.

Do you think brands/companies should use more of this type of strategy in their social networks?

Of all the respondents 15.3% (31 answers) totally disagree that entities should use more of this type of strategies in their social networks, so that, 16.3% (33 answers) disagree, 44.8% (91 answers) neither disagree nor agree, 20.2% (41 answers) agree and 3.4% (7 answers) totally agree.

If the brands/companies you follow on social networks used more of this type of strategies, would you participate?

Answering this question, of all respondents 16.7% (34 answers) totally disagree, 16.7% (34 answers) disagree, 41.9% (85 answers) neither disagree nor agree, 22.7% (46 answers) agree, and 2% (4 answers) totally agree.

These strategies used by brands/companies are reliable?

Of this statement, 6.9% (14 responses) strongly disagree, 15.8% (32 responses) disagree, 49.3% (100 responses) neither disagree nor agree, 25.1% (51 responses) agree, and 3% (6 responses) strongly agree.

Do you trust these strategies used by brands/companies?

Of this statement, 13.8% (28 responses) totally disagree, 18.2% (37 responses) disagree, 43.8% (89 responses) neither disagree nor agree, 20.7% (42 responses) agree, and 3.4% (7 responses) totally agree.

Analysis of Results

RQ1 - Does the public in this study recognize the term gamification?

In answer to question 1, 16.3% of the respondents (33 answers) who know the term gamification are between 19 and 56 years old. Of these respondents, only 3 do not have a higher education degree, and it can be assumed that 2 of these 3 respondents, since they are between the ages of 18 and 21, may still be in the process of completing a higher education degree. 90.9% of these responses are from respondents with a Higher Education Degree, which may affect why they are aware of this term, as well as their strategies and reactions in their behaviors towards it.

RQ2 - Do respondents recognize the existence of strategies on social networks with specific goals and are you aware of the strategies applied?

In answer to Q. I. 2, 16.3% (33 answers) of the respondents who know the term gamification are between 19 and 56 years old. Of these respondents, only 3 do not have a higher education degree, and it can be assumed that 2 of these 3 respondents, since they are between the ages of 18 and 21, may still be in the process of completing a higher education degree. 90.9% of these responses are from respondents with higher education qualifications, which may affect the reason why they are aware of this term, as well as their strategies and reactions in their behaviors towards it.

Of these 33 respondents who know about the term gamification and disagree or strongly disagree that these types of strategies are harmful to the reputation of brands, institutions, associations or companies, their ages range from 20 to 52, with the highest frequency range being between 20 and 30 years old. These are individuals who have followed with greater intensity the evolution of the world and the era of digital and technology, making their mentality more open in approaching this area as well as the application of it, as well as likewise no respondents Disagree or Totally Disagree that they are a good practice on the part of entities.

RQ3 - On what device are the social networks most accessed?

Regarding this question, this answer influences the whole thought of this study by the degree of recetivity of these strategies by users, depending on the functions that the devices can do. Thus, 2% (4 answers) use more the computer to access their social networks, 0% (0 answers) use the Tablet and, in turn, 98% (199 answers) use more the cell phone to access their social networks.

It is important to note that to any conclusion, all social networks, currently are available in the desktop version, and yet it is concluded that the device that most accesses social networks is the cell phone.

RQ4 - Are individuals willing to participate in gamification strategies?

Finally, it can also be concluded, in relation to P.I. 4, that respondents even not knowing the definition of gamification, nor its strategies that are applied on social networks weave their opinion and 23.65% (48 answers) agree that these strategies damage the image of the understood. Of these same respondents, more than half of the respondents usually have contact with this type of strategies, although 90.6% (184 answers) do not adhere to them frequently. This is a factor that can be easily identified due to the habits, interests, and lifestyles of users as well as the age range being heavily geared towards Millennials.

Despite their strong opinion in not participating in these strategies, almost half of the sample in this survey has no opinion if entities should use these strategies more, as well as no opinion if these types of strategies are reliable, and do not know if they would participate if entities provided more content of this kind. In this way, it can be assumed that it is a pondered society, without strong opinions about the future and with awareness of concepts such as cybersecurity.

CONCLUSION

The technological evolution and social networks are already part of the lifestyle of today's society, and it is a constant challenge for companies/brands to persist on these platforms with a high level of awareness and visibility. This social media medium has already become irreplaceable for those used and, in a business context, is where companies find potential customers, as well as analyze the choices and strategies of their competitors.

Gamification and its strategies have already been used for improvement and optimization in a wide range of challenges. Gamification can be seen as the use of game elements for user reward.

Bringing these two areas together brings us to the improvement and optimization of the strategies used in social networks for a better relationship with the user. During this chapter it was shown how this junction has already become indispensable. The main goal of gamification strategies in social networks is to improve the use and involvement of users in social networks. Thus, this chapter aimed to analyze, qualitatively and quantitatively, the behavior of users towards this type of strategies, as well as the knowledge about this area, through an online questionnaire survey.

203 answers were registered. A quite a large number, being their most frequent age between 20 and 30 years old, from the northern region of Portugal, more precisely from Viana do Castelo. After analyzing the surveys, it was possible to verify that for 83.7% of the respondents the term gamification in its pure definition is not common in the knowledge of the target audience, as well as any definition of the same area in the context of social networks, considering that 90.9% of the respondents have higher education qualifications. Although this term is not common in the knowledge of the target audience in question, a large percentage of them have observed, or even participated in strategies adopted by companies/brands without having, totally or not, notion of this area, with the precision that 90.6% of respondents do not adhere frequently to these same strategies. Even taking into account that of these respondents, most do not have any opinion on whether or not companies/brands should continue to invest in these strategies nor do they commit to their word about the future, it is possible to conclude that it is a pondered society, without assuming strong opinions about the future and with awareness of concepts such as cybersecurity and what care they should take in their exposure on the networks.

This study can benefit several areas, which are part of the business fabric, such as Marketing and Design. For managers and marketers of brands or companies, there is a huge range of needs in the strategies adopted to achieve the planned objectives and, consequently, to be closer to the consumer to understand what they want and feel about this brand-consumer relationship.

LIMITATIONS AND FUTURE RESEARCH DIRECTIONS

Following the initial research plan, a deeper analysis and study regarding the considered objectives would have been intended. Due to the time factor, it was not possible to carry them out as expected. These objectives were expanded and used for future research.
The intended objectives would be as follows:

- To identify the different types of knowledge and acceptance of strategies in gamification according to gender.

- To identify the different types of knowledge and acceptance of gamification strategies according to area of residence.
- Identify the different types of knowledge and acceptance of strategies in this area according to income.
- To identify the different types of knowledge and acceptance of strategies in this area according to the respondents' current professional situation.

For better utilization and to continue the research work, it is considered that the following research approaches can be developed.

- Apply the research to an audience with a different income level.
- Direct the study to a different target audience, in terms of age and academic qualifications.
- Use this research and drive strategy, in a specific company/brand or organization, taking advantage of the data analysis and finding positive solutions for it.
- Apply the research in different areas of residence, in large urban centers, and analyze the main differences when compared to other geographical areas.
- To extend this study to an international level.

REFERENCES

W3C. (1999). *Resource Description Framework (RDF) Model and Syntax Specification.* https://www.w3.org/TR/1999/REC-rdf-syntax-19990222/

W3C. (2004). *The Unicode Consortium.* http://www.unicode.org/

Ahlqvist, T., Bäck, A., Halonen, M., & Heinonen, S. (2008). *Social media roadmaps.* Edita Prima Oy.

Alexander, R., & Philipp, H. (2008). *Security Aspects in Web 2.0 Mashup Systems.* Technology, Altenbergerstrabe, Linz, Austria. Available at http://www.fim.uni-linz.ac.at/lva/SE_Netzwerke_und_Sicherheit_Security_Considerations_in_Intercon_Networks/semH. pdf/

Antin, J., & Churchill, E. F. (2011). Badges in social media: A social psychological perspective. *CHI, 2011 Gamification Workshop Proceedings.*

Avedon, E. M. (1971). The structural elements of games. The study of games, 419-426.

Bajdor, P., & Dragolea, L. (2011). *The gamification as a tool to improve risk management in the enterprise.* Annales Universitatis Apulensis Series Oeconomica. doi:10.29302/oeconomica.2011.13.2.38

Bista, S. K., Nepal, S., Colineau, N., & Paris, C. (2012). Using gamification in an online community. In *Collaborative Computing: Networking, Applications and Worksharing (CollaborateCom), 2012 8th International Conference.* IEEE.

Bohyun. (2013). *The Library Mobile Experience: Practices and User Expectations.* Library Technology Reports 49, no. 6. ALA TechSource.

Caillois, R. (1961). *Man, play, and games*. University of Illinois Press.

Chmielewski, D. (2013). *Nielsen Study: Social Networking Dominates Smartphone, Tablet Use*. http://articles.latimes .com/2013/jun/09/entertainment/la-et-ct-nielsenstudy-social-networking-smartphone-tablet-20130609

Christian, B., Tom, H., & Tim, B.-L. (2009). *Linked Data - The Story So Far*. Journal Semantic Web and Information Systems.

Comarch. (2014). *JetBlue Badges Deliver Gartner's Gamification*. www.comarch.us/media-resources/knowledge-articles/jetbluebadges-deliver-gartners-gamification/

Costa, C. R., Garcia, J. E., Da Fonseca, M. J. S., & Teixeira, A. (2021). Data Analysis in Content Marketing Strategies. *Iberian Conference on Information Systems and Technologies, CISTI*. 10.23919/CISTI52073.2021.9476377

Crawford, C. (2003). Chris Crawford on game design. Academic Press.

Dan, F. (2007). *From semantic Web (3.0) to the WebOS (4.0).* https://www.zdnet.com/blog/btl/from-semantic-web-30-to-the-webos-40/4499/

Deterding, S., Dixon, D., Khaled, R., & Nacke, L. (2011). From game design elements to gamefulness: defining gamification. *Proceedings of the 15th International Academic MindTrek Conference: Envisioning Future Media Environments*, 9-15. 10.1145/2181037.2181040

Facts & Figures. (2013). http://www.enterprise-gamification.com/mediawiki/index.php?title=Facts_%26_Figures

Fonseca, J. (2002). *Metodologia da pesquisa científica* [Scientific Research Methodology]. Apostila. Editora Fortaleza.

Gamification-Research.org. (2021). http://http://gamification-research.org

Gerber, Barnard, & van der Merwe. (2007). *Towards a semantic web layered architecture*. The 25th Conference on IASTED International Multi-Conference.

Groh, F. (2012). *Gamification: State of the art definition and utilization*. Institute of Media Informatics Ulm University.

Hakulinen, Auvinen, & Korhonen. (2013). Empirical Study on the Effect of Achievement Badges in TRAKLA2 Online Learning Environment. *Learning and Teaching in Computing and Engineering (LaTiCE),* 21-24.

Hamari, J., & Eranti, V. (2011). Framework for designing and evaluating game achievements. *Proceedings of DiGRA 2011: Think Design Play*, 115-134.

Hamari, J., & Koivisto, J. (2013). Social motivations to use gamification: An empirical study of gamifying exercise. *Proceedings of the 21st European Conference on Information Systems*.

Hamblen, M. (2014). *Mobile Data Traffic Is Expected to Explode 11-Fold by 2018*. www.computerworld.com/article/2487327/wireless-networking/mobile-data-traffic-is expectedto-explode-11-fold-by-2018.html

Haytham, Koutb, & Suoror. (2009). Toward An Agreement on Semantic Web Architecture. *Proceedings of World Academy of Science, Engineering And Technology, 37.*

Hemnath. (2010). *Web 4.0 - A New Web Technology*. http://website-quality.blogspot.com/2010/01/web-40-new-web-technology.html/

Huang, W. H. Y., & Soman, D. (2013). *Gamification of Education. Research Report Series: Behavioural Economics in Action*. University of Toronto.

Hunicke, R., LeBlanc, M., & Zubek, R. (2004). MDA: A formal approach to game design and game research. *Proceedings of the AAAI Workshop on Challenges in Game AI,* 4-5.

Huotari, K., & Hamari, J. (2011). Gamification from the perspective of service marketing. *Proceedings of ACM CHI 2011 Workshop on Gamification.*

Jacobs, S. (2011). *Serious Play Conference 2011: Microsoft's 'Productivity Games'*. http://www.gamasutra.com/view/news/36824/Serious_Play_Conference_2011_Mic rosofts_Productivity_Games.php

Kaplan, A. M., & Haenlein, M. (2010). Users of the world, unite! The challenges and opportunities of Social Media. *Business Horizons*, *53*(1), 53–68. doi:10.1016/j.bushor.2009.09.003

Kietzmann, J. H., Hermkens, K., McCarthy, I. P., & Silvestre, B. S. (2011). Social media? Get serious! Understanding the functional building blocks of social media. *Business Horizons*, *54*(3), 241–251. doi:10.1016/j.bushor.2011.01.005

Kotler, P., Kartajaya, H., & Setiawan, I. (2021). *Marketing 5.0: Technology for Humanity*. John Wiley & Sons Inc.

Kroeze, C., & Olivier, M. S. (2012). *Gamifying authentication. Information Security for South Africa*. ISSA.

Li, W., Grossman, T., & Fitzmaurice, G. (2012). GamiCAD: a gamified tutorial system for first time AutoCAD users. *Proceedings of the 25th annual ACM symposium on User interface software and technology*, 103-112. 10.1145/2380116.2380131

Malone, T. W. (1982). Heuristics for designing enjoyable user interfaces: Lessons from computer games. *Proceedings of the 1982 conference on Human factors in computing systems*, 63-68. 10.1145/800049.801756

Marcus, C. (2008). *Web 1.0, Web 2.0, Web 3.0 and Web 4.0 explained*. http://www.marcuscake.com/economic-development/internet-evolution/

Matyas, S., Matyas, C., Schlieder, C., Kiefer, P., Mitarai, H., & Kamata, M. (2008). Designing location-based mobile games with a purpose: collecting geospatial data with CityExplorer. Proceedings of the 2008 International Conference on Advances in Computer Entertainment Technology, 244-247. doi:10.1145/1501750.150180610.1145/1501750.1501806

Greenberg, Sutton, & Campbell. (2003). Metadata: A Fundamental Component of the Semantic Web. Bulletin of the American Society for Information Science and Technology, 29(4), 16–18.

Montola, M., Nummenmaa, T., Lucero, A., Boberg, M., & Korhonen, H. (2009). Applying game achievement systems to enhance user experience in a photo sharing service. *Proceedings of the 13th International MindTrek Conference: Everyday Life in the Ubiquitous Era*, 94-97. 10.1145/1621841.1621859

Muntean, C. I. (2011). Raising engagement in e-learning through gamification. *Proc. 6th International Conference on Virtual Learning ICVL*, 323-329.

Murugesan. (2007). Understanding Web 2.0. *Journal IT Professional.*

Neeli, B. K. (2012). A Method to Engage Employees Using Gamification in BPO Industry. *Services in Emerging Markets (ICSEM), 2012 Third International Conference*, 12-15. 10.1109/ICSEM.2012.27

OECD. (2007). *Participative web and user-created content: Web 2.0 wikis and social networking. Organization for Economic Cooperation and Development.* OECD.

Oktie, H. (2008). *Introduction to Semantic Web Technologies & Linked Data.* http://www.cs.toronto.edu/~oktie/slides/web-of-data-intro.pdf

Ortiz-Ospina. (2019). *The rise of social media.* https://ourworldindata.org/rise-of-social-media

Ossi, N. (2003). *Semantic Web: Definition.* https://www.w3c.tut.fi/talks/2003/0331umedia-on/slide6-0.html

Palmer. (2001). *The Semantic Web: An Introduction.* http://infomesh.net/2001/swintro/

Priebatsch, S. (2010). Welcome to the Decade of Games. *Harvard Business Review.* http://blogs.hbr.org/cs/2010/09/ welcome_to_the_decade_of_games.html

Rodrigues, Fonseca, & Garcia. (2022). The Use of CRM in Marketing and Communication Strategies in Portuguese Non-Profit Organizations. Navigating Digital Communication and Challenges for Organizations - Advances in E-Business Research, 223-244.

Saha, R., Manna, R., & Geetha, G. (2012). CAPTCHINO-A Gamification of Image-based CAPTCHAs to Evaluate Usability Issues. *Computing Sciences (ICCS), 2012 International Conference*, 95-99. 10.1109/ICCS.2012.18

Sandroni, A. G. (2015). *Breve Historia y Origen del Internet.* https://web.archive.org/web/20150505102148/http://www.academia.edu/5489717/BREVE_HISTORIA_Y_ORIGEN_DEL_INTERNET

Sicart, M. (2008). Defining game mechanics. *Game Studies, 8*(2).

Sudhir, B. (2006). *AJAX - Asynchronous Java Script and XML.* ITS - Information Technology and Systems Management.

Susi, T., Johannesson, M., & Backlund, P. (2007). *Serious Games: An Overview. Technical Report.* School of Humanities and Informatics, University of Skövde.

Sweetser, P., & Wyeth, P. (2005). GameFlow: A model for evaluating player enjoyment in games. *Computers in Entertainment*, *3*(3), 3–3. doi:10.1145/1077246.1077253

Boulos & Wheeler. (2007). The emerging Web 2.0 social software: An enabling suite of sociable technologies in health and health care education. *Health Information and Libraries Journal*, 2–23. PMID:17331140

Tillström, J. (2012). *Gamification in automotive marketing: A conceptual framework for implementation* [Bachelor's Thesis]. Metropolia Ammattikorkeakoulu.

Tim, B.-L., Christian, B., Tom, H., & Kingsley, I. (2008). Linked Data on the Web. *17th International World Wide Web Conference*.

Tim, B.-L., James, H., & Ora, L. (2001). The Semantic Web. *Scientific American*, *5*(1).

Velazco, C. (2012). *AT&T's Wireless Data Traffic Doubles Every Year, but Throttling Is Not the Solution.* http://techcrunch.com/ 2012/02/14/atts-wireless-data-traffic-doubles-everyyear-but-throttling-is-not-the-solution

Von Ahn, L., & Dabbish, L. (2008). Designing games with a purpose. *Communications of the ACM*, *51*(8), 58–67. doi:10.1145/1378704.1378719

Wittgenstein, L. (1953). *Philosophical investigations* (4th ed.). John Wiley & Sons.

KEY TERMS AND DEFINITIONS

Achievement: Defined as optional subgoals that are rewarded in a secondary reward system, i.e., the achievement system, separate from the primary reward system.

Badges: A commonly used game element in gamification, also known as an achievement or a trophy.

Game Mechanics: Any part of a game's rule system that covers one, and only one, type of possible interaction that takes place during the game.

Gamification: Use of game design elements in non-game contexts.

Social Networks: A set of web-based services and applications where users create and exchange content of their own creation.

Web 4.0: Known as the symbiotic web, it aims at the interaction between humans and machines in symbiosis.

Chapter 8
AI vs. Human:
Investigation of User Experience in Chat Support

Chaitra Katti
Sikkim Manipal University, India

Belem Barbosa
https://orcid.org/0000-0002-4057-360X
University of Porto, Portugal

ABSTRACT

Companies are using automated technologies customer support to elevate customer experiences. However, the question arises as to whether this strategic move is beneficial for companies. Though several studies have demonstrated the use of anthropomorphic chatbots could positively impact customers, studies still lack comparing this benefit with real-time human interaction. Thus, this study aims to examine user experience between AI and human-based chat support and understand whether AI-based chatbots necessitate replacing human interaction in customer support. The results of the study indicated that perceived benefit and relevance among human-based chats are relatively high, and perceived risk was only limited to delayed response and negative attitude. However, chatbots evidenced their perceived risk was high in anthropomorphic chatbots. Risk factors such as ethical concerns, lack of human touch were commonly seen in anthropomorphic chatbots. On the other hand. their perceived benefit and relevance were seen in chatbots acting as substitute of search queries, web navigation. or product demonstration.

INTRODUCTION

In recent years, the growing popularity of live chat has interested marketers to improve their relationships with customers. Scholars and practitioners are becoming more focussed on improving their chat services with automation such as chatbots and AI. Chatbots and AI are believed to carry huge advantages of mimicking human traits, retrieving necessary information in conversations, acting as best personal assistant (Khanna et al., 2015), reducing operation cost (Adamopoulou & Moussiades, 2020) and cater to

DOI: 10.4018/978-1-6684-5538-8.ch008

wider audience (Lu et al., 2022). Thus, this has spurred interest in among marketers to invest in chatbots and AI to elevate customer experiences during chat. A study (Grand View Research, 2022), forecasted that chatbot market size is expected to reach USD 2,485.7 million by 2028.

Despite these unquestionable advantages studies still suggest that automation like AI lacks human touch (Lu et al., 2022; Rapp et al., 2021), fails to meet customer expectations (Adam et al., 2021) and thereby highlighting the dark side of technology (Etkin, 2016; Mick & Fournier, 1998; Wilcox & Stephen, 2013). Thus, a lot of scepticism has arose among scholars and researchers about the efficiencies of chatbots.

With this growing scepticism, there is a need to understand whether chatbots can effectively replace human interaction in customer support. In other words, whether AI has the advantage over real time human communication? Although existing literature (Araujo, 2018; Chen et al., 2017; Rapp et al., 2021; Sheehan et al., 2020) argues that perfectly programmed anthropomorphic chatbots could give positive customer experiences, studies still lack comparing this benefit with real time human interaction, albeit with exceptions (Hill et al., 2015). Thus, this study aims to compare and investigate the user experience between Human and AI in chat support and propose the best communication forms to practitioners.

Consequently, this chapter adopts qualitative semi-structured interviews for its flexibility and carries the advantage to conduct in-depth examination of user experience in chat support. With the focus to investigate user perceptions between AI and human agents, the authors were inspired to use some of the approaches of Technology Acceptance Model (TAM), for its popularity to study factors influencing computer usage (Davis et al., 1989), and technology-based contexts, such as online shopping (Reimers et al., 2016). One of the crucial elements to examine a man's experiences is through perception. Rookes and Willson (2000) define perception as the process of gaining awareness or understanding of sensory information. Consequently, perceptions build a belief system of an experience. Thus, considering the importance of studying perceptions among users and guided by TAM, this study uses constructs such as perceived benefits, perceived relevance and perceived risk to examine user experiences of chatbots and human agents. The above constructs will help in understanding both positive and negative impacts of chat experience among users.

Thus, guided by TAM while adopting semi-structured interviews, the study conducts 10 interviews among participants aged 22-45 belonging to India, to meet following research objectives: (i) to explore the antecedents of chat based user experience, (ii) to analyse the perceived usefulness, relevance and risk associated with human chats and AI based chats, and (iii) to propose an appropriate communication platform for marketers and researchers.

Thus, the study makes following contributions. First, the study helps in organising the scattered contributions of user experiences in chatbots and human agents and compare the user experience between the two. Secondly, study collects empirical evidence of 10 participants from India to examine the user experience of both AI based chatbots and human agents. Thirdly, study demonstrates the user expectations and their choices (chatbot or human) of conversation in chat support and thoroughly investigate those choices based on their experiences and expectations. This study makes a useful attempt to examine user's perceptions (perceived risk, relevance and benefits) guided by TAM. Overall, the study makes contributions to practitioners and scholars about building an effective user experience on chat support.

BACKGROUND

Human communication is often complex, and more dedication is required in understanding how people relate to one another (Schram, 1954). In fact, it is through communication people form relationship with one another and ultimately a society is built (Mead, 1967). Consumers too form relationships with products or brands (Fournier, 1998) through communication. Most of the people's social and emotional needs (Burton, 1990) are fulfilled through communication. But the question arises, does this value of human interaction lie in AI based chatbots as well?

AI based Chatbots are computer programs with natural language capabilities, which can be configured to converse with human users. Tintarev et al. (2016) conceptualize chatbots as automated advice givers in that they can facilitate decision making. Firms focus on deploying chat services mainly because of the efficiencies (Brandtzaeg & Følstad, 2017) such as ability to handle pressure, decrease in operation costs, and immediate problem resolution (Adamopoulou & Moussiades, 2020) and cater to wide audience (Lu et al., 2022). However, studies have highlighted the several disadvantages of these chatbots on user experiences. Thus, there are several contradictory studies listing advantages and disadvantages of automated chats. This motivated the authors to aim at examining each of the user experiences with chatbots and human agents guided by three constructs of TAM model such as perceived relevance, perceived benefits and perceived risk.

Perceived Benefits, Relevance and Risk Associated with Human Based Chat Support

Perceived humanness is a key antecedent to customer's trust (Lu et al., 2022). Thus, both human agents and chatbots needs to have human elements during their conversation with customers (Lu et al., 2022). Lee and Lyu (2016) emphasize on the need for human interaction because of their hedonic attitudes, suggesting that consumers derive a social benefit from these interpersonal relations. Further, upon studies highlight that firm should not build customers to self-service channels especially not at the beginning of a relationship with a customer (Scherer et al., 2015).

Further, when conversing with chat support, users prefer to repair the misunderstandings with human-like agents, rather than with text-based chatbots (Corti & Gillespie, 2016). Studies suggest that there are notable differences between human-human and human-chatbot communication in the content and quality of the conversations (Hill et al., 2015). Extant literature (Adam et al., 2021; Araujo, 2018; Lu et al., 2022) highlights the importance of social and emotional contexts in humans, leading to better interpersonal relationships. Lee and Lyu (2016) demonstrate that people who are with high need of human interactions show unfavourable attitudes to self-service technologies. Text and voice-based chatbots lack current human ability such as gestures, identity-based interactions (Rapp et al., 2021), which could leave users in ambiguous state (Shi et al., 2020). The advantages of human interaction indicate social presence (Cyr et al., 2007) with their real conversation voice (Kelleher, 2009) that has positively built trust, leading the greater satisfaction with communication experience (Kaplan & Haenlein, 2010).

However, on the other hand, previous studies relate the challenges of human customer care with language barriers (Diabah, 2021; Lockwood, 2017), delayed response (Crisafulli & Singh, 2017) and negative customer attitude due to job dissatisfaction (Dean, 2004). Thus, chatbots could overcome these challenges with low operation cost and minimal time response (Adamopoulou & Moussiades, 2020). In fact solution to these problems were on using chatbots, which produces human-like cues generating stronger emotional connections with the firm (Araujo, 2018). They are programed to seek clarification and become error free, which in turn could absolutely replace human agents (Sheehan et al., 2020).

Thus, studies have listed several advantages and disadvantages of human based chats. However, the study is motivated to investigate further on user perceptions of human based chats. Thus, guided by TAM, the study uses constructs such as perceived risk, benefit, and relevance to propose following research questions:

RQ1: *What are the perceived benefits, relevance and risk associated with human based chat support?*

RQ2: *What factors (experience, emotions, efficiency) in chat support affect user's experiences?*

Perceived Risk, Relevance and Benefits Associated with AI Based Chat Support

Chatbots are equipped with natural language processing, such as, machine learning approach that enhances their ability to learn from the conversations (Adamopoulou & Moussiades, 2020). Infact the ability to respond emotionally is built through advancement in machine learning (Følstad et al., 2018). (Adamopoulou & Moussiades, 2020) study two important technology patterns of chatbots such as pattern matching approaches and machine learning and highlight their importance in various industry application, such as chatbots act as personal assistant in education sectors, provide customised health information in healthcare sectors. However, they concluded inconsistencies in communication with unpleasant experiences and data protection were major concerns in chatbots.

The risk perceptions could be high in chatbots, when chatbots fails to meet user's expectations (Adam et al., 2021) instances include lack of language skills, which could trigger negative emotion, such as frustration of not being understood (Grudin & Jacques, 2019). Informality, inconsistency and negative languages are common issues among chatbots (Adamopoulou & Moussiades, 2020; Galko et al., 2018). Furthermore, scripted responses degrade consumer engagement and content relevance (Lew et al., 2018). The majority of users showed a low degree of satisfaction with chatbots (Akhtar et al., 2019; Jain et al., 2018) due to their unsuitable responses to user's requests, leading to gap between the user's expectation and the system's performance (Luger & Sellen, 2016). Users prefers using websites rather than chatbots, which had a higher abandon rate (Beriault-Poirier et al., 2019).

On the other hand, chatbots has number of benefits. The chatbot provided a good user experience and high engagement by helping people in their productivity, and it was often perceived as humanlike (Cranshaw et al., 2017). Evidence also shows unquestionable advantages towards AI based chatbots with anthropomorphism with qualities of social presence builds emotional connection (Araujo, 2018), user compliance (Adam et al., 2021), satisfy social desires (Sheehan et al., 2020). Further upon advantages of human like chatbots builds trust (Følstad et al., 2018; Xu & Lombard, 2017), standardization (Selnes & Hansen, 2001), work efficiency (Brandtzaeg & Følstad, 2017), and contribute to satisfaction and commitment (Kelleher, 2009). According to Misischia et al. (2022), this type of chatbot carries important customer related functions such as interaction, entertainment, problem solving, trendiness, and customization.

Chatbots with high humanness can satisfy consumer's need for human interactions (Sheehan et al., 2020), thereby leading to trust (Xu & Lombard, 2017), and positive consumer responses (Toader et al., 2020). It also has been found to increase purchase intention in some situations (Barcelos et al., 2018).

Thus, chatbot anthropomorphism, could act as friend, rather than an assistant agents and this induces greater feelings of parasocial interaction with consumers, (Khanna et al., 2015), leading to more positive customer relationships (Youn & Jin, 2021). The most frequently reported motivational factor is productivity, as chatbots help users obtain efficient assistance or information (Brandtzaeg & Følstad, 2017). Consumers preferred chatbots mainly for time efficiency (Avula et al., 2018). Studies complimented anthropomorphic chatbots for its human touch such as emoji-featuring messages (Beattie et al., 2020), human like languages or names (Araujo, 2018).

Studies thus have highlighted on several advantages and disadvantages of chatbots. This motivated the authors to further examine consumer's perceptions behind these chatbot and its user experiences. Thus, the study considers consumer's perceived benefits, risk and relevance associated with chatbots and following research question is therefore proposed to examine user's chatbot perceptions:

RQ 3: *What are the perceived benefits, relevance and risk associated with AI based chat support?*

Despite the studies highlighting on advantages and disadvantages of each kind of chat experiences (AI based chatbots and human agents), studies lack in examining the user perceptions such as perceived benefits, relevance, and risks between the two. Studies have addressed the challenges of human chat experiences with anthropomorphic chatbots, however the comparison of anthropomorphic chatbots with human agents are still scarce. Considering the above contributions and the proposed research questions, the study includes semi-structured interviews in the next sections to meet its objectives.

METHOD

With the help of above-mentioned contributions presented in the previous sections and the proposed research questions, qualitative exploratory research was conducted using in-depth semi-structured interviews. Semi-structured interviews carry a huge advantage of including necessary elements of both structure and unstructured interviews such as designing set of sequential questions with a flexibility to investigate further issues brought by interviewees (Cachia & Millward, 2011) such as feelings and experiences. Furthermore, semi-structure interviews also carry additional advantage of establishing rapport and building ease (Farr, 1982) helping researchers to seek more clarification about the interviewees' experiences (Merton & Kendall, 1946; Morse & Field, 1995). In addition, it carries elements of structured interviews such as asking participants same questions in order to be compared through the detailed interview guide (McIntosh & Morse, 2015), with flexibility to probe further.

The interview outline included asking participants to narrate few of their chat experiences associated with chatbots and human agents. (e.g., Could you share a recent chat experience with chatbots). The advantages and disadvantages of each type of chat experience (Chatbot and Human) were further explored, based on the examples provided by participants.

Participants in the study were Indian citizens aged 25-45, who i) have had chat with firm on professional or personal purposes, ii) experienced at least a few times having a conversation with a AI over a chat iii) had also conversations with human based chat agents in chat. The authors identified India to conduct their study, as this country is second largest online market in the world, ranked only behind China. Currently with over 560 million internet users, it is estimated that by 2023, there would be over 650 million internet users in the country (Statista, 2021). Facilitated by the researchers' contacts and social networks, a purposive sampling method was adopted in order to select 10 participants with diverse characteristics in terms of their chat experiences (Table 1).

Table 1. Sample Characteristics. Source: Authors

Participant	Age	Education	Position
P1	25	Bachelor's in commerce	Account Manager
P2	28	Bachelor's in commerce	Sales Manager
P3	33	Master's in mass communication	Marketing Head
P4	25	Bachelor's in commerce	Clerk
P5	27	Bachelor of Arts	Broker
P6	33	Bachelor's in engineering	Trader
P7	30	Master's in business administration	Housewife
P8	35	Bachelor's in engineering	Civil Engineer
P9	25	Bachelor's in commerce	Finance Executive
P10	33	Bachelor of Arts	Business Analyst

The interview duration was 30-40 minutes, and the conversations of the participants were scripted for subsequent data analysis. Participants were briefed about the aim of the research, methodological and ethical procedures (Bryman, 2016) followed in the interviews such as informed consent, confidentiality, anonymity, and voluntary.

Thematic content analysis was independently conducted by both the authors and the contradictory opinions of the two researchers were discussed and resolved. Despite the existence of some software for content analysis (e.g., NVivo, MAXQDA), for this study the analysis was conducted without the help of specific software. Content analysis meets the goal of providing knowledge and understanding of a phenomenon in a systematic way (Downe-Wamboldt, 1992). Qualitative content analysis is particularly useful to understand subjective phenomena (Hsieh & Shannon, 2005) such as experiences associated over chat. The collected data was read multiple times and analysed word by word to derive meaning and deep insights (Morse & Field,1995) and the exact words from the data are highlighted in the results to capture the key concepts (Hsieh and Shannon, 2005).

RESULTS

The study evidenced diverse and interesting response of participants regarding chat experiences with chatbots and humans. While participants highlighted on the necessity of having a human touch during their chat interaction, they also highlighted the importance of having a chatbot to disseminate accurate information during their search queries. Consequently, perceived humanness on customer trust, became an important outcome of the results for chat conversation, thus in agreement to the studies (Lee & Lyu 2016; Lu et al., 2022). The perceived risk, benefits and relevance differed among chat experiences of human and chatbots and below mentioned are some of the experiences narrated by the participants.

Perceived Benefits, Relevance and Risk Associated with Human Based Chat Support

Participants in the study reported that there were numerous benefits experienced by them when conversing with human based chat agents during chat support. Clearly, they identified human touch, problem resolving attitude, empathy and social presences as their important factors to opt for human based chat support. Some of these factors were demonstrated in extant literature, such as human touch (Lu et al., 2022), social presence (Lee & Lyu, 2016) which plays an important role in shaping chat experiences.

Few participants made their positive assumptions based on the efficiency of human based chat agents. Participant 3 considered that "instant refunds are positive with human based chat agents, as they can understand our problems immediately". Other participants (i.e., 1, 2, 5) also used the examples of instant refund to highlight the benefits of human-based chats support.

Additionally, same participants suggested that their refunds are possible because human based chat support, understands them instantly, asks them more evidences and even apologises and narrates the reason for their mistakes. Clearly it was identified these steps were very important for consumers to build loyalty. Participant 5 highlighted "Last time, my refund experience went so smooth that I always order food from them, because I know they are responsible for their mistake."

Here are other examples provided by participants on similar experience in association with human based chats:

> I recently, received order from ecommerce website and the dress was torn. I reached customer care on chat, and he asked me to send pictures of my dress. As I did it, he asked more pictures and once they saw its their mistake, they refunded my money immediately. Additionally, I received a call from that agent personally apologising me. I felt very good about that company. (Participant 3)
>
> My food was delivered in half cooked state. I had chat with them, they asked me to send photos and they refunded me instantly after explaining, all these are not possible with chatbots. (Participant 4)
>
> One of the ecommerce firms had deducted money for the services I had not availed and when I explained them the situation they understood and instantly gave a personal call apologising me. (Participant 2)

Participants constantly repeated the importance of human presence to resolve their issues. Thus, these findings are in agreement with extant literature (e.g., Lee & Lyu 2016; Scherer et al., 2015) which attested that presence of a social actor or human is required to build interpersonal relationships.

Mostly participants reported that their problems are solved quickly when they confront human communication. Majority were happy with the experiences of human based chat communication, because of ease of communication, faster problem resolution. In addition, actions such as apology, being responsible for the mistakes, acting with guilt, regret and offering apology gift, were commonly appreciated by the participants for human based chat communication. This means that participants not only looked for problem resolution, but also expected apology with guilt, regret, personal sorry and other emotions. This eventually built loyalty and trust as evidenced in above findings.

On the other hand, these participants were asked if their cases was handled by chatbots and they mistook it to be human. And all the participants identified clear difference between humans and chatbots, they justified with answers such as:

Chatbots could not have resolved the issue so well. For instance, it did have human touch like deeply feeling sorry and then explaining their mistake on what went wrong from their side and then instantly refunding or rewards. I don't think chatbots could explain what went wrong from their side each time. (Participant 3)

No chatbot can never communicate like this, I had my food half cooked when delivered. When I had a chat, the person actually had called restaurant and then he gave his explanation about the food cooking process, but when I argued with him about it on chat, he understood and refunded me. Chatbots cannot do this. I can get to know its human. (Participant 2)

Then eventually participants narrated their experiences of human chat in comparison to chatbots. Common emotions identified by participant from human based chats were "empathy", "warmth", "compassion", "friendly" which could not be felt among chatbots. Participant 4 argued that "humans are very interactive and, that cannot be felt with chatbots."

Other arguments were:

Human are empathising. When I chat not only do my problems get resolved, but also they understand me and feel bad for my situation, with chatbots that is not possible. (Participant 5)

My problems are easily resolved when humans answer my chats. In fact, they understand faster and process it faster too. This usually happens when my food is delayed from a delivery app. (Participant 2)

Humans based agents understands me thoroughly, feels my problems, with chatbots that usually doesn't happen. (Participants 3)

Thus, participants look for emotional and social needs during chat conversation. This aligns with evidence (Adam et al., 2021; Araujo, 2018; Lu et al., 2022) highlighting the importance of social presence (Lee & Lyu, 2016) and human elements (Lu et al., 2022) during chat conversations. In addition, participants also looked for more detailed conversation. Participant 6 who mentioned "human chats give more information or try to elaborate more with additional required details, but with chatbot it is only specific accurate answer." Hence the length of conversation appears to be positively correlated with user enjoyment (Cervone et al., 2018).

Thus, the evidence demonstrate that perceived benefit and perceived relevance is relatively high in human based conversation. Participants not only looked for solution to their problems, but also wanted social and emotional presence on chat support, in agreement to the studies (Adam et al., 2021; Araujo, 2018; Cyr et al., 2007; Lee & Lyu, 2016; Lu et al., 2022) that highlighted the need of identity-based interactions (Rapp et al., 2021). Eventually this is useful in building trust (Lu et al., 2022) and greater satisfaction with communication experience (Kaplan & Haenlein, 2010).

On contrary, participants also argued about some of the drawbacks of human chat agents. The main drawback they highlighted were of delayed response (Crisafulli & Singh, 2017) and negative attitude due to job dissatisfaction (Dean, 2004). Thus, this was viewed as some of the perceived risk among human based chat communication. Though perceived risk was relatively low, there was some areas of concerns to consumers. Some examples are included below.

The only problem with human chat support is longer response, with chatbots that usually doesn't happen. (Participant 1).

Sometimes, I feel shy to ask more questions, because I feel I am wasting their time. (Participant 9)

Some of my chat experiences had arrogance problem. I personally felt they were not interested in replying. (Participant 10)

I leave chat, when the response is late, so maybe if humans are chatting, they relatively take longer time. (Participant 2)

The above-mentioned eidence clearly indicates that the perceived benefit and perceived relevance is higher in human based chat support for problem resolution and service-related queries. Along with problem resolutions consumers also look for emotional and social presences in customer care, aligning to studies (Cyr et al., 2007; Lee & Lyu, 2016; Rapp et al., 2021) which builds interpersonal relationships (Lee & Lyu, 2016). However, perceived risk such as delayed response and negative attitude were areas of concern for participants.

Differences Identified by participants

Further discussion was conducted, to understand the difference between chatbots and human. This was specifically conducted to study the evidence (e.g., Araujo, 2018; Cranshaw et al., 2017; Følstad et al.,2018; Khanna et al., 2015; Lu et al., 2022; Sheehan et al., 2020; Toader et al., 2020; Xu & Lombard, 2017) on anthromorphic chatbots to understand whether participants could identify the difference in communication between anthropomorphic chatbots or human.

While some participants could not identify the differences or had not encountered such experiences, few other participants did narrate experiences with anthropomorphic chatbots. One such incident from Participant 8 who mentioned:

Yes, a few times, I have got confused if I am chatting with a human or a chatbot. For instance, in past chatbots tried to please me (what an interesting name, you must be beautiful,) but eventually I got to know it's a chatbot, when they give inaccurate answers to my problems. (Participant 8).

Perceived Benefits, Relevance and Risk Associated with Chatbots

Perceived risk was higher in chatbots, when they tried to replace human agents, although their perceived benefits included speedy and accurate information, a substitute of search query. Chatbots demonstrated high-perceived risk, when they tried to replace human agents with superficial emotions and gestures. Thus, perceived risk was higher in anthropomorphic chatbots, not aligning to evidences (Araujo, 2018; Cranshaw et al., 2017; Khanna et al., 2015; Sheehan et al., 2020). However, their perceived benefit and relevance was high when they acted as a substitute for search queries building trust (Xu & Lombard, 2017) and satisfaction (Kelleher, 2009). Following are the responses to Chatbots in association with perceived benefits, relevance and risk.

Participants demonstrated satisfaction and ease, when chatbots gave them accurate information, during their search on websites, as such instances include. They played an important role, to ease website navigation, in bookings or government related information. "travel", "government work" or "product demonstrations" were common areas where participants highlighted their perceived usefulness for chatbots instances include:

I was flying to a country in Africa, and I wanted information on visa and covid-19 restriction of that country, and I got all the information, just a day before flying and it was useful. Chatbot is useful during the time of emergency, especially flying. (Participant 4)

Chatbot just gives you option, for example "type 8", "type 9", and when I do that, I get all required information instantly, without problems. (Participant 4)

I personally feel chatbot was useful, during the time of Covid 19, when I was looking for ventilators, I asked for the information and I got number of hospitals having ventilators. (Participant 3)

Chatbots are useful for government related queries. Most of the government information are not available on website and it takes long to get information, chatbots are useful in booking appointment, getting government services. (Participants 9)

Chatbots replies instantly, and gives me solution, if it is not a complicated problem. (Participants 1)

Chatbots are useful, to get information, I cannot find on website or google, they instantly reply with options and clicking on options gives me all answers. (Participants 2)

With chatbots, I can get information 24/7, so during emergency, say I am flying, I need some information, chatbots can help instantly, but with human that is not possible. (Participant 5)

Chatbots played an important role in demonstrating products and services, acting an user manual for participants. Thus participant 1 narrated:

My CRM had some technical glitch, and it was in mid-night, and when I started chatting, chatbot guided me step-by-step. So for product demonstrations, chatbots are very useful, but with human agents that is not possible, we have to tell them entire story and then it is time consuming. (Participant 1)

However, Participants in the study reported several areas of concerns, when chatbots tried to replace human agents. Perceived risk was relatively high in chatbots trying to replace human agents. Their major concern was on credibility and ability to understand the problems. Participants also mentioned about "superficial emotions" and "artificial gestures" of chatbots that associated to their perceived risk. Thus, the findings could not align to evidences (Araujo, 2018; Cranshaw et al., 2017; Khanna et al., 2015; Sheehan et al., 2020). Thus, anthropomorphic chatbot characteristic was not positively perceived by participants, as demonstrated by several examples provided:

Sometimes I have encountered chatbots talking just like human, trying to please me. In such situations I immediately leave, because it is scary as the company is trying to addict my emotions to chatbots. (Participant 8)

Chatbots are fake and I feel so fake, if they try to behave like human. I feel everything is fake that time. (Participant 9)

Ethical concerns were also highlighted by participants, who encountered anthropomorphic chatbots. Some recurrent concerns were on "job loss", "fear of addictions", "fear to fake emotions". For example, Participant 1 narrated

Chatbots can take our jobs, so when a chatbot start talking like human, I usually leave, with annoyance that if I encourage them in market, they might take away my job too. (Participant 1)

Thus, the above findings indicated that anthropomorphic chatbots were not positively perceived by consumers despite having its huge advantages (Araujo, 2018; Cranshaw et al., 2017; Khanna et al., 2015; Sheehan et al., 2020). On the other hand, chatbots had also major challenges such as "inaccuracy", "less problem-solving capability" "irrelevance", "lack of human touch". Some instances of participants include:

I don't feel like trusting a chatbot, because I don't know how far they understand me and solve my issue. Many times, they have been very inaccurate, when I narrate an issue. (Participant 10)

The problem with chatbots are they do not understand what I am talking in first place, they answer something irrelevant. (Participant 1)

If the issue is complicated, chatbots cannot solve it. Human are required during that time, so I prefer human chat. (Participant 4)

The above results suggests that anthromorphic chatbots negatively impacts customer experiences, there by not aligning to studies (Araujo, 2018; Cranshaw et al., 2017; Khanna et al., 2015; Sheehan et al., 2020). However, in contrast to such findings, chatbots demonstrated relatively high perceived benefits and relevance, when they tried to ease consumer's search queries or to understand product functionalities.

SOLUTIONS AND RECOMMENDATIONS

The main purpose of this paper was to investigate user experiences of chat support between human and AI based Chatbots. The chapter identified that previous authors either studied benefits of chatbots (Araujo, 2018; Avula et al., 2018; Brandtzaeg & Følstad, 2017; Chen et al., 2017; Rapp et al., 2021; Sheehan et al., 2020; Selnes & Hansen, 2001) or risk of chatbots (Galko et al., 2018; Grudin & Jacques, 2019; Lee & Lyu, 2016; Rapp et al., 2021). As far as the authors' understanding, the studies scarcely compared the chat experiences with human based chats. Thus, the authors covered this gap by demonstrating the user experiences of chatbots and human base chats with the help of constructs such as perceived risk, relevance, and benefits.

The qualitative nature of the study using some of the TAM approaches helped authors identify different types of perceptions among consumers. Some of the important ones identified; perceived benefits are high in human based chat for problem resolution, solving complex issues and queries. Perceived risk was high in anthromorphic chatbots due to ethical concerns such as fear of job loss, addiction to fake emotions and displacement of authentic human emotions permanently affected trust, satisfaction among users. Other important results aligned to the previous studies. Such that, perceived relevance in human based chat includes emotional and social presences and identity-based interactions aligning to evidences (e.g., Adam et al., 2021; Araujo, 2018; Cyr et al., 2007; Lee & Lyu, 2016; Lu et al., 2022; Scherer et al., 2015). Perceived risk in human based chat includes delayed response and negative attitude towards consumers (Crisafulli & Singh, 2017; Dean, 2004). Perceived benefits of chatbots include efficiency and standardisation (e.g., Adamopoulou & Moussiades, 2020; Avula et al., 2018; Brandtzaeg & Følstad, 2017; Selnes & Hansen, 2001). The risk factors of chatbots include lack of emotions and inaccuracy (e.g., Galko et al., 2018; Grudin & Jacques, 2019; Lee & Lyu, 2016; Rapp et al., 2021). Consequently, these above results helped authors address research questions 1 and 3.

Human based chats satisfied emotional needs such as empathy, compassion, warmth was satisfied during their conversation with the firms. Perceived humanness on customer trust, became an important outcome of the results for chat conversation, thus in agreement to the studies (Lee & Lyu 2016; Lu et al., 2022). While the challenges of human based chats such as delayed response (Crisafulli & Singh, 2017) and negative responses (Dean, 2004) were overcome by chatbots (Adamopoulou & Moussiades, 2020), still chatbots could not meet the consumers expectations, such as human touch, and emotions. Thus, chatbots negatively impacted consumers when they tried to replace human based chats. As mentioned earlier, consumers tend to discontinue their conversation with anthromorphic chatbots, due to ethical concerns such as job loss, addiction to fake emotions of chatbot or loss of authentic human emotions permanently. Thus perceived risk is high among anthromorphic chatbots. Thus, the studies limits itself to align to evidences (Araujo, 2018; Cranshaw et al., 2017; Khanna et al., 2015; Sheehan et al., 2020), and limits to demonstrate that anthropomorphic chatbots with social and emotional context could build positive customer experiences. In fact, it was identified that anger, frustration and dissatisfaction were common emotions exhibited by consumers when using chatbots for problem resolution or solving complicated queries, because such issues was expected to have human emotions such as empathy, compassion, warmth. Thus, findings are in alignment with studies (Lee & Lyu, 2016; Scherer et al., 2015) that highlighted the need for interpersonal relationship and human touch in customer support.

On the other hand, chatbots plays an important role in easing consumers' search related queries. Chatbots had relatively higher perceived benefits and relevance, during product demonstrations in chat, information related queries and search related queries. Thus, chatbots act as substitute to search queries on website and other portals. Chatbots relatively saves times and improves website experience due to their accurate information dissemination features. They also positively impacted consumers during the time of technical support of some products. Thus, chatbots builds credibility and feasibility, when they are substitute of web queries or product demonstrations. They could replace user manuals due to their efficiency. These above experiences and emotions, answers research question 2.

Hence, overall based on the findings on user experiences between chatbot and human, following contributions are made in this study:

- Perceived benefits are relatively high in human based chat for problem resolution, solving complex issues and queries
- Perceived relevance in human based chat includes emotional and social presences and identity-based interactions. These were also suggested by previous studies (e.g., Cyr et al., 2007; Lee & Lyu, 2016; Scherer et al., 2015).
- Perceived risk in human based chat is low, which includes delayed response and negative attitude towards consumers. These concerns were already evidenced in previous studies (e.g., Crisafulli & Singh, 2017; Dean, 2004).
- Perceived benefits in chatbots include timely response, accurate information dissemination, booking, and orders. These advantages were partially evidenced in studies that highlighted efficiency and standardisation in chatbots (e.g., Adamopoulou & Moussiades, 2020; Avula et al., 2018; Brandtzaeg & Følstad, 2017; Selnes & Hansen, 2001).
- Chatbots acting as substitute of web queries, user manuals, product support, and demonstrations have increased perceived relevance of chat experience among users. This is a contribution of this study.
- Chatbots, especially anthropomorphic chatbots, encountered relatively high perceived risk, when they tried to replace human agents and mimic human emotions. Inaccurate gestures, lack of emotions, lack of human touch and ethical concerns such as fear of job loss, addiction to chatbot fake

emotions and displacement of authentic human emotions permanently affected trust, satisfaction among users. Some of these concerns such as lack of emotions and inaccuracy were evidenced in studies (e.g., Galko et al., 2018; Grudin & Jacques, 2019; Lee & Lyu, 2016; Rapp et al., 2021).

- Chatbots could elevate website experiences by helping users instantly navigate to right page or save their time by giving the accurate information.

A systematization of the findings is organised in Table 2.

Table 2. Summary of the findings. Source: Authors

	Chat support-Chatbot	Chat support- Human
Perceptions		
Perceived Benefits	• Product demonstration • Website navigation • Information • User guide and product support	• Human touch • Problem resolution • Query resolution • Human emotions- empathy, compassion, warmth
Perceived relevance	• Substitute search query • Travel related information • Booking and orders	• Refunds and returns • Understanding, quick action
Perceived risk	• Ethical concerns • Inaccurate response • Lack of human touch	• Delayed response • Negative attitude

Useful Corporate Strategy

Implementing right communication platform to audience is crucial for practitioners and managers. Communication being the first stage of customer relationship needs to be well planned and designed. Managers and practitioners should thoroughly evaluate the customer's needs and wants before designing chat experiences.

Thus, this chapter helps in designing chat experiences for users using automation and human touch. Further, the study helps managers in understanding that elevating website experience with ease, convenience, effective and timely information dissemination, human agents could help companies elevate customer experiences with human touch such as empathy, warmth, compassion, thereby solving their problems and issues.

Previously from the findings of this study, it was noted that managers focussed on automation or only human chats, however this chapter has highlighted advantages and disadvantages and also has recommended their role in chat experiences. Thus this chapter will help managers design chat strategy and appropriately use each of them (human or AI) based on the requirement of consumers.

FUTURE RESEARCH DIRECTIONS

This study highlights several opportunities for future researchers. Some of the interesting concepts such as challenges of anthropomorphic chatbots or chatbots being a substitute of web queries can be examined further. Chatbots as an important navigation tool, or user guide could be studied as independent topics. Further human emotions such as empathy, warmth, compassion impacting on customer relationship can be further considered.

However, the study does come with few limitations such as limited number of participants, qualitative interviews and demographics. Additionally, limitations include region specific perceptions or inadequate chat experiences. Thus, future researchers can expand their participants number, regions with analytical approach (e.g., experimental study) that could offer many other useful insights which is beyond the scope of this study.

CONCLUSION

This chapter provides important contributions in distinguishing the different user experiences between chatbot and human based agents. The chapter highlights on both advantages and disadvantages of

chatbots and human chat experiences. Studies although have highlighted on advantages (Araujo, 2018; Cranshaw et al., 2017; Khanna et al., 2015; Sheehan et al., 2020; Toader et al., 2020; Xu & Lombard, 2017) and disadvantages (Adam et al., 2021; Galko et al., 2018; Lew et al., 2018; Akhtar et al., 2019; Jain et al., 2018) of chatbots, the comparison of these with human based chats are still limited. Thus, this study has built the understanding of User Experience by examining the chatbot and human agents and demonstrated perceived benefits, relevance and risk of each of them. The study evidenced that each of them have distinct role and cannot be replaced by one another. The findings included instances of human agents with higher perceived benefit and relevance for problem resolution, and complex queries, while chatbot's perceived benefits and relevance reflected on search queries, product support and demonstration. The study also witnessed that replacement of human chat with anthromorphic chatbots could increase perceived risk among consumers.

Previous studies on anthromorphic chatbots (Adam et al., 2021; Araujo, 2018; Cranshaw et al., 2017; Khanna et al., 2015; Sheehan et al., 2020; Toader et al., 2020; Xu & Lombard, 2017) which focusses on positive customer experiences, lacked in depth evaluation of consumer emotions and perceptions. These studies still had limitations such as unsatisfied response of consumers (e.g., Adam et al., 2021) or unclear perceptions of consumers (e.g., Araujo, 2018). Thus, this study covered such limitation and provided in-depth explanation of impacts of anthropomorphic chatbots on user experiences. Surprisingly, studies scarcely considered consumer's fears such as ethical concerns, lack of authentic human touch in chatbots. Thus, this study highlighted on challenges of anthropomorphic chatbots. Thus, the scarcity in extant literature such as comparison of the user experience between chatbot and human agents is covered in this chapter. Besides the limitations of anthropomorphic chatbots, human agents are demonstrated with recommendation and solutions.

REFERENCES

Adam, M., Wessel, M., & Benlian, A. (2021). AI-based chatbots in customer service and their effects on user compliance. *Electronic Markets*, *31*(2), 427–445. doi:10.100712525-020-00414-7

Adamopoulou, E., & Moussiades, L. (2020). Chatbots: History, technology, and applications. *Machine Learning with Applications*, *2*, 100006. doi:10.1016/j.mlwa.2020.100006

Akhtar, M., Neidhardt, J., & Werthner, H. (2019). The potential of chatbots: Analysis of Chatbot Conversations. *IEEE 21st Conference on Business Informatics (CBI),* 397-404. 10.1109/CBI.2019.00052

Araujo, T. (2018). Living up to the chatbot hype: The influence of anthropomorphic design cues and communicative agency framing on conversational agent and company perceptions. *Computers in Human Behavior*, *85*, 183–189. doi:10.1016/j.chb.2018.03.051

Avula, S., Chadwick, G., Arguello, J., & Capra, R. (2018). Searchbots: User engagement with chatbots during collaborative search. *Proceedings of the 2018 Conference on Human Information Interaction & Retrieval*, 52-61. 10.1145/3176349.3176380

Barcelos, R. H., Dantas, D. C., & Sénécal, S. (2018). Watch your tone: How a brand's tone of voice on social media influences consumer responses. *Journal of Interactive Marketing*, *41*, 60–80. doi:10.1016/j.intmar.2017.10.001

Beattie, A., Edwards, A. P., & Edwards, C. (2020). A bot and a smile: Interpersonal impressions of chatbots and humans using emoji in computer-mediated communication. *Communication Studies*, *71*(3), 409–427. doi:10.1080/10510974.2020.1725082

Beriault-Poirier, A., Prom Tep, S., & Sénécal, S. (2019). Putting chatbots to the test: Does the user experience score higher with chatbots than websites? In T. Ahram, W. Karwowski, & R. Taiar (Eds.), *Human Systems Engineering and Design* (pp. 204–212). Springer. doi:10.1007/978-3-030-02053-8_32

Brandtzaeg, P. B., & Følstad, A. (2017). Why People Use Chatbots. In I. Kompatsiaris, J. Cave, A. Satsiou, G. Carle, A. Passani, E. Kontopoulos, S. Diplaris, & D. McMillan (Eds.), *Internet Science: INSCI 2017* (pp. 377–392). Springer. doi:10.1007/978-3-319-70284-1_30

Bryman, A. (2016). *Social Research Methods*. Oxford University Press.

Burton, J. (1990). *Conflict: Human Needs Theory*. Palgrave Macmillan. doi:10.1007/978-1-349-21000-8

Cachia, M., & Millward, L. (2011). The telephone medium and semi-structured interviews: A complementary fit. *Qualitative Research in Organizations and Management*, *6*(3), 265–277. doi:10.1108/17465641111188420

Chen, R. P., Wan, E. W., & Levy, E. (2017). The effect of social exclusion on consumer preference for anthropomorphized brands. *Journal of Consumer Psychology*, *27*(1), 23–34. doi:10.1016/j.jcps.2016.05.004

Corti, K., & Gillespie, A. (2016). Co-constructing intersubjectivity with artificial conversational agents: People are more likely to initiate repairs of misunderstandings with agents represented as human. *Computers in Human Behavior*, *58*, 431–442. doi:10.1016/j.chb.2015.12.039

Cranshaw, J., Elwany, E., Newman, T., Kocielnik, R., Yu, B., Soni, S., Teevan, J., & Monroy-Hernández, A. (2017). Calendar.help: Designing a workflow-based scheduling agent with humans in the loop. *Proceedings of the 2017 CHI Conference on Human Factors in Computing Systems*, 2382–2393. 10.1145/3025453.3025780

Crisafulli, B., & Singh, J. (2017). Service failures in e-retailing: Examining the effects of response time, compensation, and service criticality. *Computers in Human Behavior*, *77*, 413–424. doi:10.1016/j.chb.2017.07.013

Cyr, D., Hassanein, K., Head, M., & Ivanov, A. (2007). The role of social presence in establishing loyalty in e-Service environments. *Interacting with Computers*, *19*(1), 43–56. doi:10.1016/j.intcom.2006.07.010

Davis, F. D., Bagozzi, R. P., & Warshaw, P. R. (1989). User acceptance of computer technology: A comparison of two theoretical models. *Management Science*, *35*(8), 982–1003. doi:10.1287/mnsc.35.8.982

Dean, A. M. (2004). Rethinking customer expectations of service quality: Are call centers different? *Journal of Services Marketing*, *18*(1), 60–78. doi:10.1108/08876040410520717

Diabah, G. (2021). Negotiating language barriers: Customer care delivery practices of a selected telecommunication company in Ghana. *Sociolinguistic Studies*, *15*(2-4), 177–199. doi:10.1558ols.41465

Downe-Wamboldt, B. (1992). Content analysis: Method, applications, and issues. *Health Care for Women International*, *13*(3), 313–321. doi:10.1080/07399339209516006 PMID:1399871

Etkin, J. (2016). The Hidden Cost of Personal Quantification. *The Journal of Consumer Research*, *42*(6), 967–984. doi:10.1093/jcr/ucv095

Farr, R. (1982). Interviewing: The social psychology of the inter-view. In A. J. Chapman & A. Gale (Eds.), *Psychology and People: A Tutorial Text* (pp. 287–305). Macmillan Education. doi:10.1007/978-1-349-16909-2_15

Følstad, A., Nordheim, C. B., & Bjørkli, C. A. (2018). What makes users trust a chatbot for customer service? An exploratory interview study. In S. S. Bodrunova (Ed.), *Internet Science* (pp. 194–208). Springer. doi:10.1007/978-3-030-01437-7_16

Fournier, S. (1998). Consumers and their brands: Developing relationship theory in consumer research. *The Journal of Consumer Research*, *24*(4), 343–353. doi:10.1086/209515

Galko, L., Porubän, J., & Senko, J. (2018). Improving the user experience of electronic university enrolment. *2018 16th International Conference on Emerging eLearning Technologies and Applications (ICETA),* 179–184. 10.1109/ICETA.2018.8572054

Grand View Research. (2022). *Chatbot Market Size Worth $3.99 Billion By 2030 | CAGR: 25.7%*. https://www.grandviewresearch.com/press-release/global-chatbot-market

Grudin, J., & Jacques, R. (2019). Chatbots, humbots, and the quest for artificial general intelligence. *Proceedings of the 2019 CHI Conference on Human Factors in Computing Systems*, 1–11. 10.1145/3290605.3300439

Hill, J., Ford, W. R., & Farreras, I. G. (2015). Real conversations with artificial intelligence: A comparison between human–human online conversations and human–chatbot conversations. *Computers in Human Behavior*, *49*, 245–250. doi:10.1016/j.chb.2015.02.026

Hsieh, H.-F., & Shannon, S. E. (2005). Three approaches to qualitative content analysis. *Qualitative Health Research*, *15*(9), 1277–1288. doi:10.1177/1049732305276687 PMID:16204405

Jain, M., Kumar, P., Kota, R., & Patel, S. N. (2018). Evaluating and informing the design of chatbots. *Proceedings of the 2018 Designing Interactive Systems Conference*, 895–906. 10.1145/3196709.3196735

Kaplan, A. M., & Haenlein, M. (2010). Users of the world, unite! The challenges and opportunities of social media. *Business Horizons*, *53*(1), 59–68. doi:10.1016/j.bushor.2009.09.003

Kelleher, T. (2009). Conversational voice, communicated commitment, and public relations outcomes in interactive online communication. *Journal of Communication*, *59*(1), 172–188. doi:10.1111/j.1460-2466.2008.01410.x

Khanna, A., Jain, M., Kumar, T., Singh, D., Pandey, B., & Jha, V. (2015). Anatomy and utilities of an artificial intelligence conversational entity. *2015 International Conference on Computational Intelligence and Communication Networks (CICN)*, 594–597. 10.1109/CICN.2015.122

Lee, H.-J., & Lyu, J. (2016). Personal values as determinants of intentions to use self-service technology in retailing. *Computers in Human Behavior*, *60*, 322–332. doi:10.1016/j.chb.2016.02.051

Lew, Z., Walther, J. B., Pang, A., & Shin, W. (2018). Interactivity in online chat: Conversational contingency and response latency in computer-mediated communication. *Journal of Computer-Mediated Communication*, *23*(4), 201–221. doi:10.1093/jcmc/zmy009

Lockwood, J. (2017). An analysis of web-chat in an outsourced customer service account in the Philippines. *English for Specific Purposes*, *47*, 26–39. doi:10.1016/j.esp.2017.04.001

Lu, L., McDonald, C., Kelleher, T., Lee, S., Chung, Y. J., Mueller, S., Vielledent, M., & Yue, C. A. (2022). Measuring consumer-perceived humanness of online organizational agents. *Computers in Human Behavior*, *128*, 107092–107092. doi:10.1016/j.chb.2021.107092

Luger, E., & Sellen, A. (2016). "Like having a really bad pA": The gulf between user expectation and experience of conversational agents. *Proceedings of the 2016 CHI Conference on Human Factors in Computing Systems*, 5286–5297. 10.1145/2858036.2858288

McIntosh, M. J., & Morse, J. M. (2015). Situating and constructing diversity in semi-structured interviews. *Global Qualitative Nursing Research*, *2*, 1–12. doi:10.1177/2333393615597674 PMID:28462313

Mead, G. (1967). *Mind, Self, & Society: From the Standpoint of a Social Behaviorist* (Vol. 1). University of Chicago Press. doi:10.7208/chicago/9780226516608.001.0001

Merton, R. K., & Kendall, P. L. (1946). The Focused Interview. *American Journal of Sociology*, *51*(6), 541–557. doi:10.1086/219886

Mick, D. G., & Fournier, S. (1998). Paradoxes of technology: Consumer cognizance, emotions, and coping strategies. *The Journal of Consumer Research*, *25*(2), 123–143. doi:10.1086/209531

Misischia, C. V., Poecze, F., & Strauss, C. (2022). Chatbots in customer service: Their relevance and impact on service quality. *The 13th International Conference on Ambient Systems, Networks and Technologies (ANT) / The 5th International Conference on Emerging Data and Industry 4.0 (EDI40), 201*, 421–428. 10.1016/j.procs.2022.03.055

Morse, J. M., & Field, P. A. (1995). *Qualitative Research Methods for Health Professionals*. SAGE Publications.

Rapp, A., Curti, L., & Boldi, A. (2021). The human side of human-chatbot interaction: A systematic literature review of ten years of research on text-based chatbots. *International Journal of Human-Computer Studies*, *151*, 102630–102630. doi:10.1016/j.ijhcs.2021.102630

Reimers, V., Chao, C.-W., & Gorman, S. (2016). Permission email marketing and its influence on online shopping. *Asia Pacific Journal of Marketing and Logistics*, *28*(2). Advance online publication. doi:10.1108/APJML-03-2015-0037

Rookes, P., & Willson, J. (2000). *Perception: Theory, Development and Organisation* (1st ed.). Routledge.

Scherer, A., Wünderlich, N. V., & von Wangenheim, F. (2015). The value of self-service: Long-term effects of technology-based self-service usage on customer retention. *Management Information Systems Quarterly*, *39*(1), 177–200. doi:10.25300/MISQ/2015/39.1.08

Schram, W. (1954). *The Process and Effects of Mass Communication*. University of Illinois Press.

Selnes, F., & Hansen, H. (2001). The potential hazard of self-service in developing customer loyalty. *Journal of Service Research*, *4*(2), 79–90. doi:10.1177/109467050142001

Sheehan, B., Jin, H. S., & Gottlieb, U. (2020). Customer service chatbots: Anthropomorphism and adoption. *Journal of Business Research*, *115*, 14–24. doi:10.1016/j.jbusres.2020.04.030

Shi, W., Wang, X., Oh, Y. J., Zhang, J., Sahay, S., & Yu, Z. (2020). Effects of persuasive dialogues: testing bot identities and inquiry strategies. In *Proceedings of the 2020 CHI Conference on Human Factors in Computing Systems* (pp. 1–13). Association for Computing Machinery. 10.1145/3313831.3376843

Statista. (2021). *Internet usage in India*. https://www.statista.com/topics/2157/internet-usage-in-india/

Tintarev, N., O'donovan, J., & Felfernig, A. (2016). Introduction to the special issue on human interaction with artificial advice givers. *ACM Transactions on Interactive Intelligent Systems*, *6*(4), 1–12. doi:10.1145/3014432

Toader, D. C., Boca, G., Toader, R., Măcelaru, M., Toader, C., Ighian, D., & Rădulescu, A. T. (2020). The effect of social presence and chatbot errors on trust. *Sustainability*, *12*(1), 256. Advance online publication. doi:10.3390u12010256

Wilcox, K., & Stephen, A. T. (2013). Are close friends the enemy? Online social networks, self-esteem, and self-control. *The Journal of Consumer Research*, *40*(1), 90–103. doi:10.1086/668794

Xu, K., & Lombard, M. (2017). Persuasive computing: Feeling peer pressure from multiple computer agents. *Computers in Human Behavior*, *74*, 152–162. doi:10.1016/j.chb.2017.04.043

Youn, S., & Jin, S. (2021). "In A.I. We Trust?" The effects of parasocial interaction and technopian versus luddite ideological views on chatbot-based customer relationship management in the emerging "feeling economy". *Computers in Human Behavior*, *119*, 106721. doi:10.1016/j.chb.2021.106721

ADDITIONAL READING

Asbjørn, F., & Brandtzæg, P. B. (2017). Chatbots and the new world of HCI. *Interaction*, *24*(4), 38–42. doi:10.1145/3085558

De Kleijn, R., Wijnen, M., & Poletiek, F. (2019). The effect of context-dependent information and sentence constructions on perceived humanness of an agent in a Turing test. *Knowledge-Based Systems*, *163*, 794–799. doi:10.1016/j.knosys.2018.10.006

Greer, S., Ramo, D., Chang, Y.-J., Fu, M., Moskowitz, J., & Haritatos, J. (2019). Use of the Chatbot "Vivibot" to Deliver Positive Psychology Skills and Promote Well-Being Among Young People After Cancer Treatment: Randomized Controlled Feasibility Trial. *JMIR mHealth and uHealth*, *7*(10), e15018. doi:10.2196/15018 PMID:31674920

Tärning, B., & Silvervarg, A. (2019). "I Didn't Understand, I'm Really Not Very Smart"—How Design of a Digital Tutee's Self-Efficacy Affects Conversation and Student Behavior in a Digital Math Game. *Education Sciences*, *9*(3), 197. doi:10.3390/educsci9030197

Trivedi, J. (2019). Examining the Customer Experience of Using Banking Chatbots and Its Impact on Brand Love: The Moderating Role of Perceived Risk. *Journal of Internet Commerce*, *18*(1), 91–111. doi:10.1080/15332861.2019.1567188

KEY TERMS AND DEFINITIONS

Anthropomorphic Chatbots: Chatbots that are designed to mimic human nature such as gestures, emotions, behaviours.

Artificial Intelligence: A technology designed to have extraordinary intelligence that do the work of several human resources bringing efficiency and standardisation.

Chatbot: Type of automated technology used in chat services to chat with customers.

Search Query: The type of search done on search engines or on websites.

User Manual: A type of document given to customers that guides how to use the products and services.

Chapter 9
The Influence of Music on Online Advertising:
The Case of YouTube

Pedro Azevedo
ISCAP, Instituto Politécnico do Porto, Portugal

Jorge Remondes
https://orcid.org/0000-0002-5259-4379
ISCAP, Instituto Politécnico do Porto, Portugal

Sandrina Teixeira
https://orcid.org/0000-0002-5859-0002
ISCAP, Instituto Politécnico do Porto, Portugal

ABSTRACT

The objective of this chapter is to understand how the use of a particular type of music may or may not capture the attention of the online user when viewing an ad and, perhaps, activate a click and an eventual conversion. An original advertisement was created to support this research and obtaining results through quantitative research in the form of an online questionnaire application. The way music can stimulate attention, memory, familiarity, the coherence of elements, rhythm, and personal tastes was analysed, as was understanding how these concepts interconnect with what the viewer visualises in an online advertising campaign. The ad was published on YouTube because in this social network there is only a short period of time of mandatory exposure of the ad to, consequently, lead to the conversion of the user to a landing page. It was concluded that the music of an online audiovisual campaign can induce the recipient's attention, confirming that the maximum optimization in an online advertising campaign occurs when the coherence of the music with the ad is prioritized.

DOI: 10.4018/978-1-6684-5538-8.ch009

INTRODUCTION

Digital advertising today plays a leading role in the dissemination of content for virtual media. However, with advertising ubiquity, we are increasingly faced with an information overload that triggers automatic conditioning on the part of the viewer, developing a stimulus predisposed to ignore almost anything that constitutes an advertising message. Therefore, companies must implement more relevant and distinctive ad strategies, placing digital advertising in constant interaction with the apogee of the audiovisual aspect (Valdellós, 2011).

According to the consultancy firm PwC, cited by Julio (2019), about 30% of advertising content produced is disseminated through the online audiovisual aspect, with an exponential growth expected over the coming years, with a view to equal or even surpass traditional advertising in the maximum period of 5 years. This trend is already predicted within the digital media itself, where there has been a considerable increase in display ads (video and banners) in favour of more traditional ads (text) present in Google's search (eMarketer, 2016). Thus, we are beginning to observe a huge trend towards the supremacy of advertising communication in the form of video, proven by recent market studies released by Cisco (2021), where it was evidenced that around 81% of internet traffic is generated through video.

However, this boom in online advertising could not be so expressive without the growth of social networks that, according to SproutSocial (2021), are the main responsible for its colossal increase. When we check the figures released by Statista (2021b), we can understand the size of this market and the existing potential in terms of advertising. In October 2020, about 4.66 billion active internet users were registered, of which 4.14 billion actively used some social network every day. Furthermore, in the study conducted by Buffer (2019), 73% of brands said that their social media ad experiences were positive or very positive from a financial return point of view. Also, every year the percentage of internet users who use social media to search for products has increased substantially, which generates great opportunities for companies in terms of advertising, culminating in a percentage of 54% in the year 2019 (GlobalWebindex, 2019).

The study presented here in this chapter was applied to Youtube, something still little explored in the literature. Consequently, it is possible to understand if the mandatory exposure time may have a different impact from the one it could have if it was displayed in the regime of "non-blank videos", techniques that are currently made available by Youtube for advertisers (Puccinelli, Wilcox, & Grewal, 2015).

However, results can only be achieved if we meet the theory of Herget, Schramm and Breves (2018), which exposes the relevance of the topic under study when they equate music to a powerful marketing tool that can influence the individual's purchasing decisions. In turn, Zoghaib (2019) points out that companies should adapt and understand this theme, as well as seek to apply this technique in their campaigns to increase the attention and predisposition of the user towards the ad. Thus, for the ad creator, the best way to avoid the Youtube ad being ignored will be to create a campaign that manages to establish a perfect audiovisual symbiosis (Tandyonomanu, 2018).

This study allows companies seeking to promote their products or services on the Youtube social network to understand whether the choice of music for one of their campaigns can substantially increase their click-through rate or not.

The aim was to understand if this element assumes a high degree of priority when carrying out an advertising campaign and, therefore, if it is a factor that pays off monetarily, since the use of music involves copyright and therefore monetary rewards for the authors of the music in question, especially in the case of very popular songs. Thus, the confirmation that music is effectively configured as a key factor capable of captivating the potential customer, may lead companies to take into account this charge when making their budget for an online advertising campaign.

The interaction that is established in virtual media presents itself as a great adversary to the more traditional audiovisual media, as proven by a study conducted by IPSOS Connect (2020), where it was concluded that the ads presented in social networks generate a greater gain of attention when compared to television advertising. In the virtual medium, as it is not possible to present pre-prepared content and programming, there is the particularity that Internet users seek content in real-time. In the case of social networks, according to Resultados Digitais (2020), one of the great benefits of this type of platform is the exposure of sponsored material to users who do not yet know they have a problem, attracting, unnoticed, potential customers who are not looking for a particular topic. Likewise, this type of platform, by allowing the investment in advertising and targeting it in a detailed way, can easily turn a user into a future customer (Parera, 2020).

It was intended to study the problem applied to the social network Youtube that, despite providing a giant business volume for all advertisers, seems to present problems in its campaign presentation, namely the recurrent use of ad blockers and a big failure in capturing the user's attention. After all, Youtube's business model is centred on the presentation of ads and that is what allows the platform to be free for users. In addition, as noted in the literature review, most ads are sold on a Cost-per-thousand impression (CPM) or Cost-per-click (CPC) system, further exacerbating this phenomenon. In this parameter, if the ad is blocked, the platform is unable to display its content and consequently collect the revenue generated. However, recent studies in the area released by eMarketer (2020) point out that most devices that access the Youtube social network cannot use this type of ad blocker, since more than half the percentage comes from the mobile and TV market, with the inclusion of smartphones, tablets and consoles / SmarthTv's, as we can see in Figure 1.

As it happens in all areas of our daily life, the effect of information overload also spreads to the social network Youtube. Although there is an increasing trend of users for the platform, the number of advertisers is also growing. Thus, it is imperative that the constitution of the advertising message is relevant, original, coherent and captivating enough to be able to retain the user's attention and, perhaps, convert him/her.

As a way to get around this natural tendency to ignore an ad, in January 2018 the YouTube platform itself began to introduce, in its advertising policy, the non-bypassable video ads that, in practice, solve this problem (YouTubeHelp, 2021a). However, if the ad does not really capture the user, at the outset, it matters little whether the user views it to the end or not. For this reason, this chapter analyzes whether one of the elements that make up the audiovisual ad (music) can induce unconsciously, not only the attention, but also the conversion of the user.

In this sense, four distinct ads were created, applied online through a questionnaire survey, in order to collect primary data, by simulating an online advertisement on the YouTube social network.

The capsta tnto(2019), suggest that the mt.LI) of higher education and in particular entrepreneurship ciurses. title is structured in four further sections:

The literature review, with views from various authors and conclusions from recent studies, which scientifically contextualises the theme of this chapter.

The methodology followed by describing the method, instrument, sample and research procedure. It also defines the objectives and problems of the study, highlighting the set of variables that influence it, according to the state of the art.

Figure 1. Percentage share on devices using Youtube. Source: eMarketer (2020)

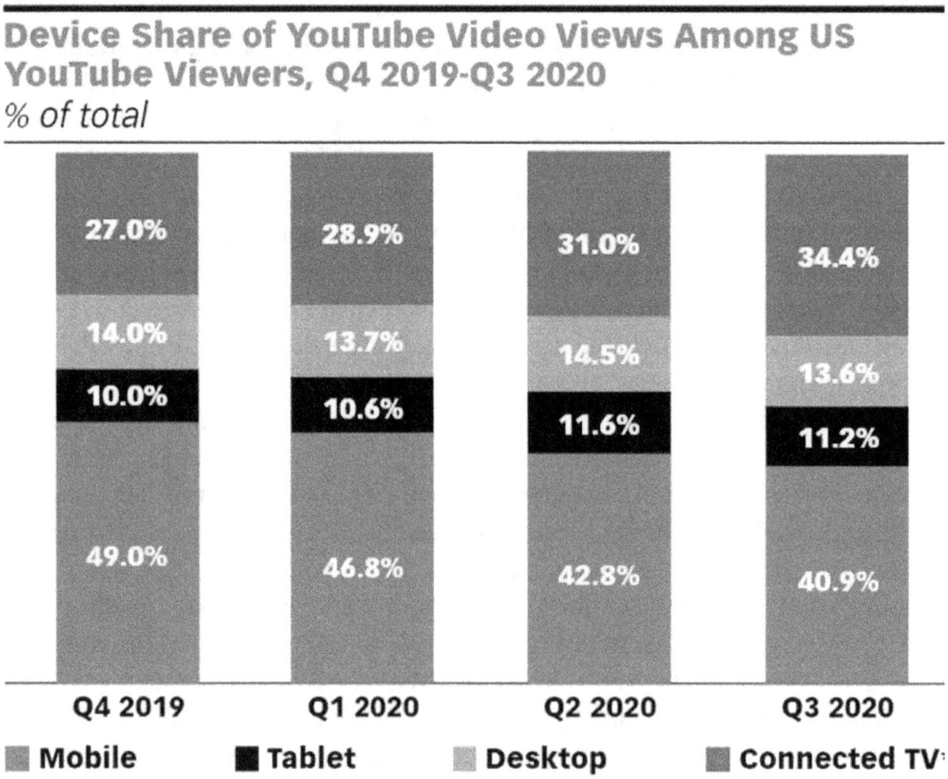

The data analysis and statistical treatment with the aid of SPSS *software* and later the validation (or not) of the research hypotheses.

The presentation of results and discussion in the light of the most relevant literature and studies.

Finally, the last section draws the main conclusions, identifies limitations and suggests future studies.

REVIEW OF LITERATURE

YouTube Advertising

Firat (2019) shows that a large part of the audiovisual ads presented on television are being transferred and modified for Youtube. According to the same author, in a survey conducted by Needle (2021), about 40% of respondents commented that, in the online medium, they preferred video-based ads due to their interactive nature. Furthermore, 83% of respondents preferred to view videos on the Youtube platform compared to any other medium, suggesting great potential for the advertising market.

In 2005, YouTube shared its first video and since then it has been growing more and more, much on account of its acquisition by the Google company. Today, according to the network's official *blog*, the number of monthly registered users has exceeded 2 billion (YouTubeOfficialBlog, 2021). Every day, YouTube users view more than 5 billion videos, and the number of videos increases by 45% every year.

Every minute, 300 hours of video are uploaded to the platform, and every day, around 30 million viewers visit the site (Biographon, 2019).

According to Zhou, Khemmarat, Gao, Wan and Zhang (2016), the success of Youtube is due not only to the huge collection of videos, but also by the discovery mechanics present in the network, such as video recommendations, search based on keywords, tab of popular videos or, if we want, "on the rise" and the instant incorporation in web pages, *blogs* and social networks. Thus, by exposing video to users, these video discovery mechanisms allow internet users to discover videos that are just of interest to them (Zhou et al., 2016). In the same line of thought, this tool is available in the *back office of* the platform, allowing an advertiser to present its advertising campaign only to the audience interested in that segment (Firat, 2019). According to data published by Biographon (2019), the number of advertisers on the network has been growing at around 40% per year, with the top 100 advertisers on the network generating around 60% of all advertising spend for Youtube. Thus, there is a balanced symbiosis between the business side and the more casual side of the platform, since the user is presented with relevant content for free and, in return, the user generates revenue for the platform through the advertising content presented.

According to Miranda (2018), the advantages of investing in Youtube advertising are several, allowing to achieve brand awareness through high percentages of segmentation, generate traffic, conversions and, at the same time, obtain *lead* generation and full knowledge of the consumer's profile, being able to resort to remarketing techniques if necessary.

In turn, in its most recent study, IPSOS Connect (2020) demonstrated the importance and potential that this social network has in the advertising landscape. Compared to television advertising, content advertised on YouTube has a much higher attention gain, reaching 83% in the *mobile* market, against only 45% provided by TV.

Also, in this research, it was clear that YouTube users have a higher tolerance for ads. Around 73% of TV users stated that they felt annoyed by the amount of ads, promoting *zapping* in their daily routine. But why does this happen? According to Resultados Digitais (2020), the answer is simple: the content advertised online takes on the particularity of being displayed when the Internet user voluntarily carries out their research for their problem. However, this does not happen with television broadcasting, largely due to the pre-programmed content that is characteristic of this medium.

Therefore, each strategy has to be thought taking into account its channel. According to Gabriel (2010), a digital presence strategy should be guided by the determination and distribution between the strategic element and the tactical element, in constant interaction with the company's ideals and its mechanisms.

YouTube Advertising Acceptance

According to Dehghani, Niaki, Ramezani and Sali (2016), there are several elements that may produce acceptance by the receiver when facing an advertisement on Youtube. The authors present a conceptual model that brings together a set of variables, statistically validated, that crosses opinions and perspectives present in the literature. This model was later validated by Firat (2019), which is also addressed in this chapter.

However, first it is important to understand why internet users tend to ignore the ads that are presented to them on Youtube. In his opinion piece for CNBC station, Handley (2017), citing data released by IPG Media Lab (2017), disclosed that about 65% of users ignore ads on Youtube, of which 76% said they do so because they think it is an intrusive behavior. To this end, he pointed out that campaigns should be emotional in nature and build a strong narrative during the mandatory exposure period of the ad. Complementarily, through their research, Shon, Shin, Hwang and Lee (2020), concluded that there are three major reasons why users develop a predisposed drive to ignore much of the advertising content. Firstly, we must take into account that the ads present in this social network "(...) cannot be easily avoided, causing consumers to need to watch the ads for a minimum period of time before consuming the specific content they wished to view (...)", causing the degree of impatience to increase and, consequently, the campaign to be ignored more easily. Interestingly, the authors demonstrate that the recurrence to ad blockers may be caused by the use of the user's personal data at the time of the ad presentation. Also, this inattention may be stimulated by advertisers' *remarketing* techniques, leading to a repeat of an already viewed ad being more easily ignored by the viewer, much encouraged by the recent update on YouTube where it is possible to show two ads in a row.

That said, and following on from the previous line of thought, Dehghani et al (2016) have gathered a total of seven elements concerning Youtube user acceptance of advertising, compiled into the following proposed model (Figure.2).

Figure 2 - Elements of advertising acceptance by Youtube users Source: Adapted from Dehghani et al (2016)

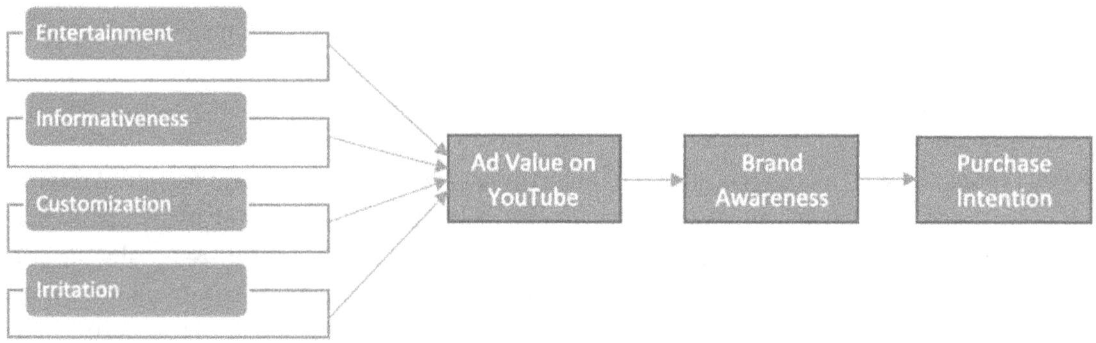

Entertainment: According to data gathered by the creators of the model (Dehghani, et al. 2016), advertising entertainment provides a greater degree of attention and encourages the receiver to consume advertising content more frequently. This concept appears to be so significant in the literature that a specific term was created to designate it, "Advertainment". In this field, several theories express the constant relation between the terms advertising and entertainment, for example, the theoretical suggestion of Russel (2007), who defines the term as the advertising practice that integrates the communicational aspects of an organization with the entertainment domain. According to Xavier (2015), taking into account new technologies and current media, there is a growing need to link advertising and entertainment, resulting in a strong storytelling appeal capable of creating and adding value to the campaign. In the particular case of YouTube, Firat (2019) determined the importance of this variable for the online medium, exposing the achievement of a positive effect on the attitude towards YouTube ads, suggesting that this is one of the main reasons why the advertising receiver assumes a positive condescension towards the ad acceptance.

Informativeness: In In general, this term can be defined as the "quality of what is informative", relating the word to the term "qualitativeness" (Priberam, 2021). According to the research of Dehghani and colleagues (2016), citing (Chen, 1999), informativeness can be characterized as a measure that places the user in front of useful and intelligent information. Thus, the advertising receiver tends to show signs of interest and acceptance towards the campaign after perceiving the unconditional exchange of interpersonal information between himself and the ad itself. Furthermore, research conducted by Clancey (1994) demonstrated the importance of the variable, suggesting that it is one of the main reasons why media users do not tend to ignore an advertisement, due to the development of the ability to absorb information during the campaign. In the most current panorama, Firat (2019) exposes that this term, when correlated with YouTube advertising, can be defined as anything capable of providing information about a company, product, service or brand. Here, it was defined that informativeness effectively produces a positive gain before the attitude and acceptance of the ad, corroborating his research with that of Yang, Huang, Yang & Yang (2017), which obtained the same positive relationship.

Customization: In the literature, this concept is very similar to the concept of segmentation when applied to the advertising industry. Thus, advertisers must follow the customer needs, modelling certain consumption:

Positive advertising value on YouTube = (Informativeness + Customization + Entertainment - Irritability)

Thus, we may conclude that a positive effect in relation to advertising value will be reached when the degree of informativeness, customization and entertainment is high and, simultaneously, when the degree of irritability is low. In turn, when the advertising message is composed of a higher level of irritability, a negative correlation will be obtained, giving rise to a diminishing effect in terms of advertising value.

Brand Awareness: The concept of *brand awareness* can be defined as the consumers' ability to recognize a brand and their consequent ability to spontaneously remember it (Betterncourt, 2017). Thus, according to the author, the ultimate goal of brands is to seek to build what would be pointed out as the *top of mind awareness, in* order to obtain a differentiation and recognition above normal or, if we want, to establish their brand as one of the best among the best. In the view of Dehghani and collaborators (2016), the new means of communication have come to expand, even more, the customer and company relationship, presenting an increase in interaction capacity and, consequently, strengthening of their communicational bonds. Applying this concept to social networks, Stephen and Toubia (2010) refer that with the increase of this type of platforms, there is a greater facility for the name of a brand to be spread throughout all networks, increasing the notification of users and leading to the creation of brand awareness. This exposure is also corroborated by the authors Barreda, Bilgihan, Nusair and Okumus (2015), evidencing that social networks contribute to the choice of a brand at the time of the purchase decision process. Thus, when this effect lasts with the user, it will be obtained a positive exposure before the brand recognition and, consequently, increase the purchase intention.

Purchase intention: Regarding the last variable of the model, we can define it as the tendency to buy a product and/or service when that same item is advertised and one has the need to obtain it (Firat, 2019). Thus, according to Dehghani and collaborators (2016), the maximum indicator of the effectiveness of a campaign can be obtained when there is purchase intention by the receiver after exposure to it. In the case of social networks, it is identified that the value of advertising assumes a positive relationship with purchase intention behaviour, much provided by the adaptation of e-commerce to these same networks

(Barreda, Bilgihan, Nusair and Okumus, 2015). However, this purchase intention, applied to social networks, is proportionally dependent according to the degree of recommendations that users make towards a brand (Dehghani, et al. 2016). More specifically, on YouTube, Yuksel (2016), cited by Firat (2019), investigated the effects of user-generated content on the platform's videos and found that the credibility and usefulness of the advertising campaign, when perceived by the receiver, positively affects purchase intention. For the creators of the model, the increase in the percentage of purchase intention can be obtained when a positive advertising value is achieved and, even higher, when it is added to a positive *brand awareness.*

That said, and by way of conclusion, Firat (2019) added in his validation paper that all variables in this model tend to produce the effects expected and hoped for by its authors, leading to its statistical feasibility.

INTERACTION BETWEEN MUSIC AND ADVERTISING

According to Belanche, Flávian and Pérez-Rueda (2017), the evidence of the relationship between music and advertising is clear: music produces a positive attention gain towards the advertising campaign. However, according to the authors, these inferences should be corroborated with more elements of interactivity, especially in online media, in order to validate their assumption.

Iyer and Aggarwal's (2019) research, suggests that music is primarily responsible for eliciting different reactions from the viewer, and can elicit extremely varied concepts and emotions, and concomitantly deliver an advertising message and even increase attention towards the ad.

However, users are increasingly demanding and selective when it comes to advertising exposure and it is up to companies to differentiate their campaigns and attract attention for possible conversion. According to Zoghaib (2019), companies understand this trend and, nowadays, more than ever, there is the production of a strong emotional appeal when transmitting values and propagative characteristics of advertising. According to the same source, one of the ways to achieve this effect is through music. This idea has been corroborated by Souza (2017) who argues that music configures itself as a communication channel of its own, capable of transmitting a message in a perceptive way, influencing the receiver to take certain attitudes intended by the advertiser. For this reason, as Valdellós (2011) mentions, about 97% of elaborated digital advertisements tend to present, in their composition, music.

Taking into account the terms of the current communicational advertising, it is safe to say that the creative industry has changed. There has been a growth and improvement in advertising expression over the years, encompassing in its conception new technological trends in the audiovisual context. However, as is known, it has not always been like this. Multiple historical stages were necessary to achieve what we know today as advertising diffusion. Despite the *layouts* and posters characteristic of the late 19th century, it was not until the appearance of television, at the beginning of the 20th century, that the advertising sector saw such a revolutionary leap. The truth, is that television provided a factor that went beyond the loudness of radio or simply reading a magazine, it provided the junction of these two worlds with the introduction of the audiovisual. On 1 July 1941, the first television advert to exploit this new broadcast medium would be created, with the Bulova brand presenting a 10-second advert of its new watch under a broadcast channel (Wikipedia, 2021). However, the ad would later come under criticism for its lack of emotion and contextualisation.

Long before the emergence of TV, music was first used in a radio advertising campaign, with the creation of the first musical jingle in 1926, for Wheaties breakfast cereal (Wikipedia, 2020). According to the same source, just 3 years after the first television advertisement (1944), the element of music was introduced in this new advertising context, with the creation of the first commercial jingle "Chiquita Banana" to be presented to the world. This time, the ad was described as catchy and informative, letting the audience know different peculiarities of how to consume a banana, in an interactive and semiotic way, assimilating that music could bring an attention gain to the campaign.

Usefulness of Music in Advertising on YouTube

In the context of music and advertising, we can point out that the theoretical approach has been moulding, according to the evolution of time, making the interaction between these two aspects studied for a long time.

This being a classic theme, when we check the existing literature on the same, we can observe a multiplicity of visions, be they congruent or disparate. Authors like Miniard, Lord and Dickson (1988), defended that music was not capable of producing sufficient persuasive effects to be considered an essential element of a campaign. In the same line of thought, Guzmán (1993) criticised the efficiency of the advertisements of his time, praising the visual elements to the detriment of background music. Also, in his more psychological current, Zajonc (1980) observed a group of people and theorized that music served only as an ornament to embellish the campaign, not producing any persuasive effect.

However, during this time, different perspectives have already emerged which are in line with the most recent literature. Hecker (1984), defended that music "(?) enhances words, colours, images and adds a form of energy that cannot be achieved in any other way". Thus, music assumes the purpose of serving as a "message intensifier" improving, in this way, the "attention - interest relationship" of the advertisement. In this framework, one of the most popular experiments in the classical literature on the subject is the study carried out by Gorn (1982). In this research, an advertisement was prepared where a light blue or beige coloured pen (neutral stimulus) was placed with pleasant or unpleasant music, according to the personal preferences of each of the viewers (not conditioned to the stimulus). As a result, about 80% of the subjects chose the pen with the more pleasant music, with the colour of the pen showing no impact on the decision - a reaction conditioned by the music.

However, with the advance of time, the visions on this theme became more and more congruent among themselves. More and more authors began to share the same vision on the relationship between music and advertising campaigns. An example of this are the theories carried out by Valdellós (2011) and by Ansani and collaborators (2020), where it is pointed out that despite its apparent little importance, music is a fundamental element in the composition of an advertisement and should be chosen quite precisely. It can enhance the expression, the mood, the situation and even the character of a campaign, culminating in the fact that, as we have seen, around 97% of elaborate digital ads tend to feature, in their composition, music (Valdellós, 2011). On the other hand, its absence will create a sense of distrust that will lead to the ineffectiveness of the campaign. Thus, music can be used to attract the attention of the ad receiver, since users consider campaigns with music more optimistic and more informative, serving this as a powerful marketing tool to replace images and words (Ng, 2011). Therefore, the decision on which music to choose for the development of a campaign, is configured as a vital part of the creation process, since if it is not consistent with the message of the ad, it can generate a certain confusion. In this way, when everything is interconnected, we obtain a strong ad message that could not be obtained without all the audiovisual symbiosis (Binet, Müllensiefen, & Edwards, 2013). To this end, Valdellós (2011) perspectives the use of music for different occasions. On one hand, it can serve only as a *background to* enhance the main message of the ad, or it can become its protagonist in order to determine the tone and communication of the campaign, whether in the form of a *jingle* or not.

Having said this, the most recent researches approach digital media. With the creation of new online platforms such as YouTube, the need has arisen to identify and implement new ways to increase viewer attention to ads, as the mandatory exposure time is relatively short.

Currently, YouTube provides various advertising policies that take into consideration the utility of creators, content providers and consumers, as we have already studied in the previous subchapters. In this medium, Tandyonomanu (2018) conducted research about ads on the YouTube platform and what would lead to their acceptance or ignoring. In this research, it was found that about 70% of the users under analysis, remained viewing the ad (which it would be possible to ignore) because they found its content interesting. Now, of this sample, about 63% of users said that the music contained in the ad was the reason why they remained viewing it, showing the importance of its existence in the composition of the campaign.

The theme of ad acceptance in relation to music was studied and deepened according to some variables, since music is a highly subjective topic and we may find divergence of opinions according to the personal tastes of each one, for example. In this context, concepts such as rhythm and coherence between the music and the ad, familiarity, personal tastes and attitude/emotions emerge in the literature and help us to better understand the theme.

Rhythm and Coherence

With his research, Ng (2011), concluded that we should analyse the intrinsic characteristics of music and adds that rhythm is the most important factor when choosing a piece of music for a campaign. Supporting his perspective, a recent study in the field by Sunaga, Meng and Zhuang (2020) concluded that the timbre and rhythm of a song stimulate greater or lesser attention to the user depending on the emotion perceived. Thus, according to the different tones of the music, different reactions and emotions arise that can influence the process of transmission and understanding of the communication and, consequently, manipulate the click on the ad. Adding to their vision, Herget, Schramm and Breves (2018), conducted a study in which they developed an instrument capable of identifying the perfect *fit of* the elements of an ad, contemplating that this "*perfect fit*" would be obtained when the coherence between the rhythm of the music and the narration of the ad were achieved, having the product and its target group in mind. We therefore entered the theme of coherence between the music and the visual elements of the ad.

Therefore, so that the user does not feel driven to ignore the ad, it will be necessary to build narratives that must be consistent with each other, involving all the congruent modalities between audio and video to obtain a greater degree of attention (Ansani, et al., 2020). When the sound and all the components of the campaign coincide, what is advertised will, at the outset, match the users' expectations, generating greater attention to the ad and, consequently, a greater chance of clicking. A high stimulation campaign performs better in congruent contexts than in incongruent contexts, in terms of ad display time, ad attitude and rhythmicity (Belanche, Flávian, & Pérez-Rueda, 2017). In this circumstance, the rhythm of the music will drive the tone of the ad depending on its tonality. Studies show that slower, more emotional music tends to generate a sense of stillness, such as the research of Sunaga, Meng and Zhuang (2020), where it was evidenced that the piano timbre provokes a sense of familiarity and security. Conversely, music with a higher tempo usually generates a higher degree of enthusiasm and energy. Ansani and colleagues (2020), therefore, argue that the music in an advertisement has to be implemented in the right measure. If, on the one hand, it is too frenetic, it may cause a discontent and a sense of alertness that will drive away any empathy with what is being advertised. On the other hand, if it is too calm, it may not spark any interest in the campaign and therefore lead to it being ignored. (Ansani, et al., 2020).

In turn, according to the literature raised by Raja, Anand and Kumar (2018), in their scale validation article, it is mentioned that when consumers evaluate the effectiveness of the music present in the advertisement, there is a tendency to analyze the compatibility of the advertisement music with the message of the advertisement and cross-reference it with other elements, such as *storytelling*. Thus, if the music is not congruent with the ad's message, it may diminish the consumers' attention towards it. For this same reason, brands seek to associate the characteristics of their product with the transmission of attributes provided by music. An example of this are the ads for energy drinks, where they are often associated with frenetic music, in order to convey the sense of energy caused by the drink itself (Blog Aquela Música do Anúncio, 2014).

Musical Familiarity

Underlying the interaction of music with advertising appears to us, in literature, the element of familiarity. Here, we understand as familiar all kinds of music, during the diffusion of the ad, are recognized by the receiver of the campaign. According to Freitas (2017), citing Zajonc (1968), according to the familiarity principle, also called the "mere exposure effect", the greater the degree of familiarity with something, the greater the level of acceptance. According to the same source, neuroscientific studies have demonstrated the importance of listening to familiar music, since they suggest that it produces a greater activation in the brain, affecting the reward and pleasure system.

Thus, the classical framework implies that the association of a product with a piece of music during a campaign (unconditioned stimulus), produces an association between the two and, therefore, a preference for the product (a conditioned response) (Zander, Apaolaza-Ibáñez, & Hartmann, 2010). In this panorama, Hahn and Hwang (1999) argue that familiarity with the music is directly associated with the emotional predisposition and predilection toward the campaign. Thus, according to the same authors, the use of unfamiliar music in a campaign may decrease the gain of attention and even stimulate the concept of information overload that we have talked about before. In this sense, when using familiar music, the number of elements to be collected before the diffusion of the campaign will be lower, and there will be a greater propensity to accept the ad.

Studies such as those of Zoghaib (2019), validated these same issues and concluded that the use of a globally known song arouses greater attention to the viewer, being more predisposed to watch the ad until the end. Taking this into account, Belanche, Flávian and Pérez-Rueda (2017) point out that campaigns that assume in their composition popular songs, provoke a high stimulation, serving to wake up viewers less engaged with the advertised item. On the other hand, users already involved with the advertised item will perceive the campaign less annoyingly, increasing the chance of clicking on the ad.

Adding to this idealization, Binet, Müllensiefen and Edward (2013) concluded that audiovisual advertisements that involve the composition of familiar songs, increase their effectiveness by about 20% to 30% compared to the use of unfamiliar songs and/or developed exclusively for the ad. Therefore, Zoghaib (2019), argues that companies, in general, perceive this trend and seek to apply this technique in their campaigns, corroborating his idea with the example of the song "I Want To Break Free" by Queen, which was present in almost all Coca-Cola ads during 2018.

Personal Tastes

According to the literature, the relationship between music and advertisement is also driven by personal tastes. According to Bourdieu (2010), the preference for specific musical genres or sub-genres can express the creation of a personal identity that will produce an association effect before various factors in our daily lives, including, for example, advertising campaigns.

Iyer and Aggarwal (2019), in their conception, argue that certain viewers can be effectively influenced by the use of sound consistent with their taste. This will lead to a positive effect on consumers' attitude and buying behaviour, as when faced with a display of music as per their tastes, there will be greater attention towards the advertisement. According to Endo and Roque (2017), citing Sousa (2006), consumers develop a search and acceptance of advertising messages when they resemble their interests, values, beliefs and expectations. On the other hand, they tend to reject, in a natural way, anything that clashes with their system.

Along the same line of thought, Vermeulen and Beukeboom (2016) validated the study carried out by Gorn in 1982, confirming that personal tastes regarding the music in an advertisement can manipulate the choice of the advertised item. Similar to Gorn's application, different music was used for each ad, and instead of using pens, mobile phone covers were applied. The result obtained was the choice of the advertisement with the most pleasant music according to the personal and individual taste of each respondent, with the mobile phone covers having no apparent persuasive effect. However, it is important to note that in this study it was evident that this conclusion may not result when a product has a high degree of involvement in the purchase. In turn, Galan (2009), through his scale validation article, showed that a musical piece with a high degree of "likeability" score or, if we wish, empathy towards musical tastes, can improve attitude towards the ad, the brand and even purchase intention. Their results indicate that music had a greater impact on attitude towards the campaign according to dimensions attributed at the personality level.

According to the analysis raised by Raja, Anand and Kumar (2018), musical appeal reflects the extent of preference for a stimulus, driven when a consumer is exposed to any advertising music. Thus, they elucidate, similar to Gorn's (1982) study, that musical appeal will influence the choices occurring in the course of the buying behaviour process.

For this reason, as we have seen previously, Dehghani and colleagues (2016) demonstrate the importance of customising advertisements to meet the personal tastes of the receiver, thus increasing the degree of acceptance and attention towards the campaign.

Emotion and Memory

The appeal to emotion and memory also stand out for their importance in developing the relationship between music and advertising. Frith (2003, p.46) suggests that "what people hear is more important to their sense of self, than what they watch or read". Music can consequently express differentiation and function as an emblem that conveys information about the person expressing a particular preference.

Indeed, for Iyer and Aggarwal (2019), the background music of a campaign can assume a highly overbearing role when affective engagement between the piece of music and the advertised item is achieved. Thus, it is a highly important concept in establishing attitudes, as the structural elements of music tend to affect an individual's mood and attitude towards a commercial (Craton, & Geoffrey, 2011).

Supporting this perspective, Simmonds and colleagues (2020) found, through their eye-tracking study, that the emotional effects of music have a greater influence on ad avoidance, reporting a high degree of correlation between emotional involvement and attention spillover to the campaign. Thus, this correlation varies, proportionally, inversely, indicating that depending on the higher degree of emotion, the dispersion of attention towards the ad will be reduced.

According to research by Rentfrow, Goldberg, Stillwell, Kosinki, Gosling and Levitin (2012), the emotional response produced by the receiver will be greater when using a musicality capable of capturing their attention, in terms of memory. Thus, there will be a change in the emotional state, which will be followed by a stimulus to ignore or accept the campaign. In this sense, Alpert and Alpert (1989) provide their view on the subject, showing that a joyful musical tone may produce a positive expression in terms of mood and, curiously, a sad sound may be able to produce influencing effects on the purchase decision process. In turn, silence or, if we wish, the absence of music, will not produce any reactional effect towards the campaign, suggesting the importance of music in advertising.

According to the conception of Abolhasani, Oakes and Oakes (2017), when a song of a campaign is memorized and memorized, it will be very difficult to remove from short-term memory and consequently lead the receiver to recall the ad. In this industry, many companies try to put their musical *jingle* in their campaign, or make use of a well-known song, so that it stays in the consumers' minds in order to activate and compenetrate their memory, in order to stimulate the intention to buy. In the authors' words, " (...) the use of extremely addictive and repetitive types of music in advertisements can have positive effects on consumer responses".

METHODOLOGY

This research aims to determine whether, solely based on music, audiovisual ads on the YouTube social network can have a circumstantial increase in capturing the user's attention, leading to an eventual improvement and effectiveness of the campaign. The specific objectives that are intended to be achieved are:

a) To understand if the complete viewing of an advertisement (30 seconds) can have a positive impact concerning the exposure of a smaller advertisement (5 seconds);
b) Understanding whether the fact that a song is coherent (or not) with what is being visually presented in the ad, has an impact on the user's action and influences him/her to click on the ad;
c) To understand if the use of a popular and familiar song can increase attention to the advertisement;
d) Understand whether the musical tastes of each individual can induce the click on the ad;
e) Understand the different reactions that may be caused by the use of different songs in the same advertisement, namely in terms of memory and emotion;
f) Consider whether the music brings any impact to the campaign.

It is thus intended to analyse and put into practice all the variables collected in the literature review which, when correlated with the collection and analysis of primary data collected, allow answering a set of questions related to the above-mentioned objectives, namely:

a) Does seeing the full version of the advertisement change the consumer's attitude compared to the shortened version?
b) Can music with different rhythms totally shape the viewer's perspective on the ad?
c) To what extent will the use of a popular song bring gain attention to the campaign?
d) Could the musical tastes of the viewer affect the effectiveness of the audiovisual campaign?
e) Could it be that if we use the same ad, but with different music, we can get extremely opposite emotional responses?
f) How the absence of music in the advertisement may (or may not) harm the campaign, using only sound effects.

Thus, our objectives can be categorised according to the variables under study in the previous chapter. More specifically, we obtain the following correspondence:

- Objective a) - Ignore or view (exposure effect);
- Objective b) - Rhythm and coherence;
- Objective c) - Familiarity;
- Objective d) - Personal (musical) tastes;
- Objective e) - Emotion and memory;
- Objective f) - Utility;

The method used comprises primary data collection, due to existing gaps in the state-of-the-art on the topic under study. Quantitative research was carried out, in the form of a survey, to obtain data through a questionnaire.

Population Definition

The sample of the present research comprises an arrangement of users of the YouTube platform, not being attributed to any specific restriction or particularity. Furthermore, it was mostly made up of young adults and its collection focused on a balanced number of responses between genders. Likewise, the instrument itself used was designed with the population in mind, presenting an advertising campaign for a fictitious *smartphone*.

Procedure

For the development of this research, advertisements were created in an authentic way and with the only purpose of application in this study. No known brand was used, so that this could not influence any user's perspective on the ad. Likewise, it was essential that the ad created had a minimally professional quality so as not to cause any diversion of attention on the part of the viewer.

So, at the beginning of the whole process, we started by making a survey of songs of different musical genres, popularity and rhythms, which could possibly lead to the stimulation of different sensations, in order to determine whether they could have an impact on the attention of the advertisement, or not.

Having said this, the development of the prototype ad began in the *After Effects* program. However, as it was necessary to test multiple variables and perspectives, it was decided to create four ads. Afterwards, three songs were selected and applied in three of the videos. In the fourth video, no music was added, leaving only the sound effects of the ad that were created.

For the finalisation of the ads, it was taken into account that all visual elements between the four videos could not have any relevant decision impact, so as not to create a diversion of attention in the study. In this way, only some subtle changes, such as changing the colour of the ad *background* and changing the colour in the presented texts, were presented between them.

Once the ads were finished, they were *uploaded* to the YouTube platform, where they were stored in a private *link*[1] .

Subsequently, a pilot questionnaire was created, via *Google Forms,* with the premise of detecting the possible and eventual existence of errors, capable of causing some confusion and misunderstanding among the respondent. Following this, six people were surveyed on 8 March 2021 and made three suggestions that were implemented in the questionnaire.

The updated version of the questionnaire was launched to the public, with its dissemination through the social networks *Facebook* and *Instagram, to* the groups that make up the institution communities of the Higher Institute of Accounting and Administration (ISCAP), of the Polytechnic Institute of Porto (IPP), inserted in the core of master's degrees and undergraduate degrees. Similarly, invitations were sent to *chat groups of the Messenger* network that fit into the definition of the population for the study.

The questionnaire was available between 15 April and 4 May 2021 and 102 responses were received.

Using the Tunebat *website* as a reference, it was decided to integrate in the ads, as shown in Table 1, songs of different genres, popularity, emotion and with different rhythms (beats per minute).

Table 1. Songs used in the advertisements

	Music	**BPM's***	**Popularity**	**Emotion**	**Musical genre**	**Source**
Announcement 1	Lemaitre - Closer ft. Jennie A.	92	62/100	Energy: 80/100 Joy: 50/100	EDM	Tunebat (2021a)
Announcement 2	Dua Lipa - Don't Start Now	124	90/100	Energy: 80/100 Joy: 79/100	Pop	Tunebat (2021b)
Announcement 3	Rousseau - Nuvole Bianche	68	48/100	Energy: 15/100 Joy: 4/100	Classic	Tunebat (2021c)
Announcement 4	n/a*	n/a*	n/a*	n/a*	n/a*	n/a*

BPM's* = beats per minute
n/a* = not applicable

Thus, it was hoped to obtain the best conclusions regarding the variables under study in this research. Finally, it is important to mention that this analysis does not apply to advertisement 4, since it does not have any associated music.

Variables Under Study

When researching for the literature review, we collected a set of variables and metrics that were compiled and validated by the authors Dessart (2018); Galan (2009); Raja, Anand, and Kumar (2018). These, served as the basis for the construction of our questionnaire, with the appropriate linguistic adaptations and, maintaining in all circumstances, the original essence of the scales. All items are gathered in Appendix I, referring to the set of questions for each variable, as well as their respective source and reliability index.

Questionnaire Online

The questionnaire was composed of 41 questions, divided into 8 sections, 40 of which were closed-ended and had to be answered.

The first section was devoted to the collection of socio-demographic data that allowed characterising the sample collected, which is described in this chapter in the results section.

Here, it is also determined which musical genre the individual usually listens to, to be able to ascertain whether personal musical tastes may interfere with the respondent's decision.

Next, the shortened version of the four adverts created (5 seconds) was presented, to simulate a situation of advertising broadcast on YouTube. At this stage, the respondent was asked whether or not he/she would ignore the ad if he/she were on the YouTube platform. Subsequently, the full versions of the ads were viewed and questions were asked based on the variables in the collected literature, to understand which variables, independently, may or may not lead to an increase in the user's attention to the ad. In this context, it is important to note that there is a strong correlation between the level of attention paid to the advertiser and the level of attention paid to the user.

Finally, in the last section, questions related to the theme were compiled, in a general way, so that it was possible to understand, once again, the effect that each of the variables under study has on the present investigation.

As for the formulation of the questions, Likert-type scales were applied in the following order: 1) I totally disagree; 2) I disagree; 3) I neither agree nor disagree; 4) I agree; 5) I totally agree; closed questions of agreement (Yes or No); multiple-choice questions and checkboxes; an open question is also included and, the only non-mandatory question in the questionnaire, regarding "Other" music genre that may not be included by the written list.

Research Hypotheses

The questionnaire was composed of 41 questions, divided into 8 sections, 40 of which were closed-ended and had to be answered.

The first section was devoted to the collection of socio-demographic data that allowed characterising the sample collected, which is described in this chapter in the results section.

Here, it is also determined which musical genre the individual usually listens to, to be able to ascertain whether personal musical tastes may interfere with the respondent's decision.

Next, the shortened version of the four adverts created (5 seconds) was presented, to simulate a situation of advertising broadcast on YouTube. At this stage, the respondent was asked whether or not he/she would ignore the ad if he/she were on the YouTube platform. Subsequently, the full versions of the ads were viewed and questions were asked based on the variables in the collected literature, to understand which variables, independently, may or may not lead to an increase in the user's attention to the ad. In this context, it is important to note that there is a strong correlation between the level of attention paid to the advertiser and the level of attention paid to the user.

Finally, in the last section, questions related to the theme were compiled, in a general way, so that it was possible to understand, once again, the effect that each of the variables under study has on the present investigation.

As for the formulation of the questions, *Likert-type* scales were applied in the following order: 1) I totally disagree; 2) I disagree; 3) I neither agree nor disagree; 4) I agree; 5) I totally agree; closed questions of agreement (Yes or No); multiple-choice questions and checkboxes; an open question is also included and, the only non-mandatory question in the questionnaire, regarding "Other" music genre that may not be included by the written list

Table 2. Research Hypotheses

H1: Full ad viewing captures more user attention in the face of reduced ad exposure (Shon, Shin, Hwang, & Lee, 2020; Puccinelli, Wilcox, & Grewal, 2015).
H2: Music with a faster rhythm and energy captures more of the viewer's attention during the advertisement (Dessarte, 2018; Galan, 2009).
H3: A popularly known song accompanying the advertisement elicits a higher degree of attention (Raja, Anand and Kumar, 2018).
H4: A song that meets the musical tastes of each consumer captures more of their interest in the campaign and consequently the click (Galan, 2009; Raja, Anand and Kumar, 2018).
H5. Songs with more upbeat melodics enhance ad recall in the future (Raja, Anand and Kumar, 2018).
H6: An ad without music does not meet its click-through target as it will be more easily ignored (Dessarte, 2018).

RESULTS

Sample and Musical Genre Heard

Data were collected through an online questionnaire with the participation of 102 respondents. The sample is mostly composed of females (58.8%), with the remaining participation of males (41.2%), and no other genders participated (see Table 3).

On the other hand, an absolute majority was recorded in the 18 to 24 age group (86.3%), with the remaining respondents in the 25 to 34 age group (13.7%), with no other age range recorded (see Table 4).

In terms of academic qualifications, it was found that the majority of the sample had a degree (64.7%), followed by a Master's degree (25.5%), with the remaining 5.9% having completed secondary education and 3.9% having a PhD (see Table 5).

With regard to the genre of music that respondents habitually listen to, 348 answers were recorded, since the content of the question allowed the respondent to answer and select more than one option.

Table 3. Gender of respondents

Gender					
		Frequency (N)	Percentage	Valid Percentage	Cumulative Percentage
Valid	Female	60	58,8	58,8	58,8
	Male	42	41,2	41,2	100,0
	Total	102	100,0	100,0	

N=102

Table 4. Age range of respondents

Age Group					
		Frequency (N)	Percentage	Valid Percentage	Cumulative Percentage
Valid	18-24	88	86,3	86,3	86,3
	25-34	14	13,7	13,7	100,0
	Total	102	100,0	100,0	

N=102

Table 5. Educational qualifications of the respondents

Educational Qualifications					
		Frequency (N)	Percentage	Valid Percentage	Cumulative percentage
Valid	Secondary Education	6	5,9	5,9	5,9
	Bachelor's Degree	66	64,7	64,7	70,6
	Master	26	25,5	25,5	96,1
	PhD	4	3,9	3,9	100,0
	Total	102	100,0	100,0	

N=102

Thus, a predominance of the Pop genre was registered (82.4%) and EDM (Electronic Music) with 51%. The remaining genres, present somewhat similar classifications between themselves, as is the case of Rock with 33.3% and Latin with 31.4%, as well as Hip-Hop/Rap with 27.7% and the R&B and Reggaeton genres sharing the same percentage of 25.5%, individually.

As it was not possible to compile all existing musical genres into a single list that was relevant to the study, the option of "Other" was made available which, when selected, questioned the respondent to specify the genre. In this option 6 answers were registered (5.9%), with the following genres: Sertanejo; J-Music; Indie; Indie folk; Jazz; Mpb (see Table 6).

Table 6. Musical genres heard by respondents

		Responses (With N=102)	Percentage
Valid	Pop	84	82,4
	Rock	34	33,3
	EDM (Electronic Music)	52	51
	Classical	14	13,7
	Hip-Hop/Rap	28	27,5
	Latin	32	31,4
	R&B	26	25,5
	Soul	10	9,8
	Reggaeton	26	25,5
	Funk	24	23,5
	Reggae	12	11,8
	Other	6	5,9
	Total	348	

N=102

Behaviour Regarding the Exposure Time of Advertisements

Simulating a skippable in-stream ad, when faced with the shortened version of the four ads (5 seconds), if they were on the YouTube platform, 70.6% of respondents would remain viewing Ad 1; 52.9% would continue to view Ad 2; 22.5% would choose not to skip Ad 3, and only 10.8% would remain to view Ad 4.

When exposed to the full version of the adverts (30 seconds) simulating a non-skippable in-stream advert, when asked if the full view of the advert captured their attention compared to the previous 5-second version, respondents answered yes, according to the following data: 73.5% for Advertisement 1; 59.8% for Advertisement 2; 32.4% for Advertisement 3; and 11.8% concerning Advertisement 4.

When we cross-reference these two data we can see if the full ad view can capture more of the user's attention compared to the minimum 5-second view.

Thus, in the case of the first ad (Ad 1), 56 of the interpolated respondents who would have remained viewing the shortened version of the ad, stated that viewing the full video further captured their attention. On the other hand, viewing the full campaign failed to capture the attention of the remaining 16 respondents who stated that they would remain to view the shortened version. Concerning the individuals who would ignore the ad in its shortened version, more than half of the respondents changed their opinion (19 out of 30), answering that viewing the ad in its full version captured their attention, while the remaining 11 did not present any change in their appreciation (see Table 7).

In Ad 2, it was possible to notice a variation in the behaviour of the data when compared to the first ad. Thus, 40 of the 54 respondents who would have remained viewing the short ad, felt that viewing the full ad brought them an increase in their attention. In turn, the full ad was not able to stimulate the attention of the remaining 14 who would primarily remain viewing the ad for more than 5 seconds. Regarding those interpolated who would ignore the ad, only 21 out of 48 of the subjects shaped their opinion and reported that viewing the full ad held more interest for them. Consequently, the remaining 27 did not change their initial viewpoint (see Table 8).

Table 7. Cross-tabulation between the variables "Ignore or view - Exposure effect" - Ad 1

Cross-tabulation: Ignore or View_Ad1 * View_Complete_Ad1				
		View Complete_Ad1		Total
		Yes	No	
Ignore or View_Ad1	View	56	16	72
	Ignore	19	11	30
Total		75	27	102

Caption:
Ignore_View_Ad = Ignore or view reduced ad (5 seconds).
Visualizar_Completo_Ad = Complete visualization of the ad (30 seconds) captured the individual's attention in relation to the reduced ad (5 seconds)

Table 8. Cross-tabulation between the variables "Ignore or view - Exposure effect" - Ad 2

Cross-tabulation: Ignore or View_Ad2 * View_complete_Ad2				
		View_Complete_Ad2		Total
		Yes	No	
Ignore or View_Ad2	View	40	14	54
	Ignore	21	27	48
Total		61	41	102

In Ad 3, it was observed that 15 out of 23 respondents who would continue to view the ad in its shortened version saw a substantial increase in attention gain towards the full campaign, with an opposite effect noted for the remaining 8. However, for those individuals who would ignore the short ad, only 18 out of 79 experienced an increase in concentration on the full campaign, with the remaining 61 showing no attention gain (see Table 9).

Table 9. Cross-tabulation between the variables "Ignore or view - Exposure effect" – Ad 3

Cross-tabulation: Ignore or View_Ad3 * View_complete_Ad3				
		View_Complete_Ad3		Total
		Yes	No	
Ignore or View_Ad3	View	15	8	23
	Ignore	18	61	79
Total		33	69	102

Finally, Ad 4 reveals the most concordant data in our study. Of the 11 individuals who would remain viewing the ad, an almost equal opinion was obtained between the two sides, with 5 responses in favour of gaining attention and 6 against. Interestingly, of the 91 individuals who would choose to ignore the ad in the first instance, only 7 would change their opinion after its full display, with the remaining 84 maintaining their initial perspective (see Table 10).

Table 10. Cross-tabulation between the variables "Ignore or view - Exposure effect" - Ad 4

Cross-tabulation: Ignore or View_Ad4 * View_complete _Ad4				
		View_Complete_Ad4		Total
		Yes	No	
Ignore or View_Ad4	View	5	6	11
	Ignore	7	84	91
Total		12	90	102

Behaviour with Regard to the Rhythm of the Music and its Consistency with the Advertisement

In other words, when asked about the pace of Advertisement 1 and whether it is appropriate for the product being advertised, the majority of respondents totally agreed with the statement.

As far as Advertisement 2 is concerned, we noticed a decline in the results, with an average of 3.29 and a median of 3, suggesting that the majority of respondents have a neutral opinion in relation to the dynamics of the product with the rhythm of the music.

In the case of Ad 3, this decay becomes even more evident, with a large proportion of those interpolated stating that they disagree with the choice of rhythmicity for the ad (mean = 2.41; median = 2).

Finally, as seen in Table 11, as ad 4 had no music, they were asked whether they would consider the absence of music appropriate for the advertised product, resulting in total disagreement with the statement (mean = 1.57; median = 1).

Table 11. Mean, median and standard deviation on the rhythm of the music in each advertisement

Rhythm of the music for each advertisement					
		Rhythm_Ad1	Rhythm_Ad2	Rhythm_Ad3	Absence_Ad4
N	Valid	102	102	102	102
	Missing	0	0	0	0
Mean		4,39	3,29	2,41	1,57
Median		5,00	3,00	2,00	1,00
Error Deviation		,798	1,279	1,410	1,076

With: 1 = Strongly Disagree; 2 = Disagree; 3 = Neither Agree Nor Disagree; 4 = Agree; 5 = Strongly Agree
Legend:
Rhythm_Ad = Rhythm of the music is suitable to the advertised product
Absence_Ad = Absence of music is adequate to the advertised product

It is also important to understand how respondents perceive the transmission of coherence to the campaign through the interaction between music and its visual elements. Thus, the data obtained are perceived in a similar way to those previously collected by the analysis of the rhythm, with Ad 1 presenting the highest degree of agreement (average = 4.27; median = 5). In this way, the individuals totally agree that the music in Ad 1 is well fit with the visual elements of the ad.

In relation to Advertisement 2, it was possible to ascertain a neutrality in the opinions gathered, reflected by an average of 3.2 and a median of 3.

For Ad 3, we again noticed decay in opinions, resulting in the majority of interpolates disagreeing with the coherence between the visual elements and the music in the ad (mean = 2.2; median = 2).

In the case of Ad 4, as seen in Table 12, individuals were asked whether the absence of music was congruent with the visual elements, resulting in an overall opinion of complete disagreement (mean = 1.75; median = 1).

Table 12. Mean, median and standard deviation on the coherence of the music in each advertisement

Consistency of the music in each advertisement					
		Coherent _Ad1	**Coherent _Ad2**	**Coherent _Ad3**	**Coherent _Ad4**
N	Valid	102	102	102	102
	Missing	0	0	0	0
Mean		4,27	3,20	2,20	1,75
Median		5,00	3,00	2,00	1,00
Error Deviation		,869	1,290	1,259	1,191

With: 1 = Strongly Disagree; 2 = Disagree; 3 = Neither Agree Nor Disagree; 4 = Agree; 5 = Strongly Agre
Legend:
Coherent_Ad = Music from the advert well fitting with the visual elements of the advert

Finally, we tried to understand if the choice of music in each advertisement was important so that the respondent would remain viewing it. As a consequence, as can be seen in Table 13, for Advertisement 1, respondents totally agreed with the statement, with a mean of 4.31 and a median of 5; for Advertisement 2, most respondents agree with the statement (mean = 3.71; median = 4); for Advertisement 3, neither agree nor disagree (mean = 2.88; median = 3) and for Advertisement 4 totally disagree (mean = 1.65; median = 1).

Table 13. Mean, median and standard deviation on the choice of music in each advertisement

Choice of music for each advertisement					
		Remain_Ad1	**Remain_Ad2**	**Remain_Ad3**	**Remain_Ad4**
N	Valid	102	102	102	102
	Missing	0	0	0	0
Mean		4,31	3,71	2,88	1,65
Median		5,00	4,00	3,00	1,00
Error Deviation		,808	1,215	1,402	1,191

With: 1 = Strongly Disagree; 2 = Disagree; 3 = Neither Agree Nor Disagree; 4 = Agree; 5 = Strongly Agree
Legend:
Stay_Ad = Choice of music is important to stay viewing the ad

However, it would be essential to ascertain whether the rhythm of a song is responsible for the attribution of coherence and the choice of the correct song for the respondent. For this purpose, an individual significance test was applied for the variables rhythm and choice of music. On the basis of Sig.< α = 0.05 one rejects H0: β 1/2/3=0, showing that in all advertisements both variables are less than 0.05. Thus, it can be concluded that both variables are statistically significant in relation to the rhythm of the music or, in the case of Ad 4, its absence.

Finally, in order to draw a conclusion about the variables analysed and if they have an impact on the attention gain, as well as on the way the receivers perceive an advertisement in its fullness, a T-test was performed for each of the variables in relation to the respondent's attention gain before the campaign. Thus, the results obtained for the first 3 ads allow us to state that all 3 elements analysed ("pace"; "remain" and "coherent"), are assumed as explanatory elements of the problem of this research, since Sig < $\alpha = 0.05$. In the case of Ad 4, the element "rhythm" is reformulated by the element "absence", however the previous conclusion is, also, obtained here.

Behaviour in the Face of Music Familiarity

As for the sample's behaviour regarding the advertisement and its connection with the familiarity of the music contained in it, it was initially sought to understand the popularity of each of the songs applied in the study. In this way, as can be seen in Table 14, for Advertisement 1, the music obtained a neutral result regarding its familiarity, with a mean of 2.69 and a median of 3. In Advertisement 2, respondents were able to recognize the music used, with the maximum degree of agreement, presenting a mean of 4.61 and a median of 5. Whereas Advertisement 3 used the music less familiar to respondents, with a mean of 2.33 and a median of 2.

Table 14. Mean, median and standard deviation on the familiarity of each song

Familiarity of each song				
		Familiar_Ad1	Familiar_Ad2	Familiar_Ad3
N	Valid	102	102	102
	Missing	0	0	0
Mean		2,69	4,61	2,33
Median		3,00	5,00	2,00
Error Deviation		1,342	,798	1,300

With: 1 = Strongly Disagree; 2 = Disagree; 3 = Neither Agree Nor Disagree; 4 = Agree; 5 = Strongly Agree
Legend:
Familiar_Ad = The music used in the advertisement is familiar to the respondente

Subsequently, it was analysed whether a popularly known and familiar song, when used in an advertisement, captivates the respondent. As result, it was obtained that 49% of the respondents totally agree with the statement; 27.5% agree; 15.7% neither agree nor disagree; 5.9% disagree and 2% totally disagree. It was also studied the individual's ability to recognize a very popular song during the showing of an advertisement, obtaining the following degrees of agreement: 60.8% totally agree; 29.4% agree and the remaining 9.8% neither agree nor disagree.

Finally, we analysed whether the use of a popularly known song arouses a gain of attention towards the campaign on the part of the receiver. Thus, we used the statistical analysis of variance through a test of individual significance for the element "Familiar". Based on Sig.$< \alpha = 0.05$ the H0 is rejected: β 1/2/3=0, showing that in all ads the variable is greater than 0.05. Thus, it will be possible to determine that familiarity is not statistically significant in relation to our problematic.

Behaviour Towards Personal Musical Tastes and the Advertisement

As we have already seen previously, the dominant personal musical tastes in our sample integrate Pop, EDM, Rock, Latin, and Hip-Hop / Rap, followed by the remaining genres in order of preference. However, according to the data collected, the music of the advertisements only integrates a small part of all possible genres determined by the respondents. For this study, it was then important to focus the attention on the songs used and the genres that transpire. In other words, it is important to analyse and cross-reference all respondents' answers and opinions that feature the following genres: EDM and Ad 1; Pop and Ad 2; Classical and Ad 3.

In this way, cross-reference tables were prepared between the music genres that individuals usually listen to and their perception of the suitability of the advertisement, to determine whether their musical preference interferes with their opinion as to whether the music has an appropriate rhythm for the advertisement and, consequently, whether it is suitable for it. As a result, it was obtained that, for Advertisement 1, around 61.5% of the respondents who habitually listen to electronic music totally agreed with the choice of music for the advertisement; 19.2% agreed with the statement and the remaining 19.2% neither agreed nor disagreed.

Regarding Advertisement 2, the data is more dispersed, with 21.2% of respondents who usually listen to Pop agreeing totally; 29.4% agreeing; 30.6% neither agree nor disagree; 9.4% disagree and the remaining 9.4% totally disagree.

With regard to Advertisement 3, for individuals who habitually listen to classical music, 28.6% totally agreed with the statement, followed by 14.28% agreeing, 28.6% giving a neutral opinion and the remaining 28.6% totally disagreed.

Finally, respondents were asked whether, during the exposure to an audiovisual advertisement, the attribution of a specific musical genre in it would help them identify with the campaign. For this question, the following levels of agreement were collected: 47.1% totally agree; 41.2% agree; 3.9% neither agree nor disagree; 2% disagree and the remaining 5.9% totally disagree.

Behaviour in the Face of Emotion and Memory

As noted earlier, most individuals tend to develop an ability to recognise and recall a popular song during the display of an advertisement. However, we also analysed whether the music of an advertisement helps the individual to remember the advertised brand or even the content of the campaign itself. Thus, in front of the question "Does the music present in an audiovisual ad help me remember the ad in the future", 66.7% of the sample agreed totally with the statement; 27.5% agreed, and the remaining 5.9% neither agreed nor disagreed.

About the rephrasing of the question before the advertised brand, 56.9% of the respondents totally agree that the music in an advertisement helps to recall a brand, followed by 31.4% agreeing; 7.8% neither agree nor disagree and 3.9% disagree.

Finally, it was analysed whether the use of inspirational music in an advertisement can transmit positive energy to the receiver. As a result, 70.6% of the respondents totally agreed; 25.5% agreed; 2% neither agreed nor disagreed and 2% totally disagreed.

Behaviour Regarding the Usefulness of Music in an Advertisement

In the approach to the effective usefulness of the music present in an advertisement on a network such as YouTube, it was necessary to confirm the degree of importance that respondents confer to the music present in advertisements on social networks. Thus, it was possible to ascertain that more than half of the respondents (55%), totally agree with the relevance of music in advertisements on social networks, with the remaining 37.3% agreeing, 3.9% neither disagree nor agree and 3.9% disagree.

At the same time, when asked whether they would more easily ignore an advertisement without music, the overwhelming majority of respondents totally agreed with the statement (80.4%), 9.8% agreed, 7.8% neither agreed nor disagreed and 2% disagreed. Interestingly, these data can be compared with the questions on the absence of music in section 4.1 of the paper.

Regarding the statement "Music adds value to an ad. Its usefulness is unquestionable", once again, a colossal majority of respondents totally agreed with the statement (78.4%), with the remaining 13.7% agreeing and 7.8% neither agreeing nor disagreeing.

Finally, the study assessed the degree to which individuals would consider the ad complete on an individual basis, with each individual's music being the only factor in the divergence of opinions. For Ad 1, 49% of respondents totally agreed that the ad is complete; 31.4% agreed; 17.6 took a neutral position and 2% disagreed.

In turn, in the data collected from Ad 2, 21.6% of the interpolated agreed totally with the statement; 33.3% agreed; 31.4% neither agreed nor disagreed; 9.8% disagreed and 3.9% totally disagreed.

As far as Advertisement 3 is concerned, the advertisement is complete according to the following degrees of agreement: 9.8% totally agreed with the statement, followed by 19.6% agreeing, 27.5% giving a neutral opinion, 31.4% disagreeing and the remaining 11.8% totally disagreeing.

Finally, and presenting data almost contrary to Ad 1, for Ad 4, the individuals considered the content of the ad according to the following percentages: 9.8% totally agree; 3.9% agreed; 13.7% neither agreed nor disagreed; 15.7% disagreed and the remaining absolute majority totally disagreed (56.9%).

Validation of the Hypotheses

For H1 (The complete viewing of the ad captures more of the user's attention when compared to the reduced exposure of the ad), the hypothesis is confirmed when the ad is perceived as complete by the individual, since for Advertisement 1 and 2 a positive permanence ratio was obtained while viewing the ad. However, for the case of Advertisement 3 and 4, H1 is not supported, based on a high number of individuals ignoring the ad, in any of its versions.

For H2 (A song with faster rhythm and energy captures more viewer's attention during the advertisement), the statement is corroborated, since songs with higher rhythm and energy, according to Table 3, are the ones that present a higher degree of coherence among respondents. Furthermore, rhythm and coherence presented a statistically significant value, with Sig $< \alpha = 0.05$.

Regarding H3 (A popularly known song accompanying the advertisement arouses a greater degree of attention), the hypothesis is not supported. Despite the data collected meeting the proposed familiarity levels, familiarity does not seem to affect the level of attention and acceptance of the campaign. Furthermore, familiarity does not have a statistically significant value, with Sig $> \alpha = 0.05$.

In what concerns H4 (A music that meets the musical tastes of each consumer captures more their interest for the campaign and, consequently, the click), it can be seen that the musical genre influences the attention to the campaign. Here, although we verify a small dispersion of data from ad 2, we can see that individuals who have the habit of listening to a certain music genre tend to agree that the music genre chosen for the ad is well fitted.

For H5 (Songs with happier melodies enhance future recall of the ad), the hypothesis is confirmed. All songs that presented a high level of joy and energy, tended to present a high agreement with the attribution of the song to the ad. Furthermore, it was confirmed that the use of a song in an ad helps the recall of the ad and the advertised brand.

Finally, as for H6 (An ad without music does not fulfil its click-through objective, as it will be more easily ignored), the hypothesis was confirmed. When we compare the opinions collected in front of the ads and if they are complete, it becomes evident that Ad 4 (the only ad without music) presents the lowest percentage of acceptance among respondents. Furthermore, according to the overwhelming majority, an ad on a social network should have music in its composition, corroborating that a campaign without music will be more easily ignored. Thus, its usefulness is, according to the same source, unquestionable. Thus, in summary form, we can summarise the validation of the hypotheses in the following table:

Table 15. Validation of hypotheses (Summary)

H1: Full ad viewing captures more user attention in the face of reduced ad exposure (Shon, Shin, Hwang, & Lee, 2020; Puccinelli, Wilcox, & Grewal, 2015).	Partially confirmed
H2: Music with a faster rhythm and energy captures more of the viewer's attention during the advertisement (Dessarte, 2018; Galan, 2009).	Confirmed
H3: A popularly known song accompanying the advertisement elicits a higher degree of attention (Raja, Anand and Kumar, 2018).	Not confirmed
H4: A song that meets the musical tastes of each consumer captures more of their interest for the campaign and consequently the click (Galan, 2009; Raja, Anand and Kumar, 2018).	Confirmed
H5: Songs with more upbeat melodies enhance ad recall in the future (Raja, Anand and Kumar, 2018).	Confirmed
H6: An ad without music does not meet its click-through target as it will be more easily ignored (Dessarte, 2018).	Confirmed

DISCUSSION

Based on the literature review, one can notice a perspective of agreement in most theories, as evidenced by Belanche, Flávian and Pérez-Rueda (2017): music produces a positive effect on the individual's attention before an advertising campaign.

Taking into account the first specific objective of the study (To understand if the complete viewing of an ad (30 seconds) can cause a positive impact on the exposure of a shorter ad (5 seconds)), the results obtained partially coincide with the research results of Shon, Shin, Hwang and Lee (2020), in which they state that 5 seconds of mandatory exposure are not effective in capturing the desired consumer attention on YouTube. We note that when the ad is perceived as complete by the viewer, they tend to increase their attention according to the exposure time before the campaign. In other words, and, obtaining results identical to those of Puccinelli, Wilcox and Grewal's theory (2015), the acceptance of an advertisement will be greater the longer the time of exposure to it. However, if the advertisement is not perceived as competent by the receiver (as is the case of Advertisement 3 and 4), the exposure time will not be a significant factor for the study, since it will be perceived as boring, favouring irritability, the only negative aspect in Dehghami et al.

Taking into account the conception of Herget, Schramm and Breves (2018), the perfect ad or "perfect fit", is obtained with the exquisite combination of the rhythm of the music and the visual elements of the ad. Thus, to understand if the fact that a song is coherent (or not) with what is being visually presented in the ad, has an impact on the user's action and influences him/her to click on the ad (Objective 2), it was possible to perceive that songs with greater rhythm and energy present a greater degree of coherence among the results obtained. In other words, and line with Ansani and collaborators (2020), when the sound and all the components of the campaign coincide with what is advertised, a greater degree of attention to the campaign is obtained. In this study, it was confirmed that Ad 1 and 2 produced the highest degree of interest in the campaign, dictated by the rhythm of the songs, fitting in the closest possible "perfect fit" perceived by the respondents. Now, these results can be explained by the theory of Sunaga, Meng and Zhuang (2020), where it is evidenced that songs with higher rhythm, generate a higher degree of enthusiasm and energy for the campaign, aspects that fit completely with the visual elements of the proposed smartphone ad. On the other hand, and still, according to the same authors, the use of slower and more emotional music tends to generate a feeling of quietness that will only produce an effect in an emotional campaign. Here, once again, the results are in line with the literature reviewed, evidenced by the lack of interest of individuals towards Advertisement 3, contained in its composition, classical and slow music.

Addressing the third objective (to understand if the use of popular and familiar music can increase the attention towards the advertisement), although the literature suggests that audiovisual advertisements that involve familiar music in their composition increase their effectiveness (Binet, Müllensiefen & Edward, 2013), the results were not in agreement with the literature. Through the data collected with the four ads, it was noted an evident tendency for the user to develop a desire to remain to view them, for a period longer than 5 seconds. However, this factor will not be driven by the familiarity with the music, but rather explained by how respondents perceived the ad to be complete. Thus, contrary to the studies of Zoghaib (2019), the use of a globally known music does not seem to arouse increased interest in the user, explained by the low levels of attention-grabbing that were obtained for Ad 2 in its full version, which undoubtedly, contains in its composition the most familiar and well-known music. Unlike the studies of Binet, Müllensiefen and Edward (2013), the familiarity of the music did not register any statistical significance during their test of variance to prove its relevance to the study.

About the fourth objective of the study (to understand whether the individual's musical tastes can induce the click on the advertisement), according to most literature, personal tastes regarding the music in an advertisement can manipulate the choice of the advertised item (Gorn, 1982; Iyer & Aggarwal, 2019; Galan, 2009). The data collected shows direct agreement with the literature, with respondents preferring advertisements that meet their musical tastes. Through the cross-tabulation between the musical genres that respondents have the habit of listening to and how they perceive the music in the advertisement as being appropriate, it is noted that most of them agree with the use of music and the dynamics that it dictates in the advertisement. Thus, as stated by Sousa (2006), the acceptance of the advertising message will be greater when it resembles the interests and personality of the receiver. Nevertheless, from Advertisement 2 we were able to perceive a small dispersion in the data. For example, for the 3 ads with music, the data obtained were in total agreement only for the first ad (EDM), with a lower degree of agreement for the remaining ads (Pop and Ad 2; Classical and Ad 3). Although the latter still shows a positive statistical significance, the literature by Vermeulen and Beukeboom (2016) can help us understand why the results show a small deviation. It is because, according to the same authors, the recipients of an advertising message can be persuaded with the use of a sound consistent with their taste, but only for products with low involvement in the purchase process. In other words, the influence of personal musical tastes in an advertisement may not be so high when an expensive product is presented, as is the case with our advertisements (smartphones).

For the fifth objective (To understand the different reactions that may be caused by the use of different songs in the same advertisement, namely in terms of memory and emotion), it was possible to determine that the songs that develop a greater emotional behaviour, manage to transmit a greater degree of attention to the campaign. Furthermore, the findings, fruit of the study developed, came to meet the literature, suggesting that the emotion established by a song can stimulate the formation of attitudes and, concomitantly, affect the mood and perception of the individual concerning an audiovisual campaign (Craton, & Geoffrey, 2011). It was also possible to perceive that the positive impact that a song can bring to the receiver of the advertisement, tends to help them remember the advertisement and brand advertised. In this sense, there is a certain consonance with the conception of Abolhasani, Oakes and Oakes (2017), which suggests that the memorization of a song from an advertisement, may originate a future direct association between the advertised brand and the content of the campaign. On the other hand, on the topic of emotion, Alpert and Alpert (1989), argue that the use of a sad sound may be able to produce effects of influence in the purchase decision process, which does not occur in this study. On the contrary, the less cheerful music (used in ad 3), was the music that least fitted the ad, according to the data collected.

Finally, it is important to understand the usefulness of music in a campaign and therefore consider if it brings any impact to the advertising campaign (objective 6). Contrary to Miniard, Lord and Dickson (1988), who considered music as a non-essential element for a campaign, the research made it clear that the importance and usefulness of music in an advertisement is unquestionable. Here, there was an absolute ineffectiveness of Ad 4 and its consequent inability to capture the receiver's attention. Although the studies by Valdellós (2011) and Ansani and collaborators (2020) are not aimed at YouTube, like much of the collected literature, it was possible to determine an agreement with their conceptions, and it can be highlighted that an ad without music will be more easily ignored by the YouTube user, not fulfilling its objective of click and subsequent conversion.

CONCLUSION

This research had as main objective to understand the importance of music in the advertising context. More specifically, it was intended to understand which intrinsic variants of music would stimulate the attention of the receiver of the advertising message, namely its rhythm and coherence; familiarity of music; personal musical tastes; emotion and memory, as well as to study its effective usefulness. Now, as we are in the digital age, it was considered relevant to apply this study to one of the largest social networks in the world, YouTube, and, at the same time, to understand if the aforementioned phenomena can induce the individual's click before the advertising campaign.

From the literature review carried out, it was possible to identify a constant mutation on the perspectives of the defined theme, starting with the more traditional media to its digital revolution. Initially, it was found that the interaction of music and the advertising medium was seen as an additional element and not very fundamental, although there were already theories that pointed to the contrary. Nowadays, in the face of the enormous overload of information that we are subjected to, it is possible to affirm that this interaction is essential to capture the user's attention.

In this sense, from the results achieved, we can see that, in general, music in an audiovisual campaign is undoubtedly useful and indispensable. However, in order to draw a conclusion about the problem and, therefore, if music can be significantly responsible for increasing the individual's attention, we must take into account the elements analysed. Thus, it was possible to confirm that the coherence between the rhythm of the music and the advertisement constitute the most important point before the theme, since these results are the most concordant of the study. Similarly, it was possible to highlight the emotional relevance that can be transmitted by the music itself and its importance in the subsequent recall of the ad. Also, it was possible to determine that the individual's personal musical tastes tend to result in a positive impact on gaining attention towards the campaign. On the other hand, it was not possible to evidence that the familiarity of a song can influence the individual's attention to the campaign in a significant way.

From a perspective outside music, and because this study is applied to YouTube, it was possible to confirm that the individual's attention to an ad will be greater the longer he/she is exposed to it. However, this factor only occurs when the ad is perceived by the receiver, at least, as reasonably complete.

Thus, taking into account what has been written and the whole content of this study, the research carried out seeks to give a contribution to all audiovisual advertising companies and to alert to the importance of music in advertisements. In this sense, it was concluded that music copyright budgets can be optimized to the maximum when they do not prioritize the use of a popular song (usually more expensive), but rather its coherence to the ad.

Study Limitations

Firstly, it is important to bear in mind that the literature analysed does not, for the most part, directly address the YouTube social network, and so the cross-referencing of data presented in the discussion may be refined as new studies are developed on the topic.

Also, in the course of the study, it was not possible to confirm whether the respondents actually viewed all the videos to the end and, therefore, provided their opinion properly grounded. Also in this parameter, we can say that the sample collected was relatively low. With 102 responses, we were not able to obtain a representation of the population.

Finally, it should be noted that the scales used for the questionnaire were not validated for the Portuguese population, so there may be a small discrepancy in the reliability indices.

Future Directions

As suggestions for future research, we recommend a possible replication of the quantitative study crossed with a qualitative eye-tracking study to understand in which direction the advert receiver directs his/her gaze and, consequently, to understand whether or not there is an increase in attention while viewing the advert.

More variation in advertisements and music genres is also suggested, to make the research more dynamic for the respondent and more complete for the researcher.

On the other hand, one could try to draw a comparison with other music-related social networks to affirm the role they could play in similar advertising contexts. Finally, it also seems opportune to elaborate on one or more interview(s) with communication and audiovisual advertisement development companies, to understand the degree of importance that they attribute to the music/soundtrack that composes their advertisements.

REFERENCES

Abolhasani, M., Oakes, S., & Oakes, H. (2017). Music in advertising and consumer identity: The search for Heideggerian authenticity. *Marketing Theory*, *17*(4), 473–490. doi:10.1177/1470593117692021

Alpert, J. I., & Alpert, M. I. (1989). Background music as an influence in consumer mood and advertising responses. *Advances in Consumer Research. Association for Consumer Research (U. S.)*, *16*, 485–491.

Ansani, A., Marini, M., D'Errico, F., & Poggi, I. (2020). How soundtracks shape what we see. Analyzing the influence of music on visual scenes through self-assessment, eye-tracking, and pupillometry. *Frontiers in Psychology*, *11*, 22–42. doi:10.3389/fpsyg.2020.02242 PMID:33117211

Arantes, M., Figueiredo, F., Almeida, J. M., & Prates, R. O. (2018, October). An investigation of user actions and experiences when exposed to *YouTube* video ads. *Proceedings of the 24th Brazilian Symposium on Multimedia and the Web*, 283-290. 10.1145/3243082.3243107

Barreda, A. A., Bilgihan, A., Nusair, K., & Okumus, F. (2015). Generating brand awareness in online social networks. *Computers in Human Behavior*, *50*, 600–609. doi:10.1016/j.chb.2015.03.023

Belanche, D., Flavián, C., & Pérez-Rueda, A. (2017). Understanding interactive online advertising: Congruence and product involvement in highly and lowly arousing, skippable video ads. *Journal of Interactive Marketing*, *37*, 75–88. doi:10.1016/j.intmar.2016.06.004

BettencourtL. A. (2017). https://www.ama.org/2017/12/20/are-you-managing-brand-equity-incorrectly/

Binet, L., Müllensiefen, D., & Edwards, P. (2013). The power of music. *Music in Advertising*, 10-13.

Biographon. (2019). https://biographon.com/youtube-stats/

Bourdieu, P. (2010). *Distinction: A Social Critique of the Judgment of Taste*. Routledge.

Buffer. (2019). https://buffer.com/state-of-social-2019

CaligiuriC.JonesB. (2019). https://www.thinkwithgoogle.com/intl/pt-br/estrategias-de-marketing/video/uncios-em-video-como-e-quando-devemos-customizar/

Cisco. (2021). https://www.cisco.com/c/en/us/solutions/collateral/executive-perspectives/annual-internet-report/white-paper-c11-741490.html

Clancey, M. (1994). The television audience examined. *Journal of Advertising Research*, *4*(34), 1–10.

Craton, L. G., & Geoffrey, P. L. (2011). Attitude toward Advertising Music: Potential Pitfalls in Commercials. *Journal of Consumer Marketing*, *28*(6), 396–411. doi:10.1108/07363761111165912

Dehghani, M., Niaki, M. K., Ramezani, I., & Sali, R. (2016). Evaluating the influence of YouTube advertising for attraction of young customers. *Computers in Human Behavior*, *59*, 165–172. doi:10.1016/j.chb.2016.01.037

Dessart, L. (2018). Do ads that tell a story always perform better? The role of character identification and character type in storytelling ads. *International Journal of Research in Marketing*, *35*(2), 289–304. doi:10.1016/j.ijresmar.2017.12.009

Digital Results. (2020). *Facebook Ads: The definitive guide*. E-book.

eMarketer. (2016). https://www.emarketer.com/Article/US-Digital-Display-Ad-Spending-Surpass-Search-Ad-Spending-2016/1013442

eMarketer. (2020). https://www.emarketer.com/chart/241048/device-share-of-youtube-video-views-among-us-youtube-viewers-q4-2019-q3-2020-of-total

Endo, A. C. B., & Roque, M. A. B. (2017). Attention, memory and perception: a conceptual analysis of Neuropsychology applied to advertising and its influence on consumer behaviour. *Revista Brasileira de Ciências da Comunicação, 40*(1), 77-96. doi:10.1590/1809-5844201715

Firat, D. (2019). YouTube adv ube advertising v tising value and its eff alue and its effects on pur ects on purchase intention chase intenti. *Journal of Global Bussiness Insights*, *4*(2), 141–155.

Freitas, C. (2017). https://www.jm-madeira.pt/opinioes/ver/741/O_poder_da_familiaridade_na_preferencia_musical

Frith, S. (2003). Music and Everyday Life. In M. Clayton, T. Herbert, & R. Middleton (Eds.), *The Cultural Study of Music* (pp. 92–101). Routledge.

Gabriel, M. (2010). *Marketing na Era Digital: Conceptos*. Platforms e Estratégias.

Galan, J. P. (2009). Music and responses to advertising: The effects of musical characteristics, likeability and congruency. *Recherche et Applications en Marketing*, *24*(4), 3–22. doi.org/10.1177/2051570902400401

Global Webindex. (2019). https://www.gwi.com/reports/social-2019

GoogleAdsHelp. (2021a). https://support.google.com/google-ads/answer/6052202?hl=pt-BR

GoogleAdsHelp. (2021b). https://support.google.com/google-ads/answer/9035882

GoogleAdsHelp. (2021c). https://support.google.com/youtube/answer/2375464?hl=pt

GoogleAdsHelp. (2021d). https://support.google.com/youtube/answer/2454017?hl=pt

Gorn, G. J. (1982). The effects of music in advertising on choice behavior: A classical conditioning approach. *Journal of Marketing*, *46*(1), 94–101. doi.org/10.1177/002224298204600109

Guzmán, J. R. S. (1993). *Teoría de la publicidad*. Tecnos.

Hahn, M., & Hwang, I. (1999). Effects of tempo and familiarity of background music on message processing in TV advertising: A resource-matching perspective. *Psychology and Marketing*, *16*(8), 659–675. doi.org/10.1002/(SICI)1520-6793

Handley, L. (2017). https://www.cnbc.com/2017/02/16/sixty-five-percent-of-people-skip-online-video-ads-heres-what-to-do.html

Hecker, S. (1984). Music for advertising effect. *Psychology and Marketing*, *1*(3-4), 3–8. doi.org/10.1002/mar.4220010303

Herget, A. K., Schramm, H., & Breves, P. (2018). Development and testing of an instrument to determine Musical Fit in audio-visual advertising. *Musicae Scientiae*, *22*(3), 362–376. doi.org/10.1177/1029864917731127

IPSOS Connect, . (2020). *Ipsos/Google Advertising Attention Research:Report*. E-book.

Iyer, S. P., & Aggarwal, A. (2019). The Role of Music in Brand Recall - A Study of Select Two-Wheeler Advertisements. *International Journal of Management, Technology And Engineering*, *4*(1), 481–503.

Julio, K. B. (2019). https://www.meioemensagem.com.br/home/midia/2019/10/08/publicidade-digital-deve-se-igualar-a-tradicional-ate-2023.html

Liu, F., Li, J., Mizerski, D., & Soh, H. (2012). Self-congruity, brand attitude, and brand loyalty: A study on luxury brands. *European Journal of Marketing*, *46*(7), 922–937. doi.org/10.1108/03090561211230098

MathiasL. (2016). https://mindminers.com/blog/pesquisa-qualitativa-quantitativa/

Miniard, P. W., Dickson, P. R., & Lord, K. R. (1988). Some central and peripheral thoughts on the routes to persuasion. *ACR North American Advances, 13*(3), 237-249. doi:10.1016/0167-8116(96)00005-5

Miranda, M. (2018). Youtube, Facebook and Instagram Ads: Get Traffic, Leads and Increase Revenue for your Business. In Marketing Digital & E-Commerce #2. Viseu: Psicosoma.

Needle, F. (2021). https://blog.hubspot.com/blog/tabid/6307/bid/23865/13-mind-bending-social-media-marketing-statistics.aspx

Ng, P. Y. (2011). *An analysis on the effectiveness of music in advertising towards consumer's buying behavior* (Doctoral dissertation). UTAR.

Ochoa, C. (2015). https://www.netquest.com/blog/br/blog/br/amostra-conveniencia

PareraE. (2020). https://postcron.com/pt/blog/como-criar-anuncios-no-facebook-ads/

Priberam. (2021). https://dicionario.priberam.org/informatividade

Puccinelli, N. M., Wilcox, K., & Grewal, D. (2015). Consumers' response to commercials: When the energy level in the commercial conflicts with the media context. *Journal of Marketing*, *79*(2), 1–18. doi.org/10.1509/jm.13.0026

Raja, M. W., Anand, S., & Kumar, I. (2018). Multi-item scale construction to measure consumers' attitude toward advertising music. *Journal of Marketing Communications*, *26*(3), 314–327. doi.org/10.1080/13527266.2018.1471615

Rentfrow, P. J., Goldberg, L. R., Stillwell, D. J., Kosinski, M., Gosling, S. D., & Levitin, D. J. (2012). The song remains the same: A replication and extension of the MUSIC model. *Music Perception*, *30*(2), 161–185. doi.org/10.1525/mp.2012.30.2.161

Russell, C. A. (2007). *Advertainment: Fusing Advertising and Entertainment*. University of Michigan: Yaffe Center.

Sahni, N. S., Stanford, G. S. B., & Zhang, C. (2020). *Are Consumers Averse to Sponsored Messages? The Role of Search Advertising in Information Discovery* [Master's thesis]. Standford University Graduate School of Bussiness.

Shon, M., Shin, J., Hwang, J., & Lee, D. (2020). Free contents vs. inconvenience costs: Two faces of online video advertising. *Telematics and Informatics*, 101476. doi:10.1016/j.tele.2020.101476

Simmonds, L., Bellman, S., Kennedy, R., Nenycz-Thiel, M., & Bogomolova, S. (2020). Moderating effects of prior brand usage on visual attention to video advertising and recall: An eye-tracking investigation. *Journal of Business Research*, *111*, 241–248. doi.org/10.1016/j.jbusres.2019.02.062

SproutSocial. (2021). https://sproutsocial.com/insights/social-media-statistics/

Statista. (2021a). https://www.statista.com/outlook/216/147/digital-advertising/portugal

Statista. (2021b). https://www.statista.com/statistics/617136/digital-population-worldwide/

Statista. (2021c). https://www.statista.com/statistics/272014/global-social-networks-ranked-by-number-of-users/

Stephen, A. T., & Toubia, O. (2010). Deriving value from social commerce networks. *JMR, Journal of Marketing Research*, *47*(2), 215–228. doi.org/10.1509/jmkr.47.2.215

Sunaga, T., Meng, Y., & Zhuang, X. (2020). Interactions between instrumental timbre and consumers' regulatory focus. *Journal of Business Research*, *121*, 1–12. doi.org/10.1016/j.jbusres.2020.08.014

Tandyonomanu, D. (2018, October). Ads on *YouTube*: Skip or Watch? In *1st International Conference on Social Sciences (ICSS 2018)*. Atlantis Press. .66 doi:10.2991/icss-18.2018

That Ad Music Blog. (2014). http://aquelamusicanuncio.blogspot.com/?view=classic

ThinkWithGoogle. (2015). https://www.thinkwithgoogle.com/intl/pt-br/estrategias-de-marketing/video/youtube-insights-segmentacao/

Tunebat. (2021a). https://tunebat.com/Info/Closer-Lemaitre-Jennie-A-/3h5IIiL9vK5aR0DZO6jD7D

Tunebat. (2021b). https://tunebat.com/Info/Don-t-Start-Now-Dua-Lipa/6WrI0LAC5M1Rw2MnX2ZvEg

Tunebat. (2021c.) https://tunebat.com/Info/Nuvole-Bianche-Rousseau/7EoBV6PM8qlIkoDms2RthA

Valdellós, A. M. S. (2011). La música en los comerciales televisivos: el arma secreto. *Sinfonía Virtual: Revista de Música Clásica y Reflexión Musical, 1*(18), 3.

Vasconcellos-Guedes, L., & Guedes, L. F. (2007). *E-surveys: Advantages and limitations of electronic questionnaires via internet in the context of scientific research.* X SemeAd-Seminário em Administração FEA/USP.

Vermeulen, I., & Beukeboom, C. J. (2016). Effects of music in advertising: Three experiments replicating single-exposure musical conditioning of consumer choice (Gorn 1982) in an individual setting. *Journal of Advertising, 45*(1), 53-61. .1088809 doi:10.1080/00913367.2015

Wikipedia. (2020). https://pt.wikipedia.org/wiki/Jingle

Wikipedia. (2021). https://en.wikipedia.org/wiki/Music_in_advertising

Xavier, A. (2015). *Storytelling: stories that leave a mark* (5th ed.). Best Seller.

Yang, K. C., Huang, C. H., Yang, C., & Yang, S. Y. (2017). Consumer attitudes toward online video advertisement: YouTube as a platform. *Kybernetes, 46*(5), 840–853. doi.org/10.1108/K-03-2016-0038

YouTubeHelp. (2021a). https://support.google.com/youtube/answer/188038?hl=pt

YouTubeHelp. (2021b). https://support.google.com/youtube/answer/2454017?hl=pt

YouTubeOfficialBlog. (2021). https://blog.youtube/press/

Zajonc, R. B. (1980). Feeling and thinking: Preferences need no inferences. *The American Psychologist, 35*(2), 151. doi.org/10.1037/0003-066X.35.2.151

Zander, M. F., Apaolaza-Ibáñez, V., & Hartmann, P. (2010). Music in Advertising: Effects on Brand and Endorser Perception. Advances in Advertising Research, 1, 127-140.

Zhou, R., Khemmarat, S., Gao, L., Wan, J., & Zhang, J. (2016). How YouTube videos are discovered and its impact on video views. *Multimedia Tools and Applications, 75*(10), 6035–6058. doi.org/10.1007/s11042-015-3206-0

Zoghaib, A. (2019). Typology of advertising music components and consumers' responses to the brand. [English Edition]. *Recherche et Applications en Marketing, 34*(3), 47–82. doi.org/10.1177/2051570718828893

KEY TERMS AND DEFINITIONS

Advertising Entertainment: Advertising with a higher degree of attention that encourages the receiver to consume advertising content more frequently.

Brand Awareness: Consumers' ability to recognize a brand and their consequent ability to remember it spontaneously.

Informativeness: Quality of what is informative.

Purchase Intention: Tendency to buy a product and/or service when that same item is advertised and one needs to obtain it.

ENDNOTES

[1] Reduced version of the adverts (5 seconds):
Announcement 1: https://youtu.be/rsa_t-Xxcx4
Announcement 2: https://youtu.be/3khttWJIzw8
Ad 3: https://youtu.be/btcc6RCw6ko
Ad 4: https://youtu.be/to5_YLPzJm0
Full version of the ads (30 seconds):
Ad 1: https://youtu.be/i_a2gsvmcFg
Announcement 2: https://youtu.be/2VVHnj8APqQ
Ad 3: https://youtu.be/XGA-t5roUok
Ad 4: https://youtu.be/Zx7T_-sKxOE

APPENDIX

Table 16

Category	Question	Authors	Cronbach's α
Rhythm and coherence	"The music of 'Ad x' and, its rhythm, are appropriate to the product being advertised."	Dessarte (2018); Galan (2009).	.92 .70
	"The choice of music for 'Ad X' is important for me to remain viewing it."	Galan (2009)	.78
	"The music in 'Ad X' seems to me to fit well with the visual elements of it."	Dessarte (2018) Galan (2009)	.93 .864
Familiarity	"The music from 'Ad X' is familiar to me"	Raja, Anand and Kumar (2018)	.71
	"The use of a popularly known song captivates me relative to the advertisement."	Raja, Anand and Kumar (2018)	.74
	"I can easily recognise a fairly popular song during the display of an advertisement."	Raja, Anand and Kumar (2018)	.71
Personal tastes	"What genre of music do you usually listen to?"	Galan (2009)	n/a
	"During the exposure of an audio-visual advertisement, a specific style of music helps me identify with the advertised item just by listening to the music."	Raja, Anand and Kumar (2018)	.66
Emotion and memory	"The use of inspirational music in an ad gives me positive energy."	Raja, Anand and Kumar (2018)	.64
	"The music present in an audio-visual ad helps me to remember the ad in the future."	Raja, Anand and Kumar (2018)	.74
	"The music in an ad helps me remember a brand."	Raja, Anand and Kumar (2018)	.69
Utility	"The 'X Advertisement' appears to be complete."	Dessarte (2018)	.81
	"Music in social media ads plays a key role in spreading your message."	Raja, Anand and Kumar (2018)	.69
	"I ignore an ad more easily without music."	Raja, Anand and Kumar (2018)	.70
	"Music adds value to an advertisement. Its usefulness is unquestionable"	Raja, Anand and Kumar (2018)	.78
Skip or view (exposure effect)	"Given 'Ad X', would you continue to view the same?"	Shon, Shin, Hwang, & Lee (2020);	n/a
	"Did the full viewing of 'Ad X' capture your attention in any way compared to its previous 5-second version?"	Puccinelli, Wilcox, & Grewal (2015)	n/a

n/a = not applicable

Chapter 10
Newspaper Digital Content During and After the Silly Season:
A Case Study

Ana Maria Lima
https://orcid.org/0000-0003-0827-392X
ISCAP, Instituto Politécnico do Porto, Portugal

Sandrina Francisca Teixeira
https://orcid.org/0000-0002-5859-0002
ISCAP, Instituto Politécnico do Porto, Portugal

Belem Barbosa
https://orcid.org/0000-0002-4057-360X
School of Economics and Management, University of Porto, Portugal

ABSTRACT

This chapter aims to understand the options for distributing newspaper content online and offline. By adopting a case study approach, it analyses the content strategy of one of the most widely read newspapers in Portugal, Jornal de Notícias (JN). Several data sources were used, including an interview with the director and a survey with 570 readers. JN shared a much larger number of news in the paper version. It is a newspaper focused on difficult news; however, it combines this strategy with the use of infotainment, especially in the online version. This chapter highlights the importance of aligning the content strategy and the preferences of readers.

DOI: 10.4018/978-1-6684-5538-8.ch010

INTRODUCTION

The days when a newspaper had only two distribution options, kiosk or bookstore sales and subscription sales, are long gone. Today, distribution options include online portal, social networks, email, e-paper (Bentley et al., 2019), just to name a few. Additionally, providing online news recommendations to users has become an important trend for news platforms, allowing them to attract more readers (Barbosa & Carvalho, 2021; Liu et al., 2019).

Considering the opportunities that are posed by digital media, this article analyses the multichannel distribution strategy of newspapers, addressing aspects such as the distribution strategy of online and offline content and the adequacy in relation to the public's information needs. The aim of this study is to understand the options for distributing content between online and offline, exploring their motivations and impacts.

A case study on one of Portuguese newspapers with a greater focus on digital is included: Jornal de Notícias (JN). JN is one of the oldest newspapers in Portugal. Apparently, it found how to position itself online and continue to be one of the most sought-after news provider online. Cherian (2015) recommends analyzing success cases in new digital business models in order to identify effective ways to integrate digital strategies. Likewise, research should also include success cases, tackling the challenges and seizing the opportunities of the digital economy (Urbach & Röglinger, 2018).

This article provides several contributions. To the authors' best of knowledge, this is the first analysis of a newspaper's content distribution strategy comparing online and offline. The article provides particularly interesting empirical evidence, which was further explored by integrating the views of both a newspaper manager and its readers. Overall, this article provides valuable insights on the integration of digital channels by newspapers and its effectiveness, which are expected to be of interest of both academics, students, and practitioners in the news sector.

The main aim of this study is to understand the options for distributing content between online and offline, exploring their motivations and impacts. As such, and considering the gaps found in the literature, five specific objectives were defined:

1. **Compare distribution:** to compare newspapers' distribution strategy online and offline.
2. **Explore:** to explore the reasons for providing different content for the paper and online versions of newspapers.
3. **Strategies:** to analyze the importance of Social Network Sites in newspapers' news development strategies.
4. **Compare success:** to compare the success of different types of news online and offline.
5. **Monetization:** to understand the monetization of different news content.

BACKGROUND

New Digital Business Models

Newspapers' traditional business model is dual: on the one hand, they raised revenue through the sale of the newspaper itself, including subscriptions, however, the main source of revenue for newspapers in recent decades has clearly been advertising.

Entering the digital context, newspapers initially tried to replicate the traditional business model, adopting the Pay Per View strategy. However, the Internet soon revealed itself as a facilitator of access to all kinds of information, in a simple, fast and, above all, free mode (Skjeret et al., 2019).

The newspaper industry is in a revenue crisis and the advertising side of the news business has virtually collapsed as a result of global competition with technology platforms (Olsen et al., 2021). This situation worsens considerably when we move towards an increasingly mobile reading (Graves & Simon, 2019).

As such, newspapers have encountered dilemmas of survival. Cook and Sirkkunen (2013) suggest that there are three possible answers: keep the sources of profit and business models intact, direct the offline business model to the online world, and find new ways to obtain financing.

Opportunities in Social Network Sites

Social network sites (SNS) have introduced a series of challenges when it comes to news gathering, verification, production, reporting, and dissemination (Zubiaga et al., 2019). According to Neuberger et al. (2018), there is still an untapped potential in SNS by increasingly meet the interests of advertisers. Thus, many newspapers and magazines included SNS in their information dissemination services, and the importance of users to actively participate in content production (Sanseverino & Santos, 2021). SNS also allow the growth of citizens' opinion, reflected in their interactions with reference newspapers' posts (Guimarães et al., 2019). Still, it is important to note that with the growth of SNS, the power went from those who had more influence to those who have more audiences. It seems to be increasingly common that the search for information, especially on mobile devices, is done on SNS (Zúñiga et al., 2017).

Online vs. Offline

Finding the right means for distributing content is an issue that concerns all media companies, not only regarding the volume of visits but also for the greater profitability that may arise. According to a recent study by Marktest, more than 5 million Portuguese people read news online, especially in the areas of communication, information, and leisure (Marktest, 2018). This figure represents about 90% of the total national inhabitants who have access to the Internet.

This leads some authors (e.g., Karimi & Walter, 2015) to suggest that priority should be given to the creation of original and appealing content on digital platforms that readers prefer. Karimi and Walter (2015) argue that this strategy results in a consistent increase in digital audiences, better financial efficiency, and more revenue opportunities.

Nowadays, the distribution of online content allows easier and immediate access to the audience (Powers & Vera-Zambrano, 2019). An exclusive no longer has the same power it once did, and within minutes, news proliferate through different channels. For the user, what is most relevant is the medium that makes the information arrive first and not necessarily the one that produced the news in the first place. As a result, more and more specialized media are trying to differentiate by the type of content, ideology, and style (Powers & Benson, 2014).

Tandoc and Vos (2015) introduced the concept of the "marketer journalist", the one who writes the news, but who also thinks and prepares it for SNS and other online media and react to readers' immediate response. Hendrickx (2021) brings out the concept of "social journalism" that demonstrates the impact that SNS had on news' design and journalists' work.

But it turns out that frequently newspapers do not know how to distribute and segment content online (Engesser & Humprecht, 2015), nor how to establish differences in content value (Halbheer et al., 2014), between paid, unpaid, and hybrid.

Content Distribution and Readers' Preferences

Boczkowski and Mitchelstein (2010) highlighted the difficulty in finding a balance between journalists' choice of and readers' demand for news. Apparently, that is still valid today. The Pew Research Center (2019) found that while individuals over 60 years of age read newspapers daily, 18-year-olds look for news in other media, including SNS. Of the younger generations, even those who used to read paper newspapers are now looking for the same information online, with time and financial gains.

In order to reach younger audiences (Thurman & Fletcher, 2017), newspapers are transferring a considerable amount of value from paper to online channels. Many media companies have implemented a 360-degree strategy that integrates content decisions based on the potential to generate value for the consumer and return through sending to multiple platforms (Waddell, 2020). While news media are multiplying, newspapers often do not know where to channel information or how to segment content (Engesser & Humprecht, 2015).

From all the literature found on this topic, two ideas stand out. The first is that the successful formulas in content distribution are yet to be found. The second, which brings us to this study, is that there is still a big gap between what are the publishers and the readers' preferences and choices, especially with regard to the digital world (Lee et al., 2012). On the other hand, although there are many forms of generating revenue, not all of them work the same way, even within similar distribution channels. It is up to the managers to find out how to adapt their business model to their distribution strategy (Berger et al., 2015).

CONTENT ANALYSIS DURING SILLY SEASON: A CASE STUDY OF JN

Methodology and Chosen Case

Method

Considering the objectives defined, this article offers evidence on a case study. Details on the methodological approach are presented in the following sections.

Case Study

The method used is the single case study (Yin, 2014), which dealt with one of the main Portuguese newspapers, with decades of history in the paper version and with a strong online presence. The selection criteria were pragmatic and sought maximum variations and not uniformity. in this case the Jornal de Notícias (JN) was chosen because it is a typical and special case, but also for reasons of convenience of access to information (Bravo & Eisman, 1998). The single case was chosen considering the relevance of analyzing a case of success in online positioning, which is worth documenting and which serves the purpose of analyzing good practice (Yin, 2014). It provides a detailed study of an object. On the other hand, it organizes and gathers numerous and detailed information, interrelating different types of data

(Goode & Hatt, 1973). It is a strategy used in different areas to collect information about a practice or phenomenon in a specific context, in order to clarify doubts and unveil future actions (Chizzotti, 2006).

By comprising the analysis of different data (e.g., interview, questionnaire, and content analysis), data triangulation reinforces the study's validity (Bruchez et al., 2015). According to Yin (2014), validity depends on four basic principles: the use of different data sources (triangulation), creating a case study database, maintaining the chain of evidence, and being careful in using data from electronic sources.

As it is a case study, and applying the usual ethical principles in social sciences, there was an embargo on the results for three years to protect the strategy of the analyzed environment. It is ethically and deontologically relevant not to disclose a company's strategy, in particular the data provided directly by those responsible for it. As we will see below, the traffic data for the newspaper is today very similar to the years of the study, that is, we can conclude that the study is current and the strategy remains the same.

Data Collection and Analysis

The analysis was carried out for 28 days (4 weeks), 14 in the month of August 2016 (15 to 28) and 14 in the month of September 2016 (5 to 18). The choice of dates took into account the concept of Silly Season (associated with the month of August in the newsrooms), associated with the idea of creating softer content during this period. In this way, it was possible to compare the two periods of analysis.

Content Analysis

The publications were categorized, typified and accounted for in the four analyzed media. The analysis used the categories defined by the newspaper: National, Local, Justice, World, Economy, Sports, People, Innovation, Culture, and Opinion. The type of content was also defined, dividing it into "soft" and "hard". According to Curran et al. (2009) and Patterson (2000), soft content is all content published in a lighter, more entertaining tone, and hard content comprises news with direct implication in the lives of readers, on politics, matters of State, but also heavier content, related to deaths, accidents or tragedies, both national and international.

On all days of analysis, the news, by category and type, published in the paper newspaper (excluding special supplements and publishing reports) and the amount of news published on the website were counted. Also counted were the number of publications on Facebook and the number of news that followed daily in the two newsletters sent by JN. Additionally, the analysis considered information provided by the newspaper's editorial staff about the most popular online news, the most successful distribution channels and the means of monetization available. In this way, the most successful online categories and typologies and distribution channels were identified. It was also possible to understand how the newspaper makes the most read news online profitable, by analyzing quantitatively the conversion points on each page. Conversion points considered were advertising banners, sponsored content, newsletter subscription form (lead capture), and e-paper subscription form (paid version of the online newspaper, with exclusive content from the paper version).

Semi Structured Interview

In a second stage of the study, an interview was conducted with the head of the newspaper's digital department. Amongst qualitative research methods, interview is a fundamental approach to access relevant information about a phenomenon. It is described as a face-to-face conversation between two people, the interviewee and the interviewer. The main objective is to uncover aspects that are difficult to explore by content analysis or another study method (Bruchez et al., 2015).

The interview outline was previously defined, but the interviewee was given freedom to explore the topic according to the newspaper's reality within each theme. The questions in the interview explored the newspaper's digital strategy and online business model and asked for information about the content creation and distribution strategy.

Survey

To substantiate data triangulation, a survey was conducted with a sample of the Portuguese population that accesses informative content through SNS, in order to understand whether the strategy analyzed in the case study corresponded to expectations and needs of the audience that reads news online.

Respondents were recruited according to the snowball method, as members of the researchers' social networks were invited to forward the link to the survey to 3 friends or acquaintances. The study was posted online on October 24, 2016, at 11:00 pm, and in less than 24 hours, 570 valid responses were reached. According to Obercom (2014), by the time of the study, around 5.5 million Portuguese people accessed the Internet. Considering a margin of error of less than 5% and a confidence level of 95%, it was defined that the study should have a minimum of 384 valid responses, which was accomplished.

The questionnaire was organized in four parts. First part comprised general questions that aim to identify reading habits of paper and online newspapers. This was followed by the section dedicated to news consumption through distribution and category, which aims to cross-check data with the analysis of content distribution in the analyzed newspaper. In the following sections, questions were asked related to the behavior of using the JN, questions only asked to respondents who said they read the JN frequently. Like Schoneville (2007), the Unified Theory of Acceptance and Use of Technology (UTAUT) model was used, using five of the factors that this model identifies to analyze users' intentions towards the particular newspaper: usage behavior, performance expectation, conditions of usability, attitude and social influence.

Chosen Case: Jornal de Notícias

Jornal de Notícias (JN) was the first generalist information newspaper to move towards a strong online presence in Portugal. It was the first publication to update its online edition daily (Martins, 2012), in 1995. More than 20 years later, JN continues to have a particularly strong online presence.

In September 2016, the JN portal was, according to the Similar Web portal (https://www.similarweb.com), in 26th place in the national ranking of most visited portals (27th in 2020) and in the 5216th position worldwide (4191st in 2020). It had an average of 9.3 million monthly visits, with an average duration of 5 minutes, 3.4 pages per visit and a bounce rate of almost 60%. These figures are particularly interesting considering the Portuguese publishing landscape.

Search engines account for around 19.73% of the total website visits, followed by referrals (references on other portals), with 6.45%. Surprisingly, email only accounted for 1.48% of the traffic, despite of the two newsletters that were sent daily. Of the traffic coming from SNS, 97.51% was from Facebook, despite the fact that JN is present on other networks, such as Twitter (which represents 1.26% of visits). In 2020, we did not see significant differences in these figures, that is, the results remain roughly the same.

RESULTS

During the four weeks of this analysis, JN published 3,723 news on paper newspaper, 2,090 insertions on the website, 1,964 posts on Facebook, and disseminated 595 linked posts in newsletters. Hence, here was a clear difference between the number of news published in the paper newspaper and online.

Content Typology and Categorization

As shown in Table 1, Sports leads the content produced, and is the dominant category in the newsletter. World occupies a primary place on Facebook (along with Sports) and on the website. The most prevalent news in the paper newspaper is Sports, followed by Local.

Table 1. Total published content by medium and category. Source: The Authors

Category / Medium	Newspaper	Website	Facebook	Newsletter	Total
Culture	258	189	147	13	616
Sports	948	377	372	213	1910
Economy	148	94	67	15	324
Innovation	8	13	14	4	39
Justice	379	141	119	45	684
Local	802	252	252	45	1351
World	331	476	477	93	1377
National	541	305	272	88	1206
Opinion	150	106	108	54	418
People	158	137	136	25	456
Total	3723	2090	1964	595	8372

Table 2 presents information on the type of news, considering the hard/soft dichotomy proposed by the literature (Curran et al., 2009; Patterson, 2000). Clearly, JN favors hard news in all distribution channels, Facebook included. The most frequent categories on paper newspaper is soft Sports news and Local hard news. People, Opinion, Economy and Innovation are the categories with less news in the paper newspaper. On the website, the hard World news type predominate, followed by soft Sports and hard National. Soft Sports predominates in the Newsletter, followed by National, World, and Opinion categories, mostly of the hard type.

Table 2. Comparison between soft and hard content published in different media. Source: The Authors

Category / Medium	Paper Newspaper		Website		Facebook		Newsletter	
	Hard	Soft	Hard	Soft	Hard	Soft	Hard	Soft
Culture	20	238	22	167	19	128	3	10
Sports	36	912	43	334	39	333	24	189
economy	147	1	90	4	61	6	15	0
Innovation	4	4	3	10	4	10	1	3
Justice	378	1	140	1	114	5	45	0
Local	507	295	210	42	205	47	38	7
World	323	8	410	66	387	90	72	21
National	486	55	259	46	220	52	74	14
Opinion	115	35	87	19	89	19	50	4
People	4	154	17	120	18	118	4	21
Total	2020	1703	1281	809	1156	808	326	269

This content analysis covered two different periods, August and September, with the aim of comparing the newspaper's activity in the period commonly known as the Silly Season and a period of normal political, economic, and social activity. The two periods were compared in order to identify possible differences both in content creation and in the type of soft/hard news (Table 3). There was no significant difference in the number of news published in the two months. In the paper newspaper, there was a slight reduction in the month of August, but during the Silly Season the website, Facebook, and Newsletter even had a slight increase in news. As demonstrated in Table 3, soft content predominated during the month of August while hard content predominated in September. The differences are greater in the hard content in the newspaper and in Facebook, and in the soft content posted in Facebook and in the Newsletter.

Table 3. Comparison between August and September news in the different media. Source: The Authors

Month / Medium / Type	August		September	
	Hard	Soft	Hard	Soft
Newspaper	973	882	1047	821
Website	615	455	666	354
Facebook	542	453	614	355
Newsletter	154	148	172	121

Hence, one conclusion drawn from this content analysis is the fact that there is a difference of almost 1700 news items published between the newspaper and the website. Arguably, this is a surprising fact, considering that the costs of publishing online are very low and the success that newspapers' online dissemination of news has had in recent years. Still, these findings are in line with Engesser and Humprecht (2015) who argue that newspapers often do not know where to channel information and how to segment content. Another finding that should be pointed out is regarding the relevance given to soft Sports in all media. Despite being a newspaper that predominantly publishes Local news on its paper version, it gives preference to the World category on the website and on Facebook. Yet JN is, in all media, a hard news newspaper. Moreover, the content strategy followed on the website and on Facebook is similar, considering the type and categories of content disseminated in both media. This result does not support the suggestions by Hermida (2012) who advocates distinct strategies for SNS. Another finding is the reduced importance given to the newsletter, with almost residual values for content distribution.

Most Read News

Based on the additional data provided by JN for this study, it was possible to conclude that all the most read news were distributed on Facebook and that that news had exactly the same conversion points (always two banners, a space for sponsored content and a space for subscribing to the newsletter).

Overall, the World category leads the ranking of the most read news, usually associated with unusual or global tragedies. Similarly, hard content leads the preference of the newspaper's online readers, as 61% of the most read news were coded as hard, while 39% were considered soft news.

Regarding its distribution, 57% of the most read news online were also published in the paper newspaper, while 43% were exclusively online. Moreover, only 11% of the most read news online were not included in the daily newsletter. In fact, the most widely read news that were not published in the paper newspaper were generally unusual world news, and 60% were soft news. Almost half of the most read news online were not published in the paper newspaper, which indicates a clear differentiation of the content that follows for each of the two media.

All the most read news have the same 4 conversion points as all news on the site. That is, the newspaper makes all its online content profitable in the same way. World hard news were the newspaper's most successful online news, despite the predominance of to the local and national news on the paper newspaper.

Cover Page Highlights

During the 28 study days, the 28 paper newspaper covers were also analyzed. Interestingly, all the news that featured on the cover on the analyzed days are hard news, even in August, in the middle of Silly Season. As point of fact, in the 28 days analyzed, only three categories were highlighted on the cover page of the paper newspaper: National, Justice, and Economy. National news lead by a wide margin. In that period, about 90% of the paper's cover highlights were published on the website, but only about 60% of were shared on Facebook. According to the information provided, 40% of the shared news comprised only an introduction and redirected the reader to the paid version. Finally, 82% of this cover news were shared in the daily newsletter, often also leading to the online newspaper version.

So, the analysis shows that the paper version of the JN maintained a hard news positioning, focusing on national themes, whether they are politics, justice (crime) or economics. Moreover, cover news were mostly featured in online channels. This idea goes against the conclusion of the previous analysis, in which online positioning was distanced, in typology and categorization, from paper positioning, betting more on world news.

Moreover, for the cover page news, JN tends to adopt a freemium version of the content, making only part available to the online reader. However, it does not always do so or with all the prominent news of the day's edition in the paper version.

Semi Structured Interview

A semi-structured interview was conducted with the Newspaper Executive Director of the Digital Department and had two general objectives: to further understand the digital strategy and the online business model of JN, and to analyze the strategy for creating and distributing content through the different media.

The interviewee acknowledged that the human resources available are scarce for the volume of work required in the digital channels. Partly due to that fact, the interviewee recognized not being aware of the fact that JN distributed much more content for the paper version than for the online version. However, he demonstrated a deep knowledge of the online readers, their profiles, schedules, and means of access: "We have tools that allow us to know in real time who is watching us, where our audience is, how they read us, for how long (…) what device they use, at what time..." referring namely to information provided by website analytics.

The interviewee explained that newsrooms are increasingly integrated, and that most journalists write for all media simultaneously. That is, they prepare news for the different means of distribution, corroborating suggestions by Tandoc and Vos (2015). As the interviewee explained:

We have a very strong team working on digital news, but the entire newsroom is sensitive [to digital channels] and, therefore, everyone participates in the elaboration of content for digital too. (…) Right now, when I edit a news item, I normally think of four different ways of editing. (Interviewee)

This conclusion is contrary to Karimi and Walter (2015) who suggest that digital is not yet thought of in the first place.

Despite being a leader in digital news, JN would not know how to survive without the paper version. However, the vision for the future is more positive: "What I can say is that the digital turnover has grown". They still bet a lot on the work of the brand that still brings them a lot of direct traffic. However, Facebook was described as a well-kept secret in the content distribution strategy for this medium. According to the interviewee, paid content does not have a predefined distribution channel, and can be distributed on paper or on the website, "Premium content is not just reserved for paper". However, the minor importance attached to the newsletter was acknowledged, not being a priority in the distribution of content: "The newsletter is automatic, it is always the same".

As for the "freemium" strategy, it was associated to brand management, namely its credibility and appreciation (in detriment to the devaluation of what is free), and not to attracting more online subscribers: "We want to let people know that paper is still a good product and that they will find exclusive information there, well-worked information".

The interviewee confirmed that JN was and specialized in hard news, some (especially those disseminated online) based on curated content from international sites: "We are very strong in hard news. It is what distinguishes us in terms of market positioning".

Readers' Perspective

Participants in this study (N = 570) were 70.7% female, and all age groups were represented (16.8% were less than 25 years old, 37.5% were 25-34 years old, 30.7% were 35-44 years old, 11.1% were 45-54 years old, and 1.2% were 55 or older). 50.7% stated that they usually read newspapers, and 49.1% did not. Moreover, 46.55% said they frequently buy paper newspapers, 52.8% of which bought JN. In addition, 94.4% said that they read news online, 37% of which read JN, and 31.4% read its competitor Jornal Público, the two most cited newspapers by the respondents.

Considering only the respondents that indicated that JN was their preferred online newspaper, Table 4 presents mean values for searches of the newspaper's categories in the different media. Clearly, the favorite channel is the website, except in the case of People category (e.g., news about celebrities), for which Facebook was the preferred medium.

Table 4. Category search by medium – mean values (response scale from 1 to 5). Source: The Authors

Categories / Medium	Paper	Website	Facebook	Newsletter
Justice	3.05	3.30	3.05	2.05
Culture	2.87	3.83	3.62	2.63
World	2.85	4.04	3.62	2.40
National	2.85	4.05	3.41	2.47
Local	2.80	3.34	3.25	2.28
Opinion	2.56	3.35	3.05	2.21
Innovation	2.54	3.61	3.47	2.39
Economy	2.43	3.29	2.84	2.12
Sports	2.36	2.96	2.75	1.91
People	1.97	2.61	2.81	1.75

Principal components exploratory factor analysis using varimax rotation was used to further explore users' preferences. The obtained solution explained 72.49% of the total variance and comprised 7 factors. The obtained factors corresponded to each medium (paper, website, Facebook, newsletter), with the exception of the Sport, People, and Opinion which formed independent categories.

Based on these results, new variables were created. Internal consistency and descriptive indicators of these new variables are presented in Table 5.

Table 5. Readers' information search preferences: factor measures statistics. Source: The Authors

Factors	Number of items	Mean	Standard deviation	Cronbach's alpha
Paper	8	2.67	1.20	0.946
Website	7	3.64	0.84	0.849
Facebook	7	3.32	1.06	0.910
Newsletter	10	2.22	1.06	0.941
Sports	3	2.69	1.24	0.803
People	3	2.47	1.11	0.778
Opinion	2	3.20	1.11	0.673

These factors enabled further exploring readers' search preferences in terms of medium. Multivariate analysis using General Linear Model revealed significant differences between each factor, indicating that the demand for information is significantly higher in the case of the website.

Additionally, significant differences were found regarding greater preference for searching in the website in the case of several news categories: Economy, Sport, Innovation, Culture, and Opinion. In accordance to previous findings, in the case of People category statistically significant differences were found regarding Facebook. In all cases, and since the Mauchy test for sphericity was violated H0 [X2 (5) = 83.95, p = 0.000], thus the sphericity of the covariance matrix could not be assumed, Huynh-Fieldt epsilon measurement (epsilon > 0.75) was used as the correction factor.

Regarding content search distribution, general linear model analysis provided interesting findings. Regarding paper newspaper, significant differences between types of content [F (7.24, 3887.34) = 57.03, p = .000] indicate that the most sought-after information corresponds to news about National, Local, World and Culture, with no statistically significant differences between them. On the website [F (7.14, 3836.23) = 111.44, p = .000], the most sought-after news belong to two categories, National and World, with no statistically significant differences between these two categories. Similarly, on Facebook [F (7.19, 3863.07) = 69.47, p = .000] the two most sought categories are World and Culture. Finally, regarding the newsletter, statistically significant differences [F (7.59, 4076.60) = 58.90, p = .000] between types of content indicate that the most sought category is Culture.

SOLUTIONS AND RECOMMENDATIONS

News media is an area in constant change, and that digital journals are now entering their third decade of evolution, alongside research on the topic (Salaverría, 2015). Journalists are expected to work on different platforms, for different audiences, with different techniques. Content production is no longer enough and the "gate watcher" replaces the "gate keeper", as nowadays content curation is one of the relevant roles of journalism, as it is also demonstrated in this study, and algorithmic curation of social media platforms is considered to create a clickbait media environment (Lischka & Garz, 2021).

Today, the distribution channels allow providing content for different audiences in different media. However, it is necessary to know how to identify readers' preferences and needs in each medium. Often, the strategy carried out by the newspaper may not be aligned with its target audiences' behavior.

Overall, this article also points out the importance of analyzing and aligning both the content distribution strategy and readers' needs and preferences. In the particular case of JN it was demonstrated that the differences were notorious. Despite the attention paid to web analytics, many aspects of the readers' behavior demand market research and informal interaction with audiences, which in turn demand resources and time that are often unavailable.

FUTURE RESEARCH DIRECTIONS

Limitations of this study should be pointed out, the most important referring to the fact that evidence presented and discussed in this article refers to one newspaper, and the data collected in 2016.

Despite the fact that data triangulation provided support to the conclusions made, future research should consider the case of other newspapers in diverse cultural contexts in order to validate these results.

Avenues for future research include both longitudinal analyses and multi-case comparison, namely in international settings. Comparing newspapers that favors their paper versions with others that only publish online may also provide interesting insights to further understand the strategies of newspapers in the digital era.

CONCLUSION

Half of the Portuguese population with Internet access still reads paper newspapers. However, they are not in the habit of buying it. Despite national statistics which indicate that JN is the most read newspaper online, this study shows that JN loses readers for its competitors. Moreover, most users have never paid for online news content or subscribed to any service of this type. Looking at the search preferences, clearly the website is the favorite medium, except for one category, People. With regard to the type of content, no significant differences were found between hard and soft.

Overall, it is important to point out that the categories sought in different media by JN readers are different from the actual strategy of the newspaper. Innovation appears in the top spots on Facebook, but JN publishes very low numbers of its own news in this category. The newsletter is the least preferred medium for JN readers and the Sports Newsletter category / medium is amongst the least sought after. Among JN online readers, there are very few who buy paper newspapers, especially among the younger age groups. Moreover, the vast majority of respondents follow JN's Facebook page. The fact that JN's Facebook page is very successful among its usual readers corroborates the point made by Casero-Ripollés (2012). However, fewer readers admit to sharing JN publications on Facebook, hence revealing little involvement with the brand.

Back in 2004, D'Haenens et al. (2004) pointed out great differences in the consumption of news on paper and online. According to this study, this is still the case. Moreover, the distribution strategy carried out by the newspaper is distinct on paper and online. This article shows that the analyzed newspaper shared a much larger amount of news in the paper version than in the online version. This is one of the most surprising conclusions of the study, which totally goes against the initial expectation.

From the different informal conversations that preceded the interview reported in this article, it was clear that the scarcity of resources prevents effectively dealing with online channels. All the little time they have available for the online channels is dedicated to analyzing, often in real time, the online readers' behavior, leaving little time to explore opportunities to create online news content. This was the main explanation for the reduced distribution of news online, be it free, freemium, or premium.

JN is a newspaper that focus on hard news, which was confirmed in this study. However, it combines this strategy with the use of Infotainment, the curation of content on international sites, looking for unusual news that creates engagement on SNS. This is a strategy often associated with content marketing strategies that consider that one piece of written content can both inform and entertain. For this newspaper, the distribution strategy is clear: the hard content goes to the cover of the newspaper, and the infotainment content (both soft and hard) goes to Facebook. Thus, there is a mixture of very different types of content created by the same medium, which is to be positioned on different channels.

In the case of the newspaper studied in this article, arguably the strategy does not intend to favor the paper newspaper for premium content. However, the probability of this happening is actually high, considering that only a small part of their news covered in the paper newspaper are also disseminated online. Evidence offered in this article provides other interesting insights on a newspaper's strategy.

Regarding a strategy for distributing different types and categories of news through different online media, it was evident in this study that some newspapers may segment categories, but share both hard and soft news in the different online media. Still, during the silly season differences between hard and soft news were noticed.

Overall, this study concluded that clearly some newspapers, JN included, do not capitalize the most popular news, as the conversion points are automatically applied to any news content. Additionally, many of the news produced were not featured online, which arguably means that newspapers such as this one may be wasting opportunities to reach, engage, and convert their readers and target audiences. Still, while incentives to subscribe the online version of the newspaper and to buy news by making freely available only a teaser or introduction to the content, evidence collected, in particular through the interview, demonstrates that newspapers' online strategy may be quite diverse from online monetization. In fact, another alternative relates to branding strategy, based on the perception that quality news should not be free.

What is certain is that, despite the growing adoption of online media, most newspapers still cannot survive exclusively from the online format. However, there are media that share informative content, in the so-called data journalism that, with far fewer human resources and far fewer ethical issues, are able to survive exclusively with online versions.

REFERENCES

Bakker, P. (2012). Aggregation, content farms and Huffinization. *Journalism Practice*, *6*(5-6), 627–637. doi:10.1080/17512786.2012.667266

Barbosa, B., & Carvalho, C. (2021). Tweet and retweet journalism during the pandemic: Dissemination of and engagement with news on twitter. *Brazilian Journalism Research*, *17*(3), 736–763. doi:10.25200/BJR.v17n3.2021.1364

Bentley, F., Quehl, K., Wirfs-Brock, J., & Bica, M. (2019). Understanding online news behaviors. *CHI '19: Proceedings of the 2019 CHI Conference on Human Factors in Computing Systems*, *590*, 1-11. 10.1145/3290605.3300820

Berger, B., Matt, C., Steininger, D. M., & Hess, T. (2015). It is not just about competition with "free": Differences between content formats in consumer preferences and willingness to pay. *Journal of Management Information Systems*, *32*(3), 105–128. doi:10.1080/07421222.2015.1095038

Boczkowski, P. J., & Mitchelstein, E. (2010). Is there a gap between the news choices of journalists and consumers? A relational and dynamic approach. *The International Journal of Press/Politics*, *15*(4), 420–440. doi:10.1177/1940161210374646

Bravo, M. P., & Eisman, L. B. (1998). *Investigación Educativa*. Ediciones Alfar.

Bruchez, A., Ciconet, B., Remussi, R., Possamai, L., Antônio, V., & Tondolo, G. (2015). análise da utilização do estudo de caso qualitativo e triangulação na brazilian business review. *Mostra de Iniciação Científica, Pós-Graduação, Pesquisa e Extensão, 4*, 1-16. doi:10.18226/35353535.v4.2015.52

Casero-Ripollés, A. (2012). Contenidos periodísticos y nuevos modelos de negocio: Evaluación de servicios digitales. *El Profesional de la Información, 21*(4), 341-346. doi:10.3145/epi.2012.jul.02

Cherian, J. (2015). Emergence of digital publishing: A great challenge to the print publications. *Procedia Economics and Finance*, *23*, 576–586. doi:10.1016/S2212-5671(15)00361-5

Chizzotti, A. (2006). *Pesquisa Qualitativa em Ciências Humanas E Sociais* [Qualitative Research in Humanities and Social Sciences]. Vozes.

Cook, C., & Sirkkunen, E. (2013). What's in a niche? exploring the business model of online journalism. *Journal of Media Business Studies*, *10*(4), 63–82. doi:10.1080/16522354.2013.11073576

Curran, J., Iyengar, S., Lund, A. B., & Salovaara-Moring, I. (2009). Media system, public knowledge and democracy: A comparative study. *European Journal of Communication*, *24*(5), 5–26. doi:10.1177/0267323108098943

D'Haenens, L., Jankowski, N., & Heuvelman, A. (2004). News in online and print newspapers: Differences in reader consumption and recall. *New Media & Society*, *6*(3), 363–382. doi:10.1177/1461444804042520

Engesser, S., & Humprecht, E. (2015). Frequency or skillfulness: How professional news media use Twitter in five western countries. *Journalism Studies*, *16*(4), 513–529. doi:10.1080/1461670X.2014.939849

Goode, W., & Hatt, P. (1973). Métodos. In *Pesquisa Social*. Editora Nacional.

Graves, D., & Simon, F. M. (2019). *Pay models for online news in the US and Europe: 2019 update*. Reuters Institute for the Study of Journalism. https://www.digitalnewsreport.org/publications/2019/pay-models-2019-update/

Guimarães, N., Miranda, F., & Figueira, A. (2019). Identifying journalistically relevant social media texts using human and automatic methodologies. *International Journal of Grid and Utility Computing*, *11*(1), 72–83. doi:10.1504/IJGUC.2020.103971

Halbheer, D., Stahl, F., Koenigsberg, O., & Lehmann, D. (2014). Choosing a digital content strategy: How much should be free? *International Journal of Research in Marketing*, *31*(2), 192–206. doi:10.1016/j.ijresmar.2013.10.004

Hendrickx, K. (2021). The rise of social journalism: An explorative case study of a youth-oriented Instagram news account. *Journalism Practice*, 1–16. Advance online publication. doi:10.1080/17512786.2021.2012500

Hermida, A. (2012). Social journalism: exploring how social media is shaping journalism. In E. Siapera & A. Veglis (Eds.), *The Handbook of Global Online Journalism* (1st ed., pp. 309–328). John Wiley & Sons, Inc. doi:10.1002/9781118313978.ch17

Karimi, J., & Walter, Z. (2015). The role of dynamic capabilities in responding to digital disruption: A factor-based study of the newspaper industry. *Journal of Management Information Systems*, *32*(1), 39–81. doi:10.1080/07421222.2015.1029380

Lee, A. M., Lewis, S. C., & Powers, M. (2012). Audience clicks and news placement: A study of time-lagged influence in online journalism. *Communication Research*, *41*(4), 1–26. doi:10.1177/0093650212467031

Lischka, J. A., & Garz, M. (2021). Clickbait news and algorithmic curation: A game theory framework of the relation between journalism, users, and platforms. *New Media & Society*. Advance online publication. doi:10.1177/14614448211027174

Liu, D. R., Liao, Y. S., & Lu, J. Y. (2019). Online news recommendations based on topic modeling and online interest adjustment. *Industrial Management & Data Systems*, *119*(8), 1802–1818. doi:10.1108/IMDS-04-2019-0251

Marktest. (2018, November 6). *Quase dois em cada três portugueses leem notícias online* [Almost two out of three Portuguese Read News Online]. https://www.marktest.com/wap/a/n/id~244d.aspx

Martins, F. (2012). Os Caminhos da Autonomia no Online [Dissertação de Mestrado, Ciências da Comunicação, Porto]. Faculdade de Letras da Universidade do Porto.

Neuberger, C., Nuernbergk, C., & Langenohl, S. (2018). Journalism as multichannel communication. *Journalism Studies*, *20*(9), 1–21. doi:10.1080/1461670X.2018.1507685

Obercom. (2014, June 22). *A Internet em Portugal – Sociedade em Rede 2014*. https://obercom.pt/a-internet-em-portugal-sociedade-em-rede-2014/

Olsen, R. K., Kalsnes, B., & Barland, J. (2021). *Do Small Streams Make a Big River? Detailing the Diversification of Revenue Streams in Newspapers' Transition to Digital Journalism Businesses*. Digital Journalism. doi:10.1080/21670811.2021.1973905

Patterson, T. E. (2000). *Doing Well and Doing Good: How Soft News and Critical Journalism Are Shrinking the News Audience and Weakening Democracy – And What News Outlets Can Do About It*. Harvard University, John F. Kennedy School of Government. https://shorensteincenter.org/how-soft-news-critical-journalism-are-shrinking-news-audience/

Pew Research Center. (2019, April 23). *The role of social media in news*. https://www.journalism.org/2019/04/23/the-role-of-social-media-in-news/

Powers, M., & Benson, R. (2014). Is the Internet homogenizing or diversifying the news? External pluralism in the U.S., Danish, and French press. *The International Journal of Press/Politics*, *19*(2), 246–265. doi:10.1177/1940161213519680

Powers, M., & Vera-Zambrano, S. (2019). Endure, Invest, ignore: How French and American journalists react to economic constraints and technological transformations. *Journal of Communication*, *69*(3), 320–343. doi:10.1093/joc/jqz015

Sanseverino, G. G., & Santos, M. F. L. (2021). Experimenting with user-generated content in journalistic practices: Adopting a user-centric storytelling approach during the covid-19 pandemic coverage in Latin America. *Brazilian Journalism Research*, *17*(2), 244–279. doi:10.25200/BJR.v17n2.2021.1369

Schoneville, S. (2007). *This just in: Analysis of factors influencing online newspaper reading behaviour* (Publication No. 140784811) [Masters dissertation, Universiteit Twente]. SemanticScholar.

Skjeret, F., Steen, F., & Wyndham, G. A. (2019). Paywalls and the demand for online news. *NHH Department of Economics Discussion Paper*, 7. doi:10.2139/ssrn.3364834

Tandoc, E. Jr, & Vos, T. (2015). The Journalist is Marketing the News. *Journalism Practice*, *10*(8), 950–966. doi:10.1080/17512786.2015.1087811

Thurman, N., & Fletcher, R. (2019). Has digital distribution rejuvenated readership? *Journalism Studies*, *20*(4), 542–562. doi:10.1080/1461670X.2017.1397532

Waddell, T. F. (2020). The authentic (and angry) audience. *Digital Journalism*, *8*(2), 249–266. doi:10.1080/21670811.2018.1490656

Yin, R. K. (2014). Case Study Research: Design and Methods (5th ed.). Academic Press.

Zubiaga, A., Heravi, B., An, J., & Kwak, H. (2019). Social media mining for journalism. *Online Information Review*, *43*(1), 2–6. doi:10.1108/OIR-02-2019-395

Zúñiga, H. G., Weeks, B., & Ardèvol-Abreu, A. (2017). Effects of the news-finds-me perception in communication: Social media use implications for news seeking and learning about politics. *Journal of Computer-Mediated Communication*, *22*(3), 105–123. doi:10.1111/jcc4.12185

ADDITIONAL READING

Chanpaneri, A., & Prachi, J. (2021). A Content Marketing as the leading technique in digital marketing: A case of Nike. *International Journal of Research and Analytical Reviews*, *8*, 226–236.

Dharmayanti, D., Bachtiar, A. M., & Fakhrul, F. N. (2021). Data visualization for content marketing domain in social media. *Journal of Engineering Science and Technology*, *16*(1), 339–349.

Júnior, M. C., & Quintas, S. D. (2014). La curaduría de contenidos y la narrativa colaborativa en el Ciberperiodismo: Estudio del caso de Storify en el diario digital elpais.com [Content curation and collaborative narrative in Cyberjournalism: a case study of Storify in the digital newspaper elpais.com]. *Estudios sobre el Mensaje Periodístico*, *1*(20), 17–31.

Lazauskas, J. (2015). *'You Need Editors, Not Brand Managers': Marketing Legend Seth Godin on the Future of Branded Content*. Contently. https://contently.com/2015/02/06/you-need-editors-not-brand-managers-marketing-legend-seth-godin-on-the-future-of-branded-content/

Mansour, D., & Barandas, H. (2017). High-tech entrepreneurial content marketing for business model innovation: A conceptual framework. *Journal of Research in Interactive Marketing*, *11*(3), 296–311. doi:10.1108/JRIM-03-2016-0022

Powers, M., & Benson, R. (2014). Is the Internet Homogenizing or Diversifying the News? External Pluralism in the U.S., Danish, and French Press. *The International Journal of Press/Politics*, *2*(19), 246–265. doi:10.1177/1940161213519680

Pullizi, J., & Rose, R. (2017). *Killing marketing: how innovative businesses are turning marketing cost into profit*. McGraw Hill.

Rose, R. (2018). *4 Business Models for Content Marketing*. Content Marketing Institute. https://contentmarketinginstitute.com/2018/10/business-model-content-marketing/

Urbach, N., & Röglinger, M. (2019). Introduction to Digitalization Cases: How Organizations Rethink Their Business for the Digital Age. In N. Urbach & M. Röglinger (Eds.), *Digitalization Cases. Management for Professionals*. Springer. doi:10.1007/978-3-319-95273-4_1

Wang, P., & McCarthy, B. (2021). What do people "like" on Facebook? Content marketing strategies used by retail bank brands in Australia and Singapore. *Australasian Marketing Journal*, *29*(2), 155–176. doi:10.1016/j.ausmj.2020.04.008

KEY TERMS AND DEFINITIONS

Content Marketing: The process of creating and distributing valuable content to engage an audience, with the objective of generating profitable customer action.

E-Paper: Digitised version of a newspaper created for printing.

Infotainment: A strategy in media that combines news and entertaining content.

Monetization: In the digital context, it represents the revenue that can be earned by publishing a piece of content.

Newsroom: Editorial office, a space where journalists work in open space.

Pay per View: Model through which we pay a certain amount to access a specific content.

Silly Season: Time of years when holidays and less intense content dominate the newsrooms.

Chapter 11
Smart Cities:
How Can Low-Density Territory Populations Be Smart Tourist Destinations? The Case of Guarda

André Alves Toste
Higher Institute of Business and Tourism, Portugal

Dulcineia Catarina Moura Catarina Moura
https://orcid.org/0000-0002-3539-4754
Universidade da Beira Interior, Portugal

Lídia Aguiar
https://orcid.org/0000-0003-1815-8546
Higher Institute of Business and Tourism, Portugal

ABSTRACT

The use of technology has been making territories smarter. This fact allows policymakers and stakeholders to develop cities in a smart, sustainable way and provide more quality of life to those who live there. Tourism destinations are beginning to be able to anticipate behaviour and customize their offer, thus becoming intelligent tourist destinations. However, it has been the large cities that have been on this journey. The qualitative methodology was used, and five structured interviews were conducted with stakeholders to understand the ability of a low-density territory, using the municipality of Guarda as the object of study. Through the identification of the essential infrastructures, the importance is given by policymakers to the theme and understanding the perception of those who influence the destination, and a proposal for a model for low-density smart tourism destinations was developed. It is concluded that low-density territories must satisfy all the associated dimensions and conceive a collective thinking that makes this goal possible: intelligent tourism destinations.

DOI: 10.4018/978-1-6684-5538-8.ch011

INTRODUCTION

In a context arising from the effects of a pandemic such as COVID-19, in which tourists' sensitivity to the identity character of the destinations they seek for their holidays differs, as does the consequent local impact of the service provided, as do their motivations, the search for smart, low-density tourism destinations is even more decisive. In other words, the search for low-density territories, as tourist destinations in which authenticity is the motivation for the tourist experience, emerges from the outset as a relevant alternative to massified territories, which in the past may have been the decision of choice and the generalized preference of tourists and visitors.

Endogeneity and the sense of place are fundamental arguments when approaching low-population density territories, since the idea of belonging to a place and transporting this image to those who visit, makes the experience unforgettable. This sense of place is shown in the relationship between people and places and in the emotional and symbolic values involved and which are conveyed by them, always related to the way people understand, experience and demonstrate their feelings. These feelings can influence the attitudes and behaviors of individuals in their relationship with place (d'Orey & Abreu, 2015).

In the early 1980s, ruralism began to be related with the concept of low-density territories. This concept does not only address territories with low population density, but also a low relational density, that is, the existence of some characteristic elements of these areas are: aging population; weak connection between people and between institutions; low level of public-private partnerships; several public institutions with little territorial dynamics, among others (Mendes, 2019).

Concerning the Portuguese territory, the theme began to gain relevance in 2015, when the application presented by the Interministerial Coordination Commission (CIC) began to define a multi-criteria approach for this type of territories. In that regard the criteria defined were based mostly on population density, population, demography, in the physical characteristics of the territory, socio-economic characteristics and in availability. Through the collaboration with the ANMP, there were classified 164 low-density cities (ANMP, 2014; CIC, 2015). Concerning the criteria for a territory to be of low density, in the Portuguese case, in 2014, the ANMP (National Association of Portuguese Cities), pointed out some extents and pointers to consider, such as: the population density; the territorial and demographic profile; the population, socio-economic and accessibility profile. Hence, and after assigning values to each of the above-mentioned indicators, the association considers low-density cities, "All cities with a density population of less than 25 inhabitants/km2 (…)" (ANMP, 2014).

Nevertheless, the difficulties that occur in this type of low-density regions, the reality is that through financial and intellectual investment and adding to the potential heritage, cultural, historical and its existence as a place, low-density territories are able to mesmerize visitors and simultaneously create a different and unforgettable image of the Portuguese countryside (Carvalho, 2018). Low-density territories have, as a tourist destination, an opportunity to redevelop the thinking of place, city and territory, despite the capacity of this type of territories that often are listed by the structural difficulties in existing as a tourist destination.

Despite being laborious and time consuming, the fact is that some projects have been designed with the aim of promoting and increasing the potential of regions and reducing the demographic desertion of territories (Ramos, 2015).

Then, it is up to the territories to aim constant development at the most varied levels. At the governance level, growth must be faced not only in the quality of life itself, but also as a tourist destination. In other words, the greater the investment, sometimes not financial, in expanding the knowledge of the territory as a place of visit by other citizens, the greater will be the clarity in the perception of itself as a tourist destination. So, regardless of its size, area or population density, the territory must seek, invest and aim to the evolution of its intellect (Pedro & Marques, 2010), and it will only be bright when it finds a balance in the various layers that take part of the building of collective intelligent thinking:

«The idea of Smart Cities means referring to a model in which, above all, the relationships between citizens and institutions, the world of the economy and, above all, between citizens themselves, are adapted in the search for a new balance. » (Aieta, 2016:1622)

Consequently, approaching the case of low density territories and their ability to adopt measures and strategies that allow them to make cities and destinations smart, for example, it is understood that, basically, governance and the way in which the territory is managed are fundamental for the formation of a common thought oriented towards a collective intelligence and methods of action that make the lives of places and visitors more effective, safe, technological and essentially intelligent (Loureiro, 2017).

In order to a territory to develop its competitiveness in the tourism sector, it is vital to take into account whether, in first place, the supporting factors and resources are strategically recognized, such as infrastructure, accessibility, facilitating resources and enterprises, and if, in a second level the main features and attractions, such as superstructures, special events, the diversity of activities, the marketing bonds, the destination's culture and history, and physiography (Boes et al., 2016). A third level is linked with the management of the destination itself, to which five basic topics are related:

1) service,
2) information,
3) organization,
4) marketing and
5) management of destination resources.

Four decisive factors are also recognized in the qualification of the destination in terms of its competitiveness (Koo et al., 2016):

1) the location of the destination,
2) its cost of living,
3) its safety and
4) its needs

Still, authors such as Brent Ritchie & Crouch (2010) argue that the destination must function as a company, that is, have its purpose and its identity well well-defined. Based on this perspective, the elements inherent to destinations are:

1) stakeholders1
2) the general public,
3) local inhabitants,
4) the media,
5) financial institutions and
6) government departments.

Despite approaches related to the management of tourism destinations, destination leadership and their governance contribute to relevant advances in tourism research, what is certain is that authors such as Volgger et al (2021), refer that the current dynamics, on a global scale, in relation to political, social, technological, and economic developments requires a review of traditional methods and instruments for developing destinations. Tourism destinations are beginning to be able to anticipate behaviour and customize their offer, thus becoming to be intelligent.

LITERATURE REVIEW

The Sustainability of the Tourism Destination

Sustainability is becoming a priority for Governments, businesses, institutions, people and it also includes cities (Baran et al., 2022). Ritchie and Crouch (2003) refer that sustainability on behalf of touristic destination managers must be based on a holistic view, stimulating the attraction of visitors; promoting local services, products and industry and improving the standard of living of local people.

Smart is generally defined as intelligent, which is the ability to learn and think quickly and show good judgement. Thus, when applying the concept of smart to devices and artifacts, smart should mean that the device is able to act in an intelligent way, learn and think quickly, and show good judgement. Most of these characteristics seem well adapted to the current capabilities of technological devices or to capabilities being developed under the artificial intelligence framework. (Bulchand-Gidumal, 2020)

Kuang and Ai (2016) mention that there is a constant concern to ensure sustainable territorial development, both with regard to Smart Tourism and Smart Tourism Destinations, namely in social, cultural, environmental and economic aspects, always supported by technology. The objective is to increase the satisfaction of tourists and visitors and improve the quality of life of resident communities (Santos Júnior et al., 2019). Thus, and reinforced by Gónzalez-Reverté (2019), Smart Tourism is a paradigm that is based on the transversal use of sustainability. In other words, there are no Smart Destinations that do not have sustainability as a priority in their designs.

The United Nations (UN), directing its efforts towards a more balanced society, developed a document that mainly aims to define objectives to create a common vision regarding the sustainable development of the World:

The 2030 Agenda is a broad and ambitious agenda that addresses various dimensions of sustainable development (socio, economic, environmental) and that promotes peace, justice and effective institutions. The Sustainable Development Goals (…) are the result of the joint work of governments and citizens around the world (…) they are the common vision for humanity, a contract between world leaders and people and "a to-do list on behalf of people and planet. (UN, 2015)

In turn, with regard to the travel experience, sustainable tourism must be included in the promotion of remarkable experiences for those who visit the place, impacting visitors and locals socially. In another way, when addressing public policies and governance, topics such as taxes, regulation, collaboration and intervention should be focused in themes such as: taxes, regulation, collaboration and intervention.

Thinking about transport, more sustainable alternatives must be created for everyone's mobility and finally communication, where the media has a fundamental role in promoting the destination's image and have the ability to influence who visits and who is visited. All this must be managed through a common vision, interconnected and, above all, with a focus on sustainability (Smith, 2017).

Sustainable tourism is as a consequence closely linked to Smart Tourism. Smart Destinations must have the ability to consider thematic outcomes such as planning, management and governance. In these outcomes, intellect matters such as urban mobility, education and research, sustainable constructions, sustainability in urban planning, energy efficiency, and everything related to Information and Communication Technologies (ICT) (Beck et al., 2018).

In short, according to López de Avila (2015):

An innovative tourism destination, built on an infrastructure of state-of-the-art technology guaranteeing the sustainable development of tourist areas, accessible to everyone, which facilitates the visitor's interaction with and integration into his or her surroundings, increases the quality of the experience at the destination, and improves residents' quality of life.

THE SMART TOURISM

With the growth of tourism in recent decades, several changes and forms of action have emerged. In parallel with this growth, technology developed and both became dominant sectors worldwide. This has given rise to a new type of tourism – Intelligent Tourism.

Regarding this topic, recent research guides research into issues such as privacy preservation, theoretical contribution, context awareness, cultural heritage, recommendation systems, social networks, the internet of things (IoT), user experience, real time, user modelling, augmented reality and big data2 (Kontogianni & Alepis, 2020).

Apropos the theoretical contribution of the theme, some authors advise interpreting Smart Tourism in three perspectives (Biloria, 2020):

1) the transformation of the tourist experience;
2) the development of bilateral strategic communication and
3) the competitiveness of the destination.

As far as context awareness is concerned, it is closely related to the technological explosion of recent decades and the influence it has had on the tourism sector. An example of this is the use of electronic devices, especially smartphones, which many of us already consider "part of the human body", that is, the amount of time that an individual spends daily with their device and through the sensors it contains, it is possible to extract data such as the date, time, location and ambient temperature, among others.

Exactly in the tourist context, these data are fundamental and allow companies in the sector to create personalized services, facilitate decision-making and even understand user preferences (Hasija et al., 2020). Also, social networks are essential in today's daily life. More than that, they are an infinite source of data3 that can be used by various sectors. According to some authors, in Smart Tourism this type of information extraction is done in two steps (Inversini et al., 2014):

1) By logging in users on social networks, such as Facebook, where they give permission for the use of their data through authentication.
2) Then, after advanced techniques of deep analysis, the systems are able to determine the preferences of the users through the published photographs and classify them, for example, in: cultural tourism, shopping tourism, among others.

From that point on, the above-mentioned recommendation systems are able to suggest, according to the tourist profile, the best attractions to be visited (Kontogianni & Alepis, 2020).

Still, in the context of tourism, the internet of things or IoT is and will be increasingly capable of originating an intelligent technological environment capable of understanding the tourist scenario and, thus, meeting their needs in a personalized way (Verma & Shukla, 2019). The influence that the IoT will have on the tourism sector, according to Verma & Shukla (2019), will be present at several levels, being categorized as: travel internet; of destinations; restaurants and kitchens; of hotels; from airports; of flights; trains and tourist attractions. For the same authors:

Positive impacts are addressed throughout the entire process: accurate data collection, acute control of user personalization, seamless travel, connectivity and smart solution for sustainability goals, energy savings, maintenance, repairs and detailed understanding of users through of real-time data analysis (Verma & Shukla, 2019:2091)

In terms of the tourist experience, it is understood that the more pleasant the moments experienced by the users, the more positive impact this will have on the evaluation made. According to Kontogianni & Alepis (2020) it is undeniable the importance of interpreting the traveler's needs and also their way of communicating and interacting with others. Moreover, the authors believe that in order to personalize an experience, three points must be taken into account:

1) the ability to adapt the experience to the tourist;
2) the motivation of the visit and
3) the expectations of the visitors.

The term real time is one of the most used when discussing the topic of Smart Tourism. Its importance becomes unquestionable, since the interconnection and synchronization of the technologies used together with the combination of collected data form a technological ecosystem, promoting an interdependence between users and devices (Vargas Sánchez, 2016).

In another sense, in the face of not being recent, the theme of increased reality has been gaining relevance with the world technological development, essentially by allowing the user to experience an environment that in reality is not real, seeing and hearing through electronic devices. In terms of tourism, this can be used, above all, by providing tourists with "contact" with fundamentally cultural, heritage or educational tourist attractions (Shih et al., 2019).

Lastly, big data is probably one of the most studied topics in the literature. Its crucial importance for the most diverse sectors of activity is indisputable and tourism has been able to take advantage of its value. For Lee et al. (2020), the amount of freely available data from social networks, sensors and mobile applications is a source of value creation and promotes the creation of a smart travel experience. However, Kontogianni & Alepis (2020:8), believe that in spite of the undeniable value of the data, the challenge goes through:

Mainly, the aim is to determine how Big Data can be most effectively sightseen in the Smart Tourism sector, as well as the challenges that arise when it comes to exploiting big data.

For that reason, the idea of Smart Tourism involves applying the concepts defined for smart territories, in a tourist context, that is, through a technological perspective, developing the destination, in order to improve the existing quality of life and capture data that allow the creation of an innovation ecosystem related to Smart Tourism (Proença Junior & Duenhas, 2020). According to Gretzel (2018), Smart Tourism consists of five categories:

1) the physical layer, where human and natural tourism resources are involved;
2) an intelligent technology layer that connects with physical infrastructures and promotes intelligent solutions and applications;
3) a data layer, that is, crucial information that can be processed and further analyzed and applied;
4) the business layer, which aims to innovate using available technologies and existing data; and finally,
5) the experience layer, which encompasses all the technological experiences of the territory.

Furthermore, governance becomes crucial, as it must serve to ensure the necessary investments, facilitate the coordination and monitoring of the process and, above all, define and promote common intelligent objectives (Inversini et al., 2014).

To Vargas Sánchez (2016), Smart Tourism focuses on the creation of experiences and business value proposals with attention to increasing efficiency, sustainability and enrichment of the experience, all of this based on the collection and creation of data derived from physical infrastructures, social connections, governmental and organizational sources, as well as the inhabitants and tourists themselves. Globally, Smart Tourism aims to improve governance, promote the emergence of innovative services and products, enhance the tourist experience and increase the competitiveness of companies and the city as a tourist destination (Lee et al., 2020). For that reason, according to Gretzel (2018:174):

Thus, while Smart Cities only consider the requirements of residents and local businesses, Smart Tourism deals with a much more dynamic ecosystem, in which inhabitants can be producers of experiences (…)

Even so, in regard to Intelligent Tourism, it is considered appropriate to record the concept of metaverse, which is associated with virtual reality and which can potentially involve socialization, involve business and economic activities, as well as signify reality itself. The outlook is not only to transport people from the real world to the virtual, but also to bring the virtual world to the real.

With the metaverse it will be possible to visit any place in the world through each one's avatar. In the final destination, a meeting with friends and family who are physically scattered to the four corners of the world, but are all gathered in the metaverse. And this reality may be much closer than it seems!

There are Startups that are already putting their gears into practice to promote travel experiences that have full potential to expand into the metaverse.

One example is CityGuyd, which combines new advancements in Amplified Reality (AR) technologies to provide travelers with a more convenient, accessible and exciting tour guide experience. No matter where the next adventure is, AR guides are on hand to show you around.

The new generations are already more than adapted to this type of technology. On that account, travel agents should also keep an eye on future updates to promote completer, attractive and even more immersive trips, even without leaving home.

Small steps in the present towards a near future full of possibilities and innovation.

in Strategic Marketing Intelligence

THE COMPETITIVENESS OF SMART DESTINATIONS

The basis of competitiveness is innovation (Baran et al., 2022). Consequently, ICT-based smart solutions are the starting point for defining Smart and Competitiveness Destinations. To Koo et al. (2016), a destination does not become intelligent without first becoming a geographic space concerned with improving the quality of life of those who inhabit it. In this sense, a Smart Destination is, then, one whose strategy throughout the tourist cycle addresses dimensions such as innovation, technology, sustainability and accessibility, consistently aiming at the social well-being of those who live and who visit. Smart Destinations can be the future of tourism, if through the integration and analysis of available data, they make decision-making more efficient, and the tourist experience unforgettable (UNTWO, 2020b). So, each destination has resources to attract visitors, however, its planning and development should focus on a detailed analysis of such attributes that offer a unique comparative advantage (Phillips & Moutinho, 2014). Such a planning process must, consequently, integrate all stakeholders, whether public or private authorities, as well as the local community (Buhalis & Amaranggana, 2015). Otowicz et al. (2022) identify the technological dimension as the most evident when analyzing the origins of Intelligent Tourism, essentially between the years 2013 to 2021. The same authors state the existence of other important dimensions in the definition of Smart Tourism Destinations, such as: the experience, sustainability and cooperation.

For Brandão et al. (2016) Smart Destinations must integrate a wide range of technology and services that serve tourism. In addition, the use of devices that monitor tourists (Susanti et al., 2016), smart transport (Deloitte, 2018), smart payment and reference applications can be fundamental (Siokas et al., 2021).

Brandão et al. (2016), based on the reference model for Smart Cities developed by Zygiaris (2013), created a reference model for Smart Tourism Destinations as presented in Table 1 - Reference model for Smart Tourism Destinations.

Table 1. Reference model for Smart Tourism Destinations. Source: Brandão et al., 2016

Order	Name	Focus
0	Tourist Destination	Fate's readiness to become intelligent
1	Green Destination	Sustainability
2	Interconnected Destination	Broadband infrastructure
3	Instrumented Destination	Real-time events
4	Integrated Destination	Open and integrated space (data)
5	Tourist Applications	Smart Services
6	Innovative Destination	Smart growth

Similar to the model by Zygiaris (2013), the base layer (zero) addresses the ability of the territory to become an intelligent destination, idealized as a tourist destination. At this stage, planning and a holistic view are crucial, as well as the appreciation and cooperation of all stakeholders involved. The creation of an identity as a destination and a strong and sustained image is essential, thus forming an intelligent vision that must be shared by all involved, not forgetting the locations (Phillips & Moutinho, 2014).

In the second layer of the model, called the green destination, the addiction that the tourism sector has on the environment is implied. In this sense, all players have been developing strategies and tools that allow, on one hand, for example, to make tourist developments more eco-friendly and, at the same time, draw tourists' attention to the importance of a responsible attitude in their travels (Weekes, 2019). Accordingly, an interconnected destination addresses, in a first stage, the connection that must exist between all those involved, when it is intended that a Smart City develops as a Smart Tourism Destination and adopts strategies according to this objective (Miskalo Cruz & Gândara, 2016), on the other hand, the crucial importance of providing the destination with fast and comprehensive internet access, facilitating the connection to Wi-Fi at various points of the destination, for locals and visitors (Brandão et al., 2016).

Based on what was stated before, an intelligent destination is reliant on the sharing of information and the creation of value from the date collected. After collecting the data, mobile applications become able to correspond to tourist operations efficiently and in real time, thus creating tourist operations (Brandão et al., 2016).

Smart Destinations use this kind of technology in order to create an intelligent context, using recommendation and decision support systems, amplified realities and even smart tools capable of monitoring tourist areas (D. Wang et al., 2013).

To conclude, the destination becomes innovative when it creates an environment favorable to innovation and capturing new business opportunities (Brandão et al., 2016). So, new technological tools can be created that promote an intelligent tourism ecosystem, favoring innovation and the creation of ideas that expand the technical and physical capabilities of tourism and synchronously develop the quality of life of those who live there and those who visit the destination (Murphy et al., 2000).

Thinking about Smart Tourism as a connection between tourism and ICT, Gretzel et al. (2015) states that for the destination to become intelligent, there must first be awareness and the objective of making the territory intelligent. The concept of Smart Tourism Destinations emerged as ICTs came to be seen as the main drivers of destination competitiveness (Buhalis and Amaranggana, 2015). However, authors such as Gelter et al. (2020) argue that there is no clear, objective definition, common to all other authors, of the concept of Smart Tourism Destination, and this can even influence the research domain.

In 2009, Barcelona City Council presented its Smart City model, which aimed to improve the quality of life of residents and ensure a more sustainable and efficient future (Mora & Bolici, 2015). The initiative intended for the city to become a world reference and an ecosystem of urban innovation, while being able to respond to future challenges that the city could face such as overtourism4. Several projects were developed with the ability to become the holistic city and open to integrate innovative ideas and create the future, according to (SEGITTUR, 2015:102):

«As a result of this important commitment, Barcelona has attracted more than 1,500 new companies and (...) has created 45,000 new jobs related to its Smart City status. This adds up to significant savings and improvements in the quality of life for residents and tourists. »

A comparable example is the city of Amsterdam. The Dutch city, though, promotes an urban innovation ecosystem that encompasses companies, authorities and smart research laboratories. Nevertheless, the approach to planning, according to Mora et al. (2015) follows a more participatory model, integrating broadband economies and smart applications. Zygiaris et al. (2013) points to the capacity of Amsterdam to become a smart tourist destination, intending to use sustainability as a fundamental requirement, reducing costs and increasing efficiency.

In addition, the city intends to implement a fiber network in all homes and easing the day-to-day life of residents and tourists. Some parts of the city are already equipped with energy meters and electrical chargers that provide data for study. Amsterdam supports its Smart City and destination plan, in projects such as EcoMaP and Onze Energy, capable of measuring energy and, through windmills, lowering energy costs (Lee et al., 2020).

Some destinations, essentially with financial power, have been investing mainly in technology at the service of tourism, examples of China and Dubai (Deloitte, 2018). In the case of Dubai, in addition to the implementation of projects such as electricity and smart water, leading to a sustainable use of resources, support programs are also being designed to make health and public and private services increasingly intelligent and sustainable in a way to improve the quality of life of the inhabitants (Virtudes et al., 2017).

In Europe, the evolution of cities and territories towards their digital transformation, the integration of the Internet of Things (IoT) and the improvement of social cohesion are promoted by Governments, namely through the application of public financial support from the European Union, more specifically, from the European Regional Development Fund (Orejon-Sanchez et al, 2022).

As far as the tourism sector is concerned, the "City of the Future" aims to be a reality and revolutionize tourism and commerce in Dubai, through blockchain. Dubai's tourism vision involves welcoming 20 million visitors and investing in Business to Business (B2B) tourism, making payments smarter and initiating a significant bet on cryptocurrencies (Bodkhe et al., 2019).

China's proposal for a Smart tourism Destination is to focus on IoT and Cloud services, focusing on visitor profile management and data collection and subsequent analysis. improvement of digital payments and the fight against overbooking (M. Wang et al., 2020). However, for D. Wang et al. (2013), China still has some limitations due to the amount of information it enhances and, despite the expensive procedures, technology can be the solution to a safer, smarter and more sustainable destination.

In short, a destination that intends to be intelligent must take into account a logical and sustained follow-up. According to Santos Junior et al. (2017), the planning of a Smart Tourism Destination should be based on four pillars:

1) technology,
2) accessibility,
3) sustainability and
4) innovation.

Yet, one of the most mentioned imperfections in Smart Cities and destinations is the lack of monitoring. To Hasija et al. (2020), territories should be concerned with both monitoring and implementation of projects, since it is through measurable metrics and strategies that the destination will understand whether a given plan is achieving the previously defined objectives. Deprived of a good monitoring, the territory is at risk of, in one hand, not taking full advantage of the information collected and, on the other hand, deviates itself from the traced route.

To understand the intelligence of tourism destinations requires the dynamic interconnection of stakeholders through a technological platform on which information related to tourism activities can be exchanged instantly. This information dynamics has been promoting extremely large data sets, known as Big Data, that can be computationally analyzed to reveal patterns and trends. In terms of Big Data, Smart Tourism Destinations must make optimal use of data, offering the right services that match the preferences of users at the right time. In this sense, it is possible to understand the existence of a relationship between the data and the tourism experience, through the guarantee of more personalized products/services to meet the unique needs and preferences of each visitor. Considering the needs, desires and desires of travelers becomes increasingly fundamental for the competitiveness of destinations (Buhalis and Amaranggana, 2015).

METHODOLOGY AND ANALYSIS OF INTERVIEWS

Methodology and Objectives

Methodology is a fundamental condition when carrying out an investigation. This must be the guiding thread in the definition of objectives and respective means to achieve them. At the same time, it allows the document to be rationally organized apart from knowledge in a dangerous way and other persecuted researchers (F. L. dos Reis, 2018).

That being the case, in the present investigation, in addition to the literature review, the descriptive qualitative analysis was adopted as a methodology, insofar as there is above all a predominance of descriptive analysis in order to offer a better perspective and understanding of the problem. Following this idea, the data collection instrument used was structured interviews. A set of standardized questions was constructed aiming at an analysis of very specific research objectives.

Stakeholders from the city of Guarda were identified who, through the response to structured interviews, provided valuable content in order to achieve the following previously defined objectives:

- To become aware of the perception of the stakeholders involved in the process, about the theme;
- Characterize and identify the infrastructure necessary for this transformation to take place;
- Understand the concern of this type of territories to develop as a city and smart destination.
- General objective: To understand how low-density territories can become a smart tourism destination.

Analysis of Interviews and Conclusions

The choice of respondents was carefully made and was essentially based on two parameters. On the one hand, individual or collective people who were present and used the analyzed territory and, on the other hand, had knowledge and influence in the city of Guarda as a Smart City.

Continuing with the analysis of the interviews, it is noted that the attention given to smart cities has gradually increased. The look at the theme has been taking different points of view. Regarding the way stakeholders perceive the city of Guarda, as a Smart City, Vitor Amaral, Vice-President of the city of Guarda admits that the territory has developed in this direction and guarantees that it is preparing itself in this regard, "The use of ICT in urban organization is a process in the implementation phase, under the of a candidacy (PEDU), aiming to provide the city with a communication axis capable of responding to the demands that this new reality brings to municipal management". In the same logic, Carlos Chaves Monteiro, Mayor of Guarda until October 2021, admits that "A smart city is a process that we can never say is finished. It is a process of continuous improvement and a governance option. In this sense, this has been the strategy adopted by the city – defining and building the foundations for the construction of a smart city, from the most basic levels of service to the citizen, monitoring and even renewed forms of promotion using ICT".

In the next segment, it was verified how the interviewees understand that the city can use its strategic location, in order to develop itself as a Smart City and later as a Smart Tourism Destination. Vitor Amaral states that "(…) in addition to the geostrategic position of the city, it is always necessary to innovate, create new products, with the aim of providing appealing tourist experiences, with the potential to attract new markets and/or improve the satisfaction levels of current tourists. (…).

The analysis shows that policy makers refer to the issue as crucial for the smart progress of the city/ destination. Taking advantage of the privileged location of the city, according to Carlos Chaves Monteiro, public policies should favor the evolution of the territory and effectively add value to the city, "(…) Guarda is the gateway to Europe (…) Hence the ability to create the conditions for the affirmation of this privileged position have to be very well thought out and strategically implemented (…)", also identifying strategies to be used, for example in the tourist case, "(…) with the support and monitoring of entrepreneurial projects, with demands in terms of innovation and product structuring (…)".

This confirms a widespread concern, with regard to the city of Guarda, to become more intelligent, particularly in relation to political actors. Vítor Amaral declares: "The city of Guarda is creating, in a guided and rational way, measures that allow the creation of bases for this implementation. (…) One of the challenges of the next decade is to apply an integrated approach methodology to the five strategic pillars – governance, technology, sustainability, accessibility and innovation". The speech is clearly balanced with that of Carlos Chaves Monteiro, who says "Guarda is a city that wants to progress (…) For the next decade the focus is on applying a methodology of integrated approach in the five strategic pillars – governance, technology, sustainability, accessibility and innovation. For this, we are working on a strategic tourism plan for our territory, with the aim of valuing Guarda as a distinctive destination, through the best use of natural and cultural attractions; innovative activities and equipment, meeting the challenges of a smart city".

Carla Basílio, from Turismo do Centro (Regional Tourism Entity), mentions that "(…) for a city to become smart "(…) a dominant factor is needed – leadership. It is necessary to organize the integration of synergies of both information and decision-makers and opinion makers, enabling new concerted ways of thinking about the city, leaving the traditional and redirecting to intelligent ways".

An opinion was sought on the possibility of low-density territories, sometimes associated with an idea of low competitiveness and with aging populations, being able to transform themselves into Smart Destinations, Carla Basílio, confidently states that "Yes, it is possible. It is not just in areas of high population concentration that structural modernization (systems related to technologies applied to urban space, information and communication) are key elements in the process of transforming conventional cities into smart cities."

Rui Ribeiro, from IAPMEI, believes that "(…) it does not seem to me that low density alone is an impediment to transforming the territory into a smart tourist destination". And then he points out a key idea, "(…) a low-density territory can have an interesting potential for the implementation of innovative projects in a test environment, thus functioning as an incubator for larger projects." This view is supported by Carlos Chaves Monteiro, "I believe and understand that it is in low-density territories that the results of public investment at this level can be shown more immediately and efficiently."

The Vice-President of the city of Guarda, reveals that "(…) due to the data of post-pandemic tourist demand, we are a territory that offers security, tranquility, landscape and pure nature, authenticity, distinct and unique heritage, gastronomy of excellence, in sum quality of life associated with principles advocated today in the goals of sustainable development".

It is possible to be concluded that although there are still constraints, it is possible and of the utmost interest to create Smart Destinations in low-density territories, such as the case of Guarda, which was taken as a case study. It was possible to verify the consonance between the interviewees, aware that through Smart Regions they can enhance the attractiveness as a Tourism Destination, simultaneously aiming at the well-being of its residents. It is clear that a territory will be all the more "intelligent", with the idea of an ideal city, if it is built as a collective political project, in the classic sense of the polis. On the other hand, low-density associated to the interior is not a casualty.

THE CASE OF GUARDA

Geographical Framework

Located next to the border with Spain, more specifically in Beira Alta or in Beiras and Serra da Estrela, being the seventh largest district capital in Portugal, the city of Guarda has 43 villages. Besides, being known for its altitude, having Torre de Menagem as its ex-libris, Guarda benefits from its location in the interior of the country, on that account becoming a unique business platform, attracting both imports and exports, having exported more of 205 million euros and imported around 220 million euros, data from 2013. Such centrality in the Iberian Peninsula, gives the region the potential for affiliations between Portugal and the rest of Europe, as it is a central point not only on road routes, but also railways. In addition, it has gained status in the connection with the main seaports, such as Aveiro and Leixões, being considered a central access corridor in this sense (Guarda, 2015).

One of the strongest projects with regard to low-density cities in Portugal was the Guarda City Council, designed in 2018 and with a long-term view, 2030. Through intelligent thinking that aims to contribute to the valorization of the interior territories, Álvaro dos Santos Amaro, at the time mayor of Guarda, stated in 2018 to the website www.smart-cities.pt, that being an intelligent territory was an "objective priority in solving local problems, in a way that sustainable and with a strategic vision for the future.", and promised to go in search of "(...) alternative paths, based not only on technology (...) but on creativity and the boldness of involving public and private partners in a survival project. territorial. A mission that demands leadership and courage." This type of vision is demonstrated at the time of application of the strategy. For Álvaro dos Santos Amaro, the Guard cannot be "afraid of trying new things and doing what no one else had the courage to do, taking risks, it is true, but guaranteeing its citizens that resignation is not an option, nor is stagnation an option. strategy. A mission that requires determination, character and conviction." (Smart Cities, 2018).

Once again, this Guarda´s project presumes a network connection between all the stakeholders involved, greater coordination in governance, greater sustainability of the place and, fundamentally, the constant improvement of the quality of life of citizens and visitors, facilitating through technology and not only, the daily life of those who use the territory, therefore creating tools and strategies to settle the population in a low-density city, but with focus and sustained thinking. From this perspective, it is understood that, although in a general panorama, low-density territories in Portugal are not usually mentioned as possible smart territories, the truth is that in a more regional governance, projects begin to emerge and several cities believe that this way can offer its inhabitants a better quality of life and make their city a better place to live (P. Reis & Baltazar, 2019).

In terms of Tourism, the city of Guarda perceives an opportunity, in the fact that it is a low-density territory. Its equidistance between Aveiro and Salamanca (159km), offers the city unique capabilities, making the territory an alliance of connection to Europe, also adding Vilar Formoso as the gateway to the European continent. Both Francisco Sá Carneiro ´s airport, in the city of Porto, and Salamanca´s airport are less than two hours away by car. According to PORDATA, in 2019 the City of Guarda had 19 tourist accommodations, 13 more than in 2010 (PORDATA, 2019).

It is a territory that presents several tourist attractions. Guarda, in addition to Torre de Menagem, has a Cathedral, built between the 14th and 16th centuries, Porta D'El Rei, and Judiaria in the historic center and which demonstrates the relationship between Christians and Jews in the past.

The existence of several ancestral trails also becomes a fundamental part of tourism in Guarda and allows tourists to visit historic villages, witnessing traditions that are still in place today. Besides to the mythical handicrafts of the City, the gastronomy usually called strong and abundant, are crucial charms in attracting new visitors and establishing residents in the territory (Portugal, 2018).

The connection between the city and the touristic sector can be demonstrated by the organization of the well-known Iberian Tourism Fair (FIT), which, presents different and innovative projects in the area, promotes the close relationship between the countries that make up the Iberian Peninsula – Portugal and Spain (Pais, 2017). So, the desire of this type of City to be able to attract more tourists and offer better living conditions to the locals has been increasing.

The Intelligence of Low-Density Destiny

Concerns and focus have changed in recent times and topics such as sustainability, carbon reduction and energy efficiency are in vogue. The achievement of these goals can be accelerated with the help of technology, which facilitates and allows any territory to become intelligent (Loureiro, 2017). But one of the biggest research questions in this area is the attempt to understand how the size or location of the city and its population density can influence the fact that a low-density territory becomes a Smart City (Lopes & Oliveira, 2017). In fact, despite being uncommon, there are some studies that point to this issue. A project that involved the 308 Portuguese cities, states that, although the territories located on the coast are most benefited due to their importance to the country and their financial power (Lopes & Oliveira, 2017:624):

(…) we can also say without a doubt that small and medium-sized cities are not behind the big ones when it comes to "Smart Cities" and that they have enormous potential to leave their mark on this concept. Living in an inner city is undoubtedly a "smart" concept. Despite all the obstacles inherent in locating these cities in sparsely populated areas, the resilient way in which these cities are overcoming all these contingencies is truly remarkable.

So, low-density territories, such as Guarda, are beginning to pay attention to this new phenomenon. In the tourism sector, the city intends to, in one hand, to appreciate Serra da Estrela as a tourist attraction to attract visitors to the region, but, at the same time, that there is no dependence in this sense (Pais, 2017). On the other hand, it aims to be a unifying pole, to the extent that, through a network of neighboring cities, that is, a synergy, taking advantage of the potential of each place, it is possible to achieve some objectives, such as increasing the length of stay, the repetition of the visit and the possible establishment of new inhabitants of the city, purposes present in the strategy of Guarda's application for European Capital of Culture 2027 (Guarda, 2027).

In this way, Guarda seems to have all the credentials to become, in the first place, a Smart City and, so, a Smart Destination, using the existing tools to achieve regional goals, as Álvaro dos Santos Amaro states, "the city of Guarda understands that being a smart city should not be a sound bite. It must be a priority goal in solving local problems, in a sustainable way and with a strategic vision for the future" (Smart Cities, 2018:1). However, it is worth noting the results of some recent research on Smart Cities that initiated a holistic dialogue for the integration of previous initiatives that had the ability to promote e-government at the European Union level (Lytras and Serban, 2020), being considered a of the great advances and indicators of an intelligent destiny.

Presentation of the Model for Low-Density Smart Tourism Destinations and Future Recommendations

The proposal is based above all on the conviction that low-density territories can and should bet on solutions that make their territory more intelligent in the various dimensions of society. Thus, an application model was created, present in Figure 2.

This model aims to create a concept – Low-Density Smart Tourism Destinations and addresses the issues that are considered most important in the transformation of a city, such as Guarda, into an intelligent destination.

It should be noted that the development of the model was based on two fundamentals: the literature review, which highlights the SCRM (Smart City Reference Model) presented by Zygiaris (2013) and the inputs provided by stakeholders in interviews.

The core point of the model is the fact that a territory, regardless of its size, must complete the following journey: City – Smart City – Smart Tourism Destination, in the sense that the creation of a collective thinking that promotes intelligent transformation becomes essential to the city/territory and this step must be taken, imperatively, before converting into a smart tourist destination. It is believed that this condition promotes a more aligned and objective trajectory.

As in the SCRM, five dimensions were defined, which are considered fundamental for this transformation, namely:

1) Governance;
2) Technology;
3) Sustainability;
4) Quality of Life and
5) Mobility.

It is understood that, by focusing on the following dimensions, the following objectives are achieved:

a) the creation of a more competitive destination;
b) the creation of a more sustainable destination;
c) the possibility of offering visitors a personalized offer and
d) better quality of life.

Figure 1. Low Density Smart Tourist Destination Model. Source: Self elaboration

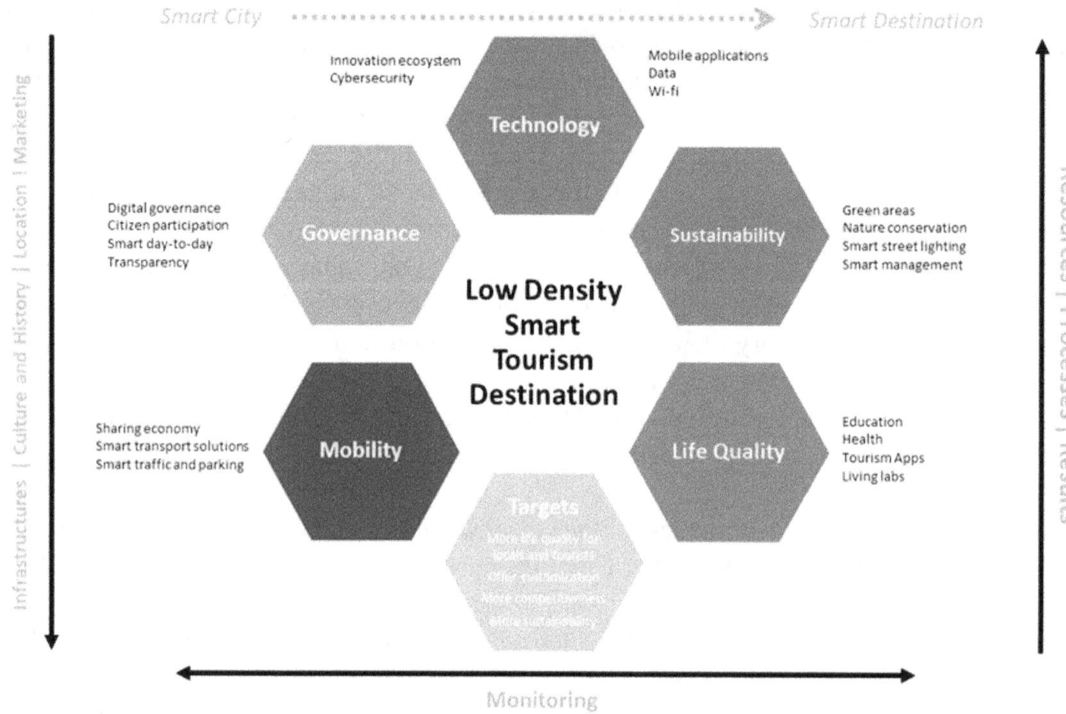

Interpretation of the Proposed Model and Conclusions

Regarding the Governance dimension, it is advisable to move towards a more digital governance, encouraging the participation of citizens in the causes of the cities, as well as the transparency of processes, which is why it is believed that it will make everyday life smarter. This progress includes actions related to city´s bill payment processes by citizens; online consultation of the various dimensions of society; promotion of a more digital health, among others, aiming to facilitate the day-to-day, both for the local inhabitants and for all those who visit the city.

This way, another dimension emerges – Technology. As this is a crucial part in the intelligent transformation of territories, it is suggested the development and creation of mobile applications that, for example, the guarantee of answers to the arguments mentioned in the Governance dimension. In addition, the availability of quality and free wi-fi seems to be a smart solution and capable of favoring locals and tourists alike. On the other hand, it is recommended, in line with one of the objectives of the model, to invest in the theme of the date, that is, to reinforce the ability to collect, preserve and process data in order to offer personalized products and services to those who want it. visit the destination and can also be used in various other matters of society.

Foreseeing a smarter territory, the desire to be more sustainable emerges as a consequence, one of the focuses of the model presented. So, it is understood that the sustainability of a smart tourist destination must go through the creation of green spaces for the enjoyment of locals and visitors, a constant concern for the preservation of nature and the promotion of the use of greener processes. It is believed that this type of procedures, such as efficient public lighting and increased recycling, favor intelligent management and the conversation of the territory.

The focus on preserving the environment must simultaneously be demonstrated in the Mobility dimension. In the model, proposals are suggested that aim to protect the environment, as well as provide resource savings for those who enjoy the territory. In this way, the sharing economy stands out, which can be applied in various areas of society, and the creation of smart transport solutions, which have the potential to emerge as a bet on promoting smart mobility. It is believed that the improvement of mobility in the territory, together with a promotion and explanation of the concept, can facilitate the daily lives of the locals and attract more visitors to the region.

All dimensions previously explored give rise to one of the main objectives of the model – greater and better Quality of Life. In this context, several themes are underlying, such as health and education, where a commitment to the development of technological solutions, such as mobile applications, capable of solving problems and valuing these areas is suggested. As an example, we suggest the design of projects that enable, through technological tools, easy contact with the health media, especially for those visiting the destination. This aspect is considered crucial, usually related to safety and trust in destinations.

In the touristic context, the development of technological materials that add value to the territory as a destination is fundamental. Thus, examples such as a complete, updated and responsive website, in addition to the creation of mobile applications capable of guiding visitors to the destination, indicating the most famous attractions, suggesting points of interest (example: gastronomy), based on tourist tastes (study date) and also reveal waiting and travel times at locations within the destination. Finally, it is suggested the creation of Living Labs, a concept of open research and where, through public-private partnerships, an ecosystem of innovation can be developed, essential in any destination that intends to be intelligent.

The model points out fundamental aspects that territories must identify and, subsequently, take advantage of, to enable the necessary transformation into smart territories.

The identification of existing infrastructures, especially those of a tourist nature; a strong commitment to the recognition of its culture, heritage and history (creation of a storytelling5) and a critical analysis of its location, it is understood as essential.

All these aspects, based on an evident commitment to the promotion and marketing of the destination, can be the key elements for evolution. Lastly, it is believed that there are three key criteria for the success of the model:

1) recognition of the resources current in the territory;
2) the subsequent definition of the processes to be developed to obtain the intended objectives and
3) the systematic analysis of the results obtained during the entire journey.

It is also argued that the application of the model requires continuous monitoring, in order to understand whether adjustments and improvements are needed in the process.

CONCLUSION, RECOMMENDATIONS FOR FUTURE INVESTIGATIONS AND LIMITATIONS

Several authors reflect in their investigations a growing trend towards the investigation of Smart Tourism and Smart Cities (Ye et al., 2020; Bastidas-Manzano et al., 2021 and Otowicz et al., 2022). Other authors have revealed the relevance of the topic by investigating the results of National governmental and European projects, or even through mere web searches (Gretzel, 2018). This fact leads us to realize that the topic is in permanent evolution and, consequently, requires careful analysis and investigation.

Understanding the intelligence of the Tourism industry and the territories themselves (not only from a local but also regional perspective) can be one of the most relevant guidelines for tourism research in the 21st century (Savić & Pavlović, 2018), thus providing a vast field of research. research – whether theoretical or practical (Ghaderi et al., 2018; Xiang, 2021).

Clearly, Smart Destinations, in a low-density context, are associated with a strategy, and which will have as a differential the diversity of sustainability and creative diversity segments, through an innovative and creative solution, and always oriented to revitalize and value the identity heritage for sustainability and environmental conservation, contributing to the increase the value of the tourism experience.

Lastly, as determined in this article, it is believed that low-density territories can become an incubator platform for projects related to Smart Cities and destinations. In this perspective, it is advisable to build projects/models that promote this type of territories in this sense, that is, identifying valid arguments that demonstrate usefulness and relevance for those who want to apply intelligent projects. By the same ideas, it is understood that the involvement of governments and the other stakeholders in territorial or regional strategies are promoters of the development of a Smart Tourism Ecosystem and that is very favorable to destinations.

Bearing in mind the enrichment of the research carried out, as future recommendations it is proposed to replicate the study in other low-density territories in Portugal or abroad, suggesting the use of the same research technique used (structured interviews with stakeholders) in order to have a similar view of the respondents' perception.

It is also suggested the application and development of the model, with a view to improving its design. Thereunto, a practical case can be used that aims to provide credibility to the investigation and from which data for possible future use in studies on the subject can be extracted, since one of the main problems in this context is the lack of information that proves the feasibility of the projects.

So far, with regard to possible future investigations, in a theme as dynamic as this one, the following is suggested:

- Examine the priorities and lines of progress of destinations for territories and Smart Cities, where there is a clear holistic approach and specialization in topics such as tourism or smart buildings, from the areas of technology, software to governance itself.
- Carry out a comparative analysis with reference to the most recognized Smart Destinations and present solutions that lead to new lines of development of tourist destinations. Evaluate the impact of public funding on the pursuit of smart tourist destinations.
- Examine the internal functioning of the cities, in terms of organizational management, with a more efficient, more technological and environmentally friendly service provision, for example, leading to the definition of a governance model 'friendly' to Smart Cities.
- To proceed with the definition of an intelligent territorial development model that is inter-municipal and that promotes notorious and beneficial effects of network and cooperation in the promotion of a resilient ecosystem of collective intelligence, of sustainable development, where intelligent citizens work and live together.

Despite the existence of various research limitations being evident and recognized by different authors (Wang et al., 2016; Bošnjak et al., 2017; Gretzel & Scarpino-John, 2018; Xiang, 2021), including the difficulty of practical application of Intelligent Tourism, as mentioned by Gretzel et al. (2015a); Xiang et al. (2015), Ballina (2020) and Lee et al. (2021), what is certain is that there is evidence of the need to deepen the themes related to destinations and smart territories (Xiang, 2021) and to recognize in them an opportunity for development and to build more resilient cities and regions (Baran et al., 2022).

In the specific case of this chapter, the main limitations were related with the response to the interviews and the absence of evidence about the understanding of the agents and stakeholders involved on the analyzed themes. In fact, it is important that Government initiatives can involve citizens in a collective intelligence strategy.

REFERENCES

ANMP. (2014) *Low Density Municipalities – criteria approved by the National Association of Portuguese Municipalities* [Municípios de Baixa Densidade - critérios aprovados pela Associação Nacional de Municípios Portugueses]. Author.

Ballina, F. J. (2020, October 5). Smart Business: The Element of Delay in the Future of Smart Tourism. *Journal of Tourism Futures*, *8*(1), 37–54. doi:10.1108/JTF-02-2020-0018

Baran, M., Kłos, M., Chodorek, M., & Marchlewska-Patyk, K. (2022, March 1). The Resilient Smart City Model–Proposal for Polish Cities. *Energies*, *15*(5), 1818. doi:10.3390/en15051818

Bastidas Manzano, A., Sánchez-Fernández, J., & Casado-Aranda, L. (2020). The Past, Present, and Future of Smart Tourism Destinations: A Bibliometric Analysis. *Journal of Hospitality & Tourism Research (Washington, D.C.)*, *45*(3), 529–552. doi:10.1177/1096348020967062

Beck, Silva Neto, & Melo Conti. (2018). *A new urban paradigm of smart and sustainable tourism* [Um novo paradigma urbano de turismo inteligente e sustentável]. XX Engema. http://engemausp.submissao.com.br/20/anais/arquivos/238.pdf

Biloria, N. (2020). From smart to empathic cities. *Frontiers of Architectural Research*, 1–14. doi:10.1016/j.foar.2020.10.001

Bodkhe, U., Bhattacharya, P., Tanwar, S., Tyagi, S., Kumar, N., & Obaidat, M. S. (2019). BloHosT: Blockchain enabled Smart Tourism and hospitality management. *CITS 2019 - Proceeding of the 2019 International Conference on Computer, Information and Telecommunication Systems*, 1–5. 10.1109/CITS.2019.8862001

Boes, K., Buhalis, D., & Inversini, A. (2016). Smart Tourism Destinations: Ecosystems for tourism destination competitiveness. *International Journal of Tourism Cities*, *2*(2), 108–124. doi:10.1108/IJTC-12-2015-0032

Bošnjak, A. (2017). *Smart Tourism Destinations: Enhancing Interconnection between Stakeholders and Tourist through Information Personalization and Feedback System.* Academic Press.

Brandao, M., Teles, A., Fluminense, U. F., & Joia, L. A. (2016). *Destino turístico inteligente : um caminho para transformação.* Academic Press.

Brent Ritchie, J. R., & Crouch, G. I. (2010). A model of competitiveness/sustainability of destinations: Brazilian perspectives [Um modelo de competitividade/sustentabilidade de destinos: Perspectivas brasileiras]. *Revista de Administração Pública, 44*(5), 1049–1066. doi:10.1590/S0034-76122010000500003

Buhalis, D., & Amaranggana, A. (2015). Smart Tourism Destinations Enhancing Tourism Experience Through Personalisation of Services. *Information and Communication Technologies in Tourism, 2015*, 377–389. doi:10.1007/978-3-319-14343-9_28

Bulchand-Gidumal, J. (2022). Post-COVID-19 Recovery of Island Tourism Using a Smart Tourism Destination Framework. *Journal of Destination Marketing & Management, 23*(Mar), 100689. Retrieved May 5, 2022, from. doi:10.1016/j.jdmm.2022.100689

Carvalho, P. (2018). Dynamics of rural low density spaces in PortugalDynamique des espaces ruraux de basse densité au Portugal. *Méditerranée, 130*(130). Advance online publication. doi:10.4000/mediterranee.10516

CIC. (2015). *Classification of Low Density Municipalities for the application of measures of positive differentiation of territories* [Classificação dos Municípios de Baixa Densidade para aplicação de medidas de diferenciação positiva dos territórios]. CIC.

Cities, S. (2018). *"Being a smart city should not be a soud bite" – Smart Cities* ["Ser uma cidade inteligente não deve ser um sound bite" - Smart Cities]. https://smart-cities.pt/opiniao-entrevista/alvaro-santosamaro3004guarda/

Costa, J., Rodrigues, D., & Gomes, J. (2019). Sustainability of tourism destinations and the importance of certification. *Worldwide Hospitality and Tourism Themes, 11*(6), 677–684. doi:10.1108/WHATT-08-2019-0050

Cronjé, D. F., & du Plessis, E. (2020). A review on tourism destination competitiveness. *Journal of Hospitality and Tourism Management, 45*, 256–265. doi:10.1016/j.jhtm.2020.06.012

Crouch, G. I., & Ritchie, J. B. (1999). Tourism, competitiveness, and societal prosperity. *Journal of Business Research, 44*(3), 137–152. doi:10.1016/S0148-2963(97)00196-3

d'Orey, F., & Abreu, R. (2015). The feeling of place and the construction of tourist destinations, proposal of a conceptual model [O Sentimento de lugar e a construção dos destinos turísticos, proposta de modelo conceptual]. *European Joural of Applied Business and Management, 1*(1), 208–223.

da Guarda, C. M. (2015). *Guia do Investidor.* http://www.guiadoinvestidor.mun-guarda.pt/en/files/assets/basic-html/index.html#1

da Guarda, C. M. (n.d.). *Guarda 2027, European Capital of Culture, Candidate City* [Guarda 2027, Capital Europeia da Cultura, Cidade candidata]. https://www.guarda2027.pt/

de Portugal, T. C. (2018). *Guarda – Tourism Center of Portugal* [Guarda – Turismo Centro Portugal]. https://turismodocentro.pt/concelho/guarda/

Deloitte. (2018). *The second wave in Smart City transformation About the Deloitte Center for Government Insights*. Deloitte.

dos Santos Júnior, A., Mendes-Filho, L., Almeida García, F., & Simões, J. (2017). Smart Tourism Destinations: a study based on the view of the stakeholders. *Revista Turismo Em Análise, 28*, 358–379.

Gelter, J., Lexhagen, M., & Fuchs, M. (2020, December 15). A Meta-Narrative Analysis of Smart Tourism Destinations: Implications for Tourism Destination Management. *Current Issues in Tourism*, *24*(20), 2860–2874. doi:10.1080/13683500.2020.1849048

Ghaderi, Z. (2021, November 4). Factors Affecting International Tourists' Intention to Use Local Mobile Apps in Online Purchase. *Asia Pacific Journal of Tourism Research*, *26*(12), 1285–1301. doi:10.1080/10941665.2021.1983626

González-Reverté. (2019, December 3). Building Sustainable Smart Destinations: An Approach Based on the Development of Spanish Smart Tourism Plans. *Sustainability*, *11*(23), 6874. doi:10.3390u11236874

Gretzel, U. (2018). From Smart Destinations to Smart Tourism regions. *Investigaciones Regionales*, *2018*(42), 171–184.

Gretzel, U. (2019). *From Smart Destinations to Smart Tourism Regions Privacy View Project Responsible Technology View Project*. Academic Press.

Gretzel, U., & Scarpino-Johns, M. (2018). Destination Resilience and Smart Tourism Destinations. *Tourism Review International*, *22*(3), 263–276. doi:10.3727/154427218X15369305779065

Gretzel, U., Sigala, M., Xiang, Z., & Koo, C. (2015a). Smart tourism: Foundations and developments. *Electronic Markets*, *25*(3), 179–188. doi:10.100712525-015-0196-8

Gretzel, U., Werthner, H., Koo, C., & Lamsfus, C. (2015b). Conceptual foundations for understanding smart tourism ecosystems. *Computers in Human Behavior*, *50*, 558–563. doi:10.1016/j.chb.2015.03.043

Hasija, S., Shen, Z. J. M., & Teo, C. P. (2020). Smart City operations: Modeling challenges and opportunities. *Manufacturing & Service Operations Management*, *22*(1), 203–213. doi:10.1287/msom.2019.0823

Inversini, A., Cantoni, L., & De Pietro, M. (2014). Destination Online Communication: Why Less is Sometimes More. A Study of Online Communications of English Destinations. *Journal of Travel & Tourism Marketing*, *31*(5), 563–575. doi:10.1080/10548408.2014.883949

Kontogianni, A., & Alepis, E. (2020). Smart tourism: State of the art and literature review for the last six years. *Array*, *6*(January), 100020. doi:10.1016/j.array.2020.100020

Koo, C., Shin, S., Gretzel, U., Hunter, W. C., & Chung, N. (2016). Conceptualization of Smart Tourism Destination Competitiveness. *Asia Pacific Journal of Information Systems*, *26*(4), 561–576. doi:10.14329/apjis.2016.26.4.561

Lee, P., Hunter, W. C., & Chung, N. (2020). Smart tourism city: Developments and transformations. *Sustainability (Switzerland)*, *12*(10), 1–15. doi:10.3390u12103958

Lopes, I. M., & Oliveira, P. (2017). Can a small city be considered a Smart City? *Procedia Computer Science*, *121*, 617–624. doi:10.1016/j.procs.2017.11.081

López De Á vila, A., Lancis, E., García, S., Alcantud, A., García, B., & Muñoz, N. (2015, September). *Smart tourism destinations report: Building the future* [Informe destinos turísticos inteligentes: Construyendo el future]. Madrid, Spain: SEGITTUR. https://www.thinktur.org/media/Libro-Blanco-Destinos-Tursticos-Inteligentes-construyendo-el-futuro.pdf

López de Avila, A. (2015). *Smart destinations: XXI century tourism.* ENTER2015 Conference on Information and Communication TechNologies in Tourism, LugaNo, Switzerland.

Loureiro, B. M. P. (2017). *Smart Cities na Política Europeia e o Smart Living em Portugal.* Academic Press.

Lytras, M. D., & Serban, A. C. (2020). E-Government Insights to Smart Cities Research: European Union (EU) Study and the Role of Regulations. *IEEE Access: Practical Innovations, Open Solutions, 8*, 65313–65326. doi:10.1109/ACCESS.2020.2982737

Marques, R. P., Melo, A. I., Natário, M. M., & Biscaia, R. (2021). *Impact of Tourist Activities on Low-Density Territories.* Springer International Publishing. doi:10.1007/978-3-030-65524-2

Mendes, B. (2019). *The Problem of Low Density Territories: Four Cases Studies* [A Problemática dos Territórios de Baixa Densidade : Quatro Estudos de Caso]. Academic Press.

Miskalo Cruz, M., & Gândara, J. (2016). Indicators for monitoring and managing smart tourist destinations [Indicadores para monitoramento e gestão de destinos turísticos inteligentes]. *CULTUR - Revista de Cultura e Turismo, 10*(2), 3–22.

Mora, L., & Bolici, R. (2015). The development process of Smart City strategies: the case of Barcelona. *1st International City Regeneration Congress.*

Murphy, P., Pritchard, M. P., & Smith, B. (2000). The destination product and its impact on traveller perceptions. *Tourism Management, 21*(1), 43–52. doi:10.1016/S0261-5177(99)00080-1

Ödemiş, M. (2022). Smart Tourism Destinations: A Literature Review on Applications in Turkey's Touristic Destinations. *Optimizing Digital Solutions for Hyper-Personalization in Tourism and Hospitality*, 131-153.

ONU. (2015). *Sustainable Development Goals* [Objetivos de Desenvolvimento Sustentável]. https://unric.org/pt/objetivos-de-desenvolvimento-sustentavel/

Orejon-Sanchez, R. D., Crespo-Garcia, D., Andres-Diaz, J. R., & Gago-Calderon, A. (2022). Smart Cities' Development in Spain: A Comparison of Technical and Social Indicators with Reference to European Cities. *Sustainable Cities and Society, 81*(June), 103828. Retrieved May 5, 2022, from. doi:10.1016/j.scs.2022.103828

Otowicz, M. (2022). Dimensions of Smart Tourism and Its Levels: An Integrative Literature Review. *Journal of Smart Tourism, 2*(1), 5–19. doi:10.52255marttourism

Pais, I. (2017). *Territorial Brand: The residents' view of their city (Cidade da Guarda)* [Master's dissertation]. Faculty of Economics of Porto.

Pedro, A., & Marques, S. (2010). *From Space Construction to Territory Construction* [Da Construção do Espaço à Construção do Território]. Academic Press.

Phillips, P., & Moutinho, L. (2014). Critical review of strategic planning research in hospitality and tourism. *Annals of Tourism Research*, *48*, 96–120. doi:10.1016/j.annals.2014.05.013

PORDATA. (2019). *Your municipality in numbers!* [O seu município em números!]. Guarda.

Priano, F. H., Armas, R. L., & Guerra, C. F. (2016). A model for the smart development of island territories. In Y. Kim, & M. Liu (Eds.), *Proceedings of the 17th International Digital Government Research Conference on Digital Government Research* (pp. 465–474). New York: Association for Computing Machinery. 10.1145/2912160.2912187

Proença, M. Jr, & Duenhas, R. A. (2020). Smart cities and sustainable cities: Convergence of actions mere advertising? [Cidades inteligentes e cidades sustentáveis: convergência de ações ou mera publicidade?]. *Revista Brasileira de Planejamento e Desenvolvimento*, *9*(2), 317. doi:10.3895/rbpd.v9n2.10234

Ramos, G. M. de A. (2015). *Institutional Innovation, Tourism and Development in Low Density Territories: the case of Naturtejo Geopark* [Inovação Institucional, Turismo e Desenvolvimento em Territórios de Baixa Densidade: o caso do Geopark Naturtejo]. Academic Press.

Reis, F. L. dos. (2018). *Scientific Research and Academic Works - Practical Guide* [Investigação Científica e Trabalhos Académicos - Guia Prático]. Academic Press.

Reis, P., & Baltazar, M. da S. (2019). Low-density rural territories as a space for leisure and tourism. The tourist destination Historic Villages of Portugal [Os territórios rurais de baixa densidade como espaço de lazer e de turismo. O destino turístico Aldeias Históricas de Portugal]. *Sociologia on Line*, *21*, 141–166. doi:10.30553ociologiaonline.2019.21.6

Ritchie, J. R. B., & Crouch, G. I. (2003). *The Competitive Destination: A Sustainable Tourism Perspective*. https://books.google.pt/books?hl=pt-PT&lr=&id=yvydAwAAQBAJ&oi=fnd&pg=PR5&dq=Ritchie,+J.+R.+B.,+%26+Crouch,+G.+I.+(2003).+The+competitive+destination:+a+sustainable+tourism+perspective.+Wallingford,+Reino+Unido:+CAB+International.&ots=lSIZ6ph9SO&sig=qvdIW5

Santos-Júnior, A. (2019). Entendiendo La Gobernanza de Los Destinos Turísticos Inteligentes: El Caso de Florianópolis -Brasil [Understanding Smart Tourism Destinations' Governance: The Case of Florianópolis – Brazil]. *International Journal of Information Systems and Tourism (IJIST).*

Savić, J., & Pavlović, G. (2018). Analysis of Factors of Smart Tourism Development in Serbia. *Menadzment U Hotelijerstvu I Turizmu*, *6*(1), 81–91. doi:10.5937/menhottur1801081S

SEGITTUR. (2015). *Smart Destinations Report: building the future*. https://www.segittur.es/opencms/export/sites/segitur/.content/galerias/descargas/documentos/Libro-Destinos-Inteligentes-en-Ingls.pdf

Shih, N. J., Diao, P. H., & Chen, Y. (2019). ARTS, an AR tourism system, for the integration of 3D scanning and smartphone AR in cultural heritage tourism and pedagogy. *Sensors (Switzerland)*, *19*(17), 3725. Advance online publication. doi:10.339019173725 PMID:31466354

Smith, J. (2017). *Transforming Travel: Realising the Potential of Sustainable Tourism*. https://books.google.pt/books?id=eQJDDwAAQBAJ&printsec=frontcover&dq=Transforming+Travel+Realising+the+potential+of+sustainable+tourism&hl=pt-PT&sa=X&ved=2ahUKEwjR4NWE0qXvAhXF8OAKHZiPApoQ6AEwAHoECAMQAg#v=onepage&q=Transforming%20Travel%20Realising%20the%20potenti&f=false

Susanti, R., Soetomo, S., Buchori, I., & Brotosunaryo, P. M. (2016). Smart Growth, Smart City and Density: In Search of The Appropriate Indicator for Residential Density in Indonesia. Procedia - Social and Behavioral Sciences, 227, 194–201. doi:10.1016/j.sbspro.2016.06.062

UNTWO. (2020b). *Digital Transformation*. UNWTO. https://www.unwto.org/digital-transformation

Vargas Sánchez, A. (2016). Exploring the concept of smart tourist destination. *Enlightening Tourism: A Pathmaking Journal, 6*(2), 178–196. doi:10.33776/et.v6i2.2913

Verma, A., & Shukla, V. (2019). Analyzing the Influence of IoT in Tourism Industry. SSRN *Electronic Journal*. doi:10.2139/ssrn.3358168

Virtudes, A., Abbara, A., & Sá, J. (2017). Dubai: A Pioneer Smart City in the Arabian Territory. *IOP Conference Series. Materials Science and Engineering*, *245*(5), 052071. Advance online publication. doi:10.1088/1757-899X/245/5/052071

Volgger, Erschbamer, G., & Pechlaner, H. (2021, January). Destination Design: New Perspectives for Tourism Destination Development. *Journal of Destination Marketing & Management, 19*, 100561. doi:10.1016/j.jdmm.2021.100561

Wang, Li, X. R., Zhen, F., & Zhang, J. H. (2016). How Smart Is Your Tourist Attraction?: Measuring Tourist Preferences of Smart Tourism Attractions via a FCEM-AHP and IPA Approach. *Tourism Management*, *54*(June), 309–320. doi:10.1016/j.tourman.2015.12.003

Wang, D., Li, X., & Li, Y. (2013). China's "Smart Tourism Destination" initiative: A taste of the service-dominant logic. *Journal of Destination Marketing & Management*, *2*(2), 59–61. doi:10.1016/j.jdmm.2013.05.004

Wang, M., Zhou, T., & Wang, D. (2020). Tracking the evolution processes of Smart Cities in China by assessing performance and efficiency. *Technology in Society*, *63*(June), 101353. doi:10.1016/j.techsoc.2020.101353

Weekes, S. (2019). *Insight Report Smart waste, an opportunity cities should not trash*. Academic Press.

Xiang, Z. (2021, March 31). A New Platform to Support and Define an Emerging Field. *Journal of Smart Tourism*, *1*(1), 1–2. doi:10.52255marttourism.2021.1.1.1

Ye, Ye, H., & Law, R. (2020, April 22). Systematic Review of Smart Tourism Research. *Sustainability*, *12*(8), 3401. doi:10.3390u12083401

Zygiaris, S. (2013). Smart City Reference Model: Assisting Planners to Conceptualize the Building of Smart City Innovation Ecosystems. *Journal of the Knowledge Economy*, *4*(2), 217–231. doi:10.100713132-012-0089-4

KEY TERMS AND DEFINITIONS

Competitiveness: According to Cronjé and Plessis (2020) "most of the research conducted on tourism destination competitiveness are based on the work of Porter, but prior to this, the issue of competitiveness was developed as a scientific discipline by Smith in the late 70's. Due to the reason that tourism is primarily a service driven industry, tourism researchers had to adjust definitions, develop new models and identify factors that would be applicable to the tourism industry. Crouch and Ritchie (1999) developed the concept of destination competitiveness and, according to the authors, it consists of the ability of a given destination to provide higher quality tourism experiences compared to other destinations.

Low Density Territories: In Portugal, are considered low density territories: "All cities with a density population of less than 25 inhabitants/km2" (ANMP, 2014). In low density territories tourism is emerging as an excellent opportunity to promote innovative dynamics, to lure investment, and to attract new people. These territories represent an economic challenge, especially because most of them are home to an increasingly aging population unable to pursue economic development. However, in these territories it is important to promote sustainable tourism, which preserves the environment and communities' quality of life (Marques et al., 2021).

Smart Destinations: It represents a consolidated innovative space, based on the territory and a technological infrastructure. Furthermore, it is a territory committed to environmental and cultural preservation and to the socioeconomic factors related to its identity and habitat. In these territories there are intelligence systems that capture and process information at real time, with the aim of facilitating the interaction of tourist and visitor with the environment and facilitating the acquisition of intelligence by the managers of these destinations, increasing their efficiency and improving the quality of tourist experiences (de Avila Muñoz & Sánchez, 2013; Otowicz et al., 2022).

Smart Tourism: It is tourism supported by the use of technology combined with integrated efforts by destinations to collect, aggregate and harness data from physical infrastructure, social connections, governmental and organizational sources, human bodies and minds, with the aim of transforming this data into on-site experiences and business value propositions focused on efficiency, sustainability and differentiating experiences (Gretzel et al., 2015a). Conceptualized by Ödemiş (2022):

The concept of smart tourism is one of the tourism-based concepts and practices used for making tourism activities much more effective: making cities and destinations smart, and increasing tourist satisfaction, and has emerged with the need to ensure the adaptation and integration of internet and mobile technologies, which are among the technological developments of today, to tourism activities.

Smart Tourism City: According to Lee et al. (2020) "more over cities around the world are increasingly recognizing the smart tourism city concept and related strategies as means of optimizing sustainable environments. Particularly for cities facing emerging issues of residents' negative perceptions towards tourism, smart tourism city empowers a city to rise to this challenge by creating urban spaces that residents and visitors can enjoy together."

Smart Tourism Ecosystem: Gretzel et al. (2015b) defines smart tourism ecosystem as a system related with intense knowledge sharing and value creation, that depends on technology in the creation, management and provision of smart tourism services and experiences, and monitor technological advances.

Smart Tourism Region: According to Priano et al. (2016) a smart tourism region it´s that to identify its strengths and opportunities and, in addition, adequately coordinates available - and generally limited - resources to generate maximum productivity in the areas that comprise it.

Sustainability: At the level of tourism activity and tourism experience, considering sustainability is an efficient fit between human activity and development and environmental protection, which requires a clear and equitable distribution of responsibilities, with notorious respect for resources, consumption and behavior in relation to resources. natural and cultural features of tourism destinations (Costa, Rodrigues & Gomes, 2019).

ENDNOTES

1 A group of people who have common interests in the outcome of destiny.

2 Great number of data that is, structured or not, which are generated every second and which can allow users to anticipate and get to know them.

3 User information that can be used for different purposes in tourism and beyond.

4 When there are too many tourists visiting the same place at the same time.

5 Tool used in the tourist sector, which promotes the attraction of visitors, through the creation of a plot around the destination.

Compilation of References

Abel, S. M., & Wong, A. (2010). Is There A Second Life For Trademarks. In *Second Life*. Fenwick & West LLP.

Abolhasani, M., Oakes, S., & Oakes, H. (2017). Music in advertising and consumer identity: The search for Heideggerian authenticity. *Marketing Theory*, *17*(4), 473–490. doi:10.1177/1470593117692021

Adam, M., Wessel, M., & Benlian, A. (2021). AI-based chatbots in customer service and their effects on user compliance. *Electronic Markets*, *31*(2), 427–445. doi:10.100712525-020-00414-7

Adamopoulou, E., & Moussiades, L. (2020). Chatbots: History, technology, and applications. *Machine Learning with Applications*, *2*, 100006. doi:10.1016/j.mlwa.2020.100006

Ahlqvist, T., Bäck, A., Halonen, M., & Heinonen, S. (2008). *Social media roadmaps*. Edita Prima Oy.

Ailawadi, K. L., & Farris, P. W. (2017). Managing Multi- and Omni-Channel Distribution: Metrics and Research Directions. *Journal of Retailing*, *93*(1), 120–135. doi:10.1016/j.jretai.2016.12.003

Akhtar, M., Neidhardt, J., & Werthner, H. (2019). The potential of chatbots: Analysis of Chatbot Conversations. *IEEE 21st Conference on Business Informatics (CBI),* 397-404. 10.1109/CBI.2019.00052

Akter, S., Hossain, T. M. T., & Strong, C. (2021). What Omnichannel Really Means? *Journal of Strategic Marketing*, *29*(7), 567–573. doi:10.1080/0965254X.2021.1937284

Alafeef, M., Singh, D., & Ahmad, K. (2012). The Influence of Demographic Factors and User Interface on Mobile Banking Adoption: A Review. *Journal of Applied Sciences*, *12*(20), 2082–2095. doi:10.3923/jas.2012.2082.2095

Alexander, R., & Philipp, H. (2008). *Security Aspects in Web 2.0 Mashup Systems*. Technology, Altenbergerstrabe, Linz, Austria. Available at http://www.fim.uni-linz.ac.at/lva/SE_Netzwerke_und_Sicherheit_Security_Considerations_in_Intercon_Networks/semH. pdf/

Alexander, B., & Cano, M. B. (2020a). Store of the Future: Towards a (re) Invention and (re) Imagination of Physical Store Space in an Omni-Channel Context. *Journal of Retailing and Consumer Services*, *55*, 101913. doi:10.1016/j.jretconser.2019.101913

Alexander, B., & Kent, A. (2020b). Change in Technology-Enabled Omnichannel Customer Experiences in-store. *Journal of Retailing and Consumer Services*, *65*, 102338. Advance online publication. doi:10.1016/j.jretconser.2020.102338

AlNuaimi, B. K., Singh, S. K., Ren, S., Budhwar, P., & Vorobyev, D. (2022). Mastering digital transformation: The nexus between leadership, agility, and digital strategy. *Journal of Business Research*, *145*, 636–648. doi:10.1016/j.jbusres.2022.03.038

Alpert, J. I., & Alpert, M. I. (1989). Background music as an influence in consumer mood and advertising responses. *Advances in Consumer Research. Association for Consumer Research (U. S.)*, *16*, 485–491.

Al-Saedi, K., & Al-Emran, M. (2021). A systematic review of mobile payment studies from the lens of the UTAUT model. *Recent Advances in Technology Acceptance Models and Theories*, 79-106.

Amankwah-Amoah, J., Khan, Z., Wood, G., & Knight, G. (2021). COVID-19 and digitalization: The great acceleration. *Journal of Business Research*, *136*, 602–611. doi:10.1016/j.jbusres.2021.08.011 PMID:34538980

Ameen, N., Tarhini, A., Shah, M., & Madichie, N. O. (2021). Going with the flow: Smart shopping malls and omnichannel retailing. *Journal of Services Marketing*, *35*(3), 325–348. doi:10.1108/JSM-02-2020-0066

Ammanath, B. (2022, March 21). An Ethical Metaverse Takes Trustworthy AI. *The Wall Street Journal*. https://deloitte.wsj.com/articles/an-ethical-metaverse-takes-trustworthy-ai-01647884164

ANACOM. (2021). *Pandemic COVID-19 - Impacto na utilização dos serviços de comunicações*. ANACOM.

Anderson, T., & Rivera-Vargas, P. (2020). A Critical Look at Educational Technology from a Distance Education Perspective. *Digital Education Review*, *37*(37), 208–229. doi:10.1344/der.2020.37.208-229

ANMP. (2014) *Low Density Municipalities – criteria approved by the National Association of Portuguese Municipalities* [Municípios de Baixa Densidade - critérios aprovados pela Associação Nacional de Municípios Portugueses]. Author.

Ansani, A., Marini, M., D'Errico, F., & Poggi, I. (2020). How soundtracks shape what we see. Analyzing the influence of music on visual scenes through self-assessment, eye-tracking, and pupillometry. *Frontiers in Psychology*, *11*, 22–42. doi:10.3389/fpsyg.2020.02242 PMID:33117211

Antin, J., & Churchill, E. F. (2011). Badges in social media: A social psychological perspective. *CHI, 2011 Gamification Workshop Proceedings.*

Antunes, J., Pinto, A., Reis, P., & Henriques, C. (2019). A Indústria 4.0 e o seu Impacto no Tecido Económico Empresarial Português: O Caso da Região Dão-Lafões [Industry 4.0 and its Impact on the Portuguese Business Economic Fabric: The Case of the Dão-Lafões Region]. *Fronteiras: Journal of Social, Technological and Environmental Science*, *8*, 263-291. https://repositorio.ipv.pt/handle/10400.19/5654

Arantes, M., Figueiredo, F., Almeida, J. M., & Prates, R. O. (2018, October). An investigation of user actions and experiences when exposed to *YouTube* video ads. *Proceedings of the 24th Brazilian Symposium on Multimedia and the Web*, 283-290. 10.1145/3243082.3243107

Araujo, T. (2018). Living up to the chatbot hype: The influence of anthropomorphic design cues and communicative agency framing on conversational agent and company perceptions. *Computers in Human Behavior*, *85*, 183–189. doi:10.1016/j.chb.2018.03.051

Ardito, L., Simon, R., Vito, A., & Bernardo, B. (2021). The Duality of Digital and Environmental Orientations in the Context of SMEs: Implications for Innovation Performance. *Journal of Business Research*, *123*, 44–56. doi:10.1016/j.jbusres.2020.09.022

Arora, S., Sahney, S., & Parida, R. R. (2022). Drivers of showrooming behaviour: insights from integrated perspectives. *International Journal of Retail & Distribution Management*, *50*(3), 398-413. doi: 10.1108/IJRDM-09-2020-0374/

Arora, S., Parida, R. R., & Sahney, S. (2020). Understanding Consumers' Showrooming Behaviour: A Stimulus–Organism–Response (SOR) Perspective. *International Journal of Retail & Distribution Management*, *48*(11), 1157–1176. doi:10.1108/IJRDM-01-2020-0033

Ashton, K. (2009). That 'Internet of Things' Thing. *RFID Journal*. https://www.rfidjournal.com/articles/view?4986

Associação Portuguesa de Bancos. (2020). *Boletim Informativo APB*. Associação Portuguesa de Bancos.

Avedon, E. M. (1971). The structural elements of games. The study of games, 419-426.

Avula, S., Chadwick, G., Arguello, J., & Capra, R. (2018). Searchbots: User engagement with chatbots during collaborative search. *Proceedings of the 2018 Conference on Human Information Interaction & Retrieval*, 52-61. 10.1145/3176349.3176380

Aw, E. C. X. (2019). Understanding the webrooming phenomenon: Shopping motivation, channel-related benefits and costs. *International Journal of Retail & Distribution Management, 47*(10), 1074-1092. doi: 10.1108/IJRDM-01-2019-0026/

Bailenson, J. (2018). *Experience on demand: What virtual reality is, how it works, and what it can do*. WW Norton & Company.

Bajdor, P., & Dragolea, L. (2011). *The gamification as a tool to improve risk management in the enterprise*. Annales Universitatis Apulensis Series Oeconomica. doi:10.29302/oeconomica.2011.13.2.38

Bakker, P. (2012). Aggregation, content farms and Huffinization. *Journalism Practice*, *6*(5-6), 627–637. doi:10.1080/17512786.2012.667266

Balkan, B. (2021). Impacts of Digitalization on Banks and Banking. In The Impact of Artificial Intelligence on Governance, Economics and Finance, Volume 1 (pp. 33-50). Springer. doi:10.1007/978-981-33-6811-8_3

Ballina, F. J. (2020, October 5). Smart Business: The Element of Delay in the Future of Smart Tourism. *Journal of Tourism Futures*, *8*(1), 37–54. doi:10.1108/JTF-02-2020-0018

Banco de Portugal. (2018). *Aviso n.º 2/2018*. Retrieved from October 9, 2021, from https://www.bportugal.pt/aviso/22018

Banco de Portugal. (2020). *Inclusão Financeira e Digital e Escolha de Produtos Bancários em Portugal*. Author.

Banik, S. (2021). Exploring the involvement-patronage link in the phygital retail experiences. *Journal of Retailing and Consumer Services*, *63*, 102739. doi:10.1016/j.jretconser.2021.102739

Bankewitz, M., Åberg, C., & Teuchert, C. (2016). Digitalization and boards of directors: A new era of corporate governance? *Business and Management Research*, *5*(2), 58–69. doi:10.5430/bmr.v5n2p58

Baran, M., Kłos, M., Chodorek, M., & Marchlewska-Patyk, K. (2022, March 1). The Resilient Smart City Model–Proposal for Polish Cities. *Energies*, *15*(5), 1818. doi:10.3390/en15051818

Barann, B., Hermann, A., Heuchert, M., & Becker, J. A. (2022). Can't touch this? Conceptualizing the Customer Touchpoint in the Context of Omni-Channel Retailing. *Journal of Retailing and Consumer Services*, *65*, 102269. Advance online publication. doi:10.1016/j.jretconser.2020.102269

Barbosa, B., & Carvalho, C. (2021). Tweet and retweet journalism during the pandemic: Dissemination of and engagement with news on twitter. *Brazilian Journalism Research*, *17*(3), 736–763. doi:10.25200/BJR.v17n3.2021.1364

Barcelos, R. H., Dantas, D. C., & Sénécal, S. (2018). Watch your tone: How a brand's tone of voice on social media influences consumer responses. *Journal of Interactive Marketing*, *41*, 60–80. doi:10.1016/j.intmar.2017.10.001

Bardzell, S., & Shankar, K. (2007). Video game technologies and virtual design: A study of virtual design teams in a metaverse. Lecture Notes in Computer Science, 4563, 607–616. doi:10.1007/978-3-540-73335-5_65

Barreda, A. A., Bilgihan, A., Nusair, K., & Okumus, F. (2015). Generating brand awareness in online social networks. *Computers in Human Behavior*, *50*, 600–609. doi:10.1016/j.chb.2015.03.023

Barwitz, N., & Maas, P. (2018). Understanding the Omnichannel Customer Journey: Determinants of Interaction Choice. *Journal of Interactive Marketing*, *43*, 116–133. doi:10.1016/j.intmar.2018.02.001

Bastidas-Manzano, A., Sánchez-Fernández, J., & Casado-Aranda, L. (2020). The Past, Present, and Future of Smart Tourism Destinations: A Bibliometric Analysis. *Journal of Hospitality & Tourism Research (Washington, D.C.), 45*(3), 529–552. doi:10.1177/1096348020967062

Batat, W. (2022). What does phygital really mean? A Conceptual Introduction to the Phygital Customer Experience (Ph-Cx) Framework. *Journal of Strategic Marketing*, •••, 1–24. doi:10.1080/0965254X.2022.2059775

BBC. (2021). *Metaverso: o que é a economia do mundo paralelo e como ela pode ser explorada nos próximos anos* [Metaverse:What is the shadow world economy and how it can be exploited in the coming years]. https://www.bbc.com/portuguese/internacional-59431866

Beattie, A., Edwards, A. P., & Edwards, C. (2020). A bot and a smile: Interpersonal impressions of chatbots and humans using emoji in computer-mediated communication. *Communication Studies, 71*(3), 409–427. doi:10.1080/10510974.2020.1725082

Beck, Silva Neto, & Melo Conti. (2018). *A new urban paradigm of smart and sustainable tourism* [Um novo paradigma urbano de turismo inteligente e sustentável]. XX Engema. http://engemausp.submissao.com.br/20/anais/arquivos/238.pdf

Beck, N., & Rygl, D. (2015). Categorization of Multiple Channel Retailing in Multi-, Cross-, and Omni-Channel Retailing for Retailers and Retailing. *Journal of Retailing and Consumer Services, 27*, 170–178. doi:10.1016/j.jretconser.2015.08.001

Belanche, D., Flavián, C., & Pérez-Rueda, A. (2017). Understanding interactive online advertising: Congruence and product involvement in highly and lowly arousing, skippable video ads. *Journal of Interactive Marketing, 37*, 75–88. doi:10.1016/j.intmar.2016.06.004

Bentley, F., Quehl, K., Wirfs-Brock, J., & Bica, M. (2019). Understanding online news behaviors. *CHI '19: Proceedings of the 2019 CHI Conference on Human Factors in Computing Systems, 590*, 1-11. 10.1145/3290605.3300820

Berger, B., Matt, C., Steininger, D. M., & Hess, T. (2015). It is not just about competition with "free": Differences between content formats in consumer preferences and willingness to pay. *Journal of Management Information Systems, 32*(3), 105–128. doi:10.1080/07421222.2015.1095038

Beriault-Poirier, A., Prom Tep, S., & Sénécal, S. (2019). Putting chatbots to the test: Does the user experience score higher with chatbots than websites? In T. Ahram, W. Karwowski, & R. Taiar (Eds.), *Human Systems Engineering and Design* (pp. 204–212). Springer. doi:10.1007/978-3-030-02053-8_32

Berman, B., & Thelen, S. (2018). Planning and implementing an effective omnichannel marketing program. *International Journal of Retail & Distribution Management, 46*(7), 598-614. doi: 10.1108/IJRDM-08-2016-0131/

BettencourtL. A. (2017). https://www.ama.org/2017/12/20/are-you-managing-brand-equity-incorrectly/

Bican, P. M., & Brem, A. (2020). Digital business model, digital transformation, digital entrepreneurship: Is there a sustainable "digital?" *Sustainability, 12*(13), 5239. doi:10.3390u12135239

Biloria, N. (2020). From smart to empathic cities. *Frontiers of Architectural Research*, 1–14. doi:10.1016/j.foar.2020.10.001

Binance Academy. (2022). *Retrospectiva 2021*. https://academy.binance.com/pt/articles/binance-academy-2021-year-in-review

Binet, L., Müllensiefen, D., & Edwards, P. (2013). The power of music. *Music in Advertising*, 10-13.

Biographon. (2019). https://biographon.com/youtube-stats/

Birkinshaw, J., & Ansari, S. (2015). Understanding management models: going beyond "what" and "why" to "how" work gets done in organizations. In Business model innovation: The organizational dimension (pp. 85-103). University Press.

Bista, S. K., Nepal, S., Colineau, N., & Paris, C. (2012). Using gamification in an online community. In *Collaborative Computing: Networking, Applications and Worksharing (CollaborateCom), 2012 8th International Conference*. IEEE.

Boczkowski, P. J., & Mitchelstein, E. (2010). Is there a gap between the news choices of journalists and consumers? A relational and dynamic approach. *The International Journal of Press/Politics*, *15*(4), 420–440. doi:10.1177/1940161210374646

Bodkhe, U., Bhattacharya, P., Tanwar, S., Tyagi, S., Kumar, N., & Obaidat, M. S. (2019). BloHosT: Blockchain enabled Smart Tourism and hospitality management. *CITS 2019 - Proceeding of the 2019 International Conference on Computer, Information and Telecommunication Systems*, 1–5. 10.1109/CITS.2019.8862001

Boes, K., Buhalis, D., & Inversini, A. (2016). Smart Tourism Destinations: Ecosystems for tourism destination competitiveness. *International Journal of Tourism Cities*, *2*(2), 108–124. doi:10.1108/IJTC-12-2015-0032

Bohyun. (2013). *The Library Mobile Experience: Practices and User Expectations*. Library Technology Reports 49, no. 6. ALA TechSource.

Bolton, R. N., Gustafsson, A., Tarasi, C. O., & Witell, L. (2022). Designing Satisfying Service Encounters: Website Versus Store Touchpoints. *Journal of the Academy of Marketing Science, 50*(1), 85-107. doi: 10.1007/s11747-021-00808-9

Boltz, L. O., Yadav, A., Dillman, B., & Robertson, C. (2021). Transitioning to remote learning: Lessons from supporting K-12 teachers through a MOOC. *British Journal of Educational Technology*, *52*(4), 1377–1393. doi:10.1111/bjet.13075

Bošnjak, A. (2017). *Smart Tourism Destinations: Enhancing Interconnection between Stakeholders and Tourist through Information Personalization and Feedback System*. Academic Press.

Boulos & Wheeler. (2007). The emerging Web 2.0 social software: An enabling suite of sociable technologies in health and health care education. *Health Information and Libraries Journal*, 2–23. PMID:17331140

Bourbon, R., & Lamas, M. (2021). Modelos de Negócio e Internacionalização: O caso da Jump Willy. *Portuguese Journal of Finance, Management and Accounting*, *7*(13), 47-69. http://u3isjournal.isvouga.pt/index.php/PJFM

Bourdieu, P. (2010). *Distinction: A Social Critique of the Judgment of Taste*. Routledge.

Bourlakis, M., Papagiannidis, S., & Li, F. (2009). Retail spatial evolution: Paving the way from traditional to metaverse retailing. *Electronic Commerce Research*, *9*(1–2), 135–148. doi:10.100710660-009-9030-8

Brandao, M., Teles, A., Fluminense, U. F., & Joia, L. A. (2016). *Destino turístico inteligente : um caminho para transformação*. Academic Press.

Brandtzaeg, P. B., & Følstad, A. (2017). Why People Use Chatbots. In I. Kompatsiaris, J. Cave, A. Satsiou, G. Carle, A. Passani, E. Kontopoulos, S. Diplaris, & D. McMillan (Eds.), *Internet Science: INSCI 2017* (pp. 377–392). Springer. doi:10.1007/978-3-319-70284-1_30

Bravo, M. P., & Eisman, L. B. (1998). *Investigación Educativa*. Ediciones Alfar.

Brent Ritchie, J. R., & Crouch, G. I. (2010). A model of competitiveness/sustainability of destinations: Brazilian perspectives [Um modelo de competitividade/sustentabilidade de destinos: Perspectivas brasileiras]. *Revista de Administração Pública*, *44*(5), 1049–1066. doi:10.1590/S0034-76122010000500003

Bruchez, A., Ciconet, B., Remussi, R., Possamai, L., Antônio, V., & Tondolo, G. (2015). análise da utilização do estudo de caso qualitativo e triangulação na brazilian business review. *Mostra de Iniciação Científica, Pós-Graduação, Pesquisa e Extensão, 4*, 1-16. doi:10.18226/35353535.v4.2015.52

Brummer, C., & Yadav, Y. (2019). Fintech and the Innovation Trilemma. *The Georgetown Law Journal*, *107*(235), 235–307.

Bryman, A. (2016). *Social Research Methods*. Oxford University Press.

Buffer. (2019). https://buffer.com/state-of-social-2019

Buhalis, D., & Amaranggana, A. (2015). Smart Tourism Destinations Enhancing Tourism Experience Through Personalisation of Services. *Information and Communication Technologies in Tourism, 2015*, 377–389. doi:10.1007/978-3-319-14343-9_28

Bulchand-Gidumal, J. (2022). Post-COVID-19 Recovery of Island Tourism Using a Smart Tourism Destination Framework. *Journal of Destination Marketing & Management, 23*(Mar), 100689. Retrieved May 5, 2022, from. doi:10.1016/j.jdmm.2022.100689

Burton, J. (1990). *Conflict: Human Needs Theory*. Palgrave Macmillan. doi:10.1007/978-1-349-21000-8

Cachia, M., & Millward, L. (2011). The telephone medium and semi-structured interviews: A complementary fit. *Qualitative Research in Organizations and Management, 6*(3), 265–277. doi:10.1108/17465641111188420

Caillois, R. (1961). *Man, play, and games*. University of Illinois Press.

Caixa Geral de Depósitos. (2019). *Missão e Estratégia do Grupo CGD*. Retrieved October 9, 2021, from https://www.cgd.pt/Institucional/Governo-Sociedade-CGD/Pages/Missao-Estrategia.aspx

Caixa Geral de Depósitos. (2020). *Grupo CGD*. Retrieved October 10, 2021, from https://www.cgd.pt/Institucional/Grupo-CGD/Documents/Organograma-do-Grupo-CGD.pdf

Caixa Geral de Depósitos. (2021). *CGD*. Retrieved October 10, 2021, from https://www.cgd.pt/Institucional/Patrimonio-Historico-CGD/Historia/Pages/Historia-CGD.aspx

Caixa Geral de Depósitos. (2021). *EUSOUDIGITAL - Programa de Capacitação Digital de Adultos*. Retrieved September 29, 2021, from https://www.cgd.pt/Institucional/eusoudigital/Pages/eusoudigital.aspx

Caixa Geral de Depósitos. (2021). *Institucional: 9M2021 - Resultados*. Retrieved November 5, 2021 from https://www.cgd.pt/institucional/sala-de-imprensa/2021/pages/atividade-consolidada-30-setembro-2021.aspx

Cai, Y. J., & Lo, C. K. (2020). Omni-channel management in the new retailing era: A systematic review and future research agenda. *International Journal of Production Economics, 229*, 107729. doi:10.1016/j.ijpe.2020.107729

CaligiuriC.JonesB. (2019). https://www.thinkwithgoogle.com/intl/pt-br/estrategias-de-marketing/video/uncios-em-video-como-e-quando-devemos-customizar/

Cao, L., & Li, L. (2018). Determinants of Retailers' Cross-Channel Integration: An Innovation Diffusion Perspective on Omni-Channel Retailing. *Journal of Interactive Marketing, 44*, 1–16. doi:10.1016/j.intmar.2018.04.003

Carvalho, P. (2018). Dynamics of rural low density spaces in PortugalDynamique des espaces ruraux de basse densité au Portugal. *Méditerranée, 130*(130). Advance online publication. doi:10.4000/mediterranee.10516

Casero-Ripollés, A. (2012). Contenidos periodísticos y nuevos modelos de negocio: Evaluación de servicios digitales. *El Profesional de la Información, 21*(4), 341-346. doi:10.3145/epi.2012.jul.02

Case, S. (2017). *A Terceira Onda da Internet*. HSM.

Catanzaro, M. (1988). Using Qualitative Analytical Techniques. In N. Woods & M. Catanzaro (Eds.), *Nursing Research: Theory and Practice* (pp. 437–456). Mosby Incorporated.

Cateora, P., Gilly, M., & Graham, J. (2011). *International Marketing* (15th ed.). McGraw-Hill Companies, Inc.

Cavusgil, T., Knight, G., & Riesenberger, R. (2008). *International Business: Strategy, Management, and the New Realities*. Pearson Prentice Hall.

Cerqueira, C., Lamas, M., Baranovskiy, S. (2021). Interculturality Role in Startups Internationalization: Eligent Club's Virtual Accelerator Program Case. *E-Revista de Estudos Interculturais do CEI–ISCAP*, *3*(9), 1–22. doi:10.34630/erei.v3i9.4217

Certificado de Maturidade Digital. (2022). https://selosmaturidadedigital.incm.pt/SMD/

Chang, Y. P., & Li, J. (2022). Seamless Experience in the Context of Omnichannel Shopping: Scale Development and Empirical Validation. *Journal of Retailing and Consumer Services*, *64*, 102800. doi:10.1016/j.jretconser.2021.102800

Chatterjee, S., Chaudhuri, R., & Vrontis, D. (2021). Examining the global retail apocalypse during the COVID-19 pandemic using strategic omnichannel management: A consumers' data privacy and data security perspective. *Journal of Strategic Marketing*, *29*(7), 617–632. doi:10.1080/0965254X.2021.1936132

Chaudhary, P., Singh, A., & Sharma, S. (2021). Understanding the antecedents of omni-channel shopping by customers with reference to fashion category: The Indian millennials' perspective. *Young Consumers*, *23*(2), 304–320. doi:10.1108/YC-05-2021-1327

Chen, C., & Kim, C. (Eds.). (2018). *The digitization of business in China: Exploring the transformation from manufacturing to a digital service hub*. Springer International Publishing. doi:10.1007/978-3-319-79048-0

Chen, C., & Yao, M. Z. (2022). Strategic use of immersive media and narrative message in virtual marketing: Understanding the roles of telepresence and transportation. *Psychology and Marketing*, *39*(3), 524–542. doi:10.1002/mar.21630

Chen, R. P., Wan, E. W., & Levy, E. (2017). The effect of social exclusion on consumer preference for anthropomorphized brands. *Journal of Consumer Psychology*, *27*(1), 23–34. doi:10.1016/j.jcps.2016.05.004

Chen, T. Y., Yeh, T. L., Wu, H. L., & Deng, S. (2022). Effect of Channel Integration Quality on Consumer Responses within Omni-channel Retailing. *Asia Pacific Journal of Marketing and Logistics*. Advance online publication. doi:10.1108/APJML-04-2021-0270

Cherian, J. (2015). Emergence of digital publishing: A great challenge to the print publications. *Procedia Economics and Finance*, *23*, 576–586. doi:10.1016/S2212-5671(15)00361-5

Child, J., Karmowska, J., & Shenkar, O. (2022). The role of context in SME internationalization – A review. *Journal of World Business*, *57*(1), 101267. doi:10.1016/j.jwb.2021.101267

Chimborazo-Azogue, L. E., Mollá-Descals, A., Miquel-Romero, M. J., & Frasquet, M. (2022). Mobile dependency and uncertainty reduction: Influence on showrooming behaviours and user-generated content creation. *International Journal of Retail & Distribution Management*, *50*(8/9), 996–1014. Advance online publication. doi:10.1108/IJRDM-10-2021-0487

Chizzotti, A. (2006). *Pesquisa Qualitativa em Ciências Humanas E Sociais* [Qualitative Research in Humanities and Social Sciences]. Vozes.

Chmielewski, D. (2013). *Nielsen Study: Social Networking Dominates Smartphone, Tablet Use*. http://articles.latimes.com/2013/jun/09/entertainment/la-et-ct-nielsenstudy-social-networking-smartphone-tablet-20130609

Chou, S. Y., Shen, G. C., Chiu, H. C., & Chou, Y. T. (2016). Multichannel service providers' strategy: Understanding customers' switching and free-riding behavior. *Journal of Business Research*, *69*(6), 2226–2232. doi:10.1016/j.jbusres.2015.12.034

Christian, B., Tom, H., & Tim, B.-L. (2009). *Linked Data - The Story So Far.* Journal Semantic Web and Information Systems.

CIC. (2015).*Classification of Low Density Municipalities for the application of measures of positive differentiation of territories* [Classificação dos Municípios de Baixa Densidade para aplicação de medidas de diferenciação positiva dos territórios]. CIC.

Cisco. (2021). https://www.cisco.com/c/en/us/solutions/collateral/executive-perspectives/annual-internet-report/white-paper-c11-741490.html

Cities, S. (2018). *"Being a smart city should not be a soud bite" – Smart Cities* ["Ser uma cidade inteligente não deve ser um sound bite" - Smart Cities]. https://smart-cities.pt/opiniao-entrevista/alvaro-santosamaro3004guarda/

Clancey, M. (1994). The television audience examined. *Journal of Advertising Research, 4*(34), 1–10.

Comarch. (2014). *JetBlue Badges Deliver Gartner's Gamification.* www.comarch.us/media-resources/knowledge-articles/jetbluebadges-deliver-gartners-gamification/

Cook, C., & Sirkkunen, E. (2013). What's in a niche? exploring the business model of online journalism. *Journal of Media Business Studies, 10*(4), 63–82. doi:10.1080/16522354.2013.11073576

Corti, K., & Gillespie, A. (2016). Co-constructing intersubjectivity with artificial conversational agents: People are more likely to initiate repairs of misunderstandings with agents represented as human. *Computers in Human Behavior, 58*, 431–442. doi:10.1016/j.chb.2015.12.039

Costa, C. R., Garcia, J. E., Da Fonseca, M. J. S., & Teixeira, A. (2021). Data Analysis in Content Marketing Strategies. *Iberian Conference on Information Systems and Technologies, CISTI.* 10.23919/CISTI52073.2021.9476377

Costa, J., Rodrigues, D., & Gomes, J. (2019). Sustainability of tourism destinations and the importance of certification. *Worldwide Hospitality and Tourism Themes, 11*(6), 677–684. doi:10.1108/WHATT-08-2019-0050

Cranshaw, J., Elwany, E., Newman, T., Kocielnik, R., Yu, B., Soni, S., Teevan, J., & Monroy-Hernández, A. (2017). Calendar.help: Designing a workflow-based scheduling agent with humans in the loop. *Proceedings of the 2017 CHI Conference on Human Factors in Computing Systems*, 2382–2393. 10.1145/3025453.3025780

Craton, L. G., & Geoffrey, P. L. (2011). Attitude toward Advertising Music: Potential Pitfalls in Commercials. *Journal of Consumer Marketing, 28*(6), 396–411. doi:10.1108/07363761111165912

Crawford, C. (2003). Chris Crawford on game design. Academic Press.

Crisafulli, B., & Singh, J. (2017). Service failures in e-retailing: Examining the effects of response time, compensation, and service criticality. *Computers in Human Behavior, 77*, 413–424. doi:10.1016/j.chb.2017.07.013

Cronjé, D. F., & du Plessis, E. (2020). A review on tourism destination competitiveness. *Journal of Hospitality and Tourism Management, 45*, 256–265. doi:10.1016/j.jhtm.2020.06.012

Cropanzano, R., & Mitchell, M. S. (2005). Social exchange theory: An interdisciplinary review. *Journal of Management, 31*(6), 874–900. doi:10.1177/0149206305279602

Crouch, G. I., & Ritchie, J. B. (1999). Tourism, competitiveness, and societal prosperity. *Journal of Business Research, 44*(3), 137–152. doi:10.1016/S0148-2963(97)00196-3

Cui, T. H., Ghose, A., Halaburda, H., Iyengar, R., Pauwels, K., Sriram, S., Tucker, C., & Venkataraman, S. (2021). Informational Challenges in Omnichannel Marketing: Remedies and Future Research. *Journal of Marketing, 85*(1), 103–120. doi:10.1177/0022242920968810

Curran, J. M., & Meuter, M. L. (2005). Self-service technology adoption: Comparing three technologies. *Journal of Services Marketing*, *19*(2), 103–113. doi:10.1108/08876040510591411

Curran, J., Iyengar, S., Lund, A. B., & Salovaara-Moring, I. (2009). Media system, public knowledge and democracy: A comparative study. *European Journal of Communication*, *24*(5), 5–26. doi:10.1177/0267323108098943

Cyr, D., Hassanein, K., Head, M., & Ivanov, A. (2007). The role of social presence in establishing loyalty in e-Service environments. *Interacting with Computers*, *19*(1), 43–56. doi:10.1016/j.intcom.2006.07.010

D'Haenens, L., Jankowski, N., & Heuvelman, A. (2004). News in online and print newspapers: Differences in reader consumption and recall. *New Media & Society*, *6*(3), 363–382. doi:10.1177/1461444804042520

d'Orey, F., & Abreu, R. (2015). The feeling of place and the construction of tourist destinations, proposal of a conceptual model [O Sentimento de lugar e a construção dos destinos turísticos, proposta de modelo conceptual]. *European Joural of Applied Business and Management*, *1*(1), 208–223.

da Guarda, C. M. (2015). *Guia do Investidor*. http://www.guiadoinvestidor.mun-guarda.pt/en/files/assets/basic-html/index.html#1

da Guarda, C. M. (n.d.). *Guarda 2027, European Capital of Culture, Candidate City* [Guarda 2027, Capital Europeia da Cultura, Cidade candidata]. https://www.guarda2027.pt/

Daehn, C., Costa, A., & Pereira, R. (2021) Digital transformation and sustainability: Challenges and trends. In *ENGEMA 2021 - XXIII International Meeting on Business Management and the Environment*. FEA/USPAt.

Dahana, W. D., Shin, H., & Katsumata, S. (2018). Influence of individual characteristics on whether and how much consumers engage in showrooming behavior. *Electronic Commerce Research*, *18*(4), 665–692. doi:10.100710660-017-9277-4

Damar, M. (2021). Metaverse Shape of Your Life for Future : A bibliometric snapshot. *Journal of Metaverse*, *1*(1), 1–8.

Dan, F. (2007). *From semantic Web (3.0) to the WebOS (4.0)*. https://www.zdnet.com/blog/btl/from-semantic-web-30-to-the-webos-40/4499/

Davis, F. D. (1989). Perceived Usefulness, Perceived Ease of Use, and User Acceptance of Information Technology. *Management Information Systems Research Center, 13*(3), 319-340.

Davis, F. D. (1989). Perceived usefulness, perceived ease of use and user acceptance of information technology. *Management Information Systems Quarterly*, *13*(3), 319–339. doi:10.2307/249008

Davis, F. D., Bagozzi, R. P., & Warshaw, P. R. (1989). User Acceptance of Computer Technology: A Comparison of Two Theoretical Models. *Management Science*, *35*(8), 982–1003. doi:10.1287/mnsc.35.8.982

de Alvarenga, A. (2019). *Digital Transformation in Public Administration: Case Study* [Master's Dissertation]. ISCTE – Lisbon University Institute, Department of Information Sciences and Technologies.

De Keyser, A., Lemon, K. N., Klaus, P., & Keiningham, T. L. (2015). A framework for understanding and managing the customer experience. *Marketing Science Institute working paper series, 85*(1), 15-121

De Keyser, A., Verleye, K., Lemon, K. N., Keiningham, T. L., & Klaus, P. (2020). Moving the Customer Experience Field Forward: Introducing the Touchpoints, Context, Qualities (TCQ) Nomenclature. *Journal of Service Research*, *23*(4), 433–455. doi:10.1177/1094670520928390

De Marchi, V., Di Maria, E., & Gereffi, G. (2018). *Local clusters in global value chain: Linking actors and territories through manufacturing and innovation*. Routledge. doi:10.4324/9781315182049 de Portugal, T. C. (2018). *Guarda – Tourism Center of Portugal* [Guarda – Turismo Centro Portugal]. https://turismo-docentro.pt/concelho/guarda/

Dean, A. M. (2004). Rethinking customer expectations of service quality: Are call centers different? *Journal of Services Marketing, 18*(1), 60–78. doi:10.1108/08876040410520717

Dehghani, M., Niaki, M. K., Ramezani, I., & Sali, R. (2016). Evaluating the influence of YouTube advertising for attraction of young customers. *Computers in Human Behavior, 59*, 165–172. doi:10.1016/j.chb.2016.01.037

Deloitte. (2015). *Millenials and wealth management - Trends and challenges of the new clientele.* Retrieved October 16, 2021 from https://www2.deloitte.com/content/dam/Deloitte/lu/Documents/about-deloitte/Inside/lu-inside9-full.pdf

Deloitte. (2018). *Digital Maturity Model.* https://www2.deloitte.com/content/dam/Deloitte/global/Documents/Technology-Media-Telecommunications/deloitte-digital-maturity-model.pdf

Deloitte. (2018). *The second wave in Smart City transformation About the Deloitte Center for Government Insights.* Deloitte.

Denicolai, S., Zucchella, A., & Magnani, G. (2021). Internationalization, digitalization, and sustainability: Are SMEs ready? A survey on synergies and substituting effects among growth paths. *Technological Forecasting and Social Change, 166*(12), 120650. doi:10.1016/j.techfore.2021.120650

Dessart, L. (2018). Do ads that tell a story always perform better? The role of character identification and character type in storytelling ads. *International Journal of Research in Marketing, 35*(2), 289–304. doi:10.1016/j.ijresmar.2017.12.009

Deterding, S., Dixon, D., Khaled, R., & Nacke, L. (2011). From game design elements to gamefulness: defining gamification. *Proceedings of the 15th International Academic MindTrek Conference: Envisioning Future Media Environments*, 9-15. 10.1145/2181037.2181040

Dethine, B., Enjolras, M., & Monticolo, D. (2020). Digitalization and SME's export Management: Impacts on Resources and Capabilities. *Technology Innovation Management Review, 10*(4), 18–34. doi:10.22215/timreview/1344

Devlin, J. F. (1995). Technology and innovation in retail banking distribution. *International Journal of Bank Marketing, 13*(4), 19–25. doi:10.1108/02652329510082915

Diabah, G. (2021). Negotiating language barriers: Customer care delivery practices of a selected telecommunication company in Ghana. *Sociolinguistic Studies, 15*(2-4), 177–199. doi:10.1558ols.41465

Digital Results. (2020). *Facebook Ads: The definitive guide.* E-book.

Diogo, R., Junior, A., & dos Santos, N. (2019). Digital transformation and knowledge management: contributions to the improvement of productive and organizational processes. P2P & Innovation, 5(2), 154-175. doi:10.21721/p2p.2019v5n2.p154-175

Djelassi, S., Diallo, M. F., & Zielke, S. (2018). How self-service technology experience evaluation affects waiting time and customer satisfaction? A moderated mediation model. *Decision Support Systems, 111*, 38–47. doi:10.1016/j.dss.2018.04.004

Dominguinhos, P. (2000). *A Internacionalização das Empresas Portuguesas: Balanço da Evidência Empírica* [The Internationalization of Portuguese Companies: Balance of Empirical Evidence]. ESCE-IPS.

dos Santos Júnior, A., Mendes-Filho, L., Almeida García, F., & Simões, J. (2017). Smart Tourism Destinations: a study based on the view of the stakeholders. *Revista Turismo Em Análise, 28*, 358–379.

Downe-Wamboldt, B. (1992). Content analysis: Method, applications, and issues. *Health Care for Women International, 13*(3), 313–321. doi:10.1080/07399339209516006 PMID:1399871

Duričin, D., & Herceg, I. V. (2018). Industry 4.0 and Paradigm Change in Economics and Business Management. In AMP 2018, LNME (pp. 37–56). Academic Press.

Eden, R. E., Jones, A. B., Casey, V., & Draheim, M. (2019). Digital Transformation Requires Workforce Transformation. *MIS Quarterly Executive, 18*(1), 1–17. doi:10.17705/2msqe.00005

Eisenhardt, K. (1989). Building theories from case study research. *Academy of Management Review, 14*(4), 532–550. doi:10.2307/258557

eMarketer. (2016). https://www.emarketer.com/Article/US-Digital-Display-Ad-Spending-Surpass-Search-Ad-Spending-2016/1013442

eMarketer. (2020). https://www.emarketer.com/chart/241048/device-share-of-youtube-video-views-among-us-youtube-viewers-q4-2019-q3-2020-of-total

Endo, A. C. B., & Roque, M. A. B. (2017). Attention, memory and perception: a conceptual analysis of Neuropsychology applied to advertising and its influence on consumer behaviour. *Revista Brasileira de Ciências da Comunicação, 40*(1), 77-96. doi:10.1590/1809-5844201715

Engesser, S., & Humprecht, E. (2015). Frequency or skillfulness: How professional news media use Twitter in five western countries. *Journalism Studies, 16*(4), 513–529. doi:10.1080/1461670X.2014.939849

Etkin, J. (2016). The Hidden Cost of Personal Quantification. *The Journal of Consumer Research, 42*(6), 967–984. doi:10.1093/jcr/ucv095

EUR-Lex. (2015). *Payment services (PSD 2) - Directive (EU) 2015/2366.* Retrieved October 27, 2021 from https://eur-lex.europa.eu/legal-content/EN/TXT/HTML/?uri=LEGISSUM:2404020302_1

European Central Bank. (2021). *EU structural financial indicators: end of 2020.* Retrieved November 20, 2021 from https://www.ecb.europa.eu/press/pr/date/2021/html/ecb.pr210526~7469dedaaf.en.html

European Commission. (2021). *Annual Report on European SMEs 2020/2021.* https://ec.europa.eu/docsroom/documents/46062

Every Realm. (2022). *Top Metaverses o Follow in 2022.* https://everyrealm.com/blog/education/top-metaverses-to-follow-in-2022

Facts & Figures. (2013). http://www.enterprise-gamification.com/mediawiki/index.php?title=Facts_%26_Figures

Fang, J., Liu, H., Li, Y., & Cai, Z. (2021). Retaining customers with in-store mobile usage experience in omni-channel retailing: The moderating effects of product information overload and alternative attractiveness. *Electronic Commerce Research and Applications, 46*, 101028. doi:10.1016/j.elerap.2020.101028

Farah, M. F., Hasni, M. J., & Abbas, A. K. (2018). Mobile-banking adoption: Empirical evidence from the banking sector in Pakistan. *International Journal of Bank Marketing, 36*(7), 1386–1413. doi:10.1108/IJBM-10-2017-0215

Farah, M. F., Ramadan, Z. B., & Harb, D. H. (2019). The examination of virtual reality at the intersection of consumer experience, shopping journey and physical retailing. *Journal of Retailing and Consumer Services, 48*, 136–143. doi:10.1016/j.jretconser.2019.02.016

Farr, R. (1982). Interviewing: The social psychology of the inter-view. In A. J. Chapman & A. Gale (Eds.), *Psychology and People: A Tutorial Text* (pp. 287–305). Macmillan Education. doi:10.1007/978-1-349-16909-2_15

Ferm, L. E. C., & Thaichon, P. (2021). Customer pre-participatory social media drivers and their influence on attitudinal loyalty within the retail banking industry: A multi-group analysis utilizing social exchange theory. *Journal of Retailing and Consumer Services*, *61*, 102584. doi:10.1016/j.jretconser.2021.102584

Fernandes, C., Ferreira, J. J., Veiga, P. M., Kraus, S., & Dabić, M. (2022). Digital entrepreneurship platforms: Mapping the field and looking towards a holistic approach. *Technology in Society*, *70*, 101979. Advance online publication. doi:10.1016/j.techsoc.2022.101979

Fernández, N. V., Pérez, M. J. S., & Vázquez-Casielles, R. (2018). Webroomers versus showroomers: Are they the same? *Journal of Business Research*, *92*, 300–320. doi:10.1016/j.jbusres.2018.08.004

Ferreira, J. J., Fernandes, C. I., Veiga, P. M., & Hughes, M. (2022). Prevailing theoretical approaches predicting sustainable business models: A systematic review. *International Journal of Productivity and Performance Management*, *71*(3), 790–813. doi:10.1108/IJPPM-12-2020-0653

Fiestas, J. C., & Tuzovic, S. (2021). Mobile-assisted showroomers: Understanding their purchase journey and personalities. *Journal of Retailing and Consumer Services*, *58*, 102280. doi:10.1016/j.jretconser.2020.102280

Filev, A. (2020, March 30). *C*OVID-19 Is A Before-And-After Moment In The Digital Transformation. *Forbes*. https://www.forbes.com/sites/andrewfilev/2020/03/30/covid-19-is-a-before-and-after-moment-in-the-digital-transformation/

Firat, D. (2019). YouTube adv ube advertising v tising value and its eff alue and its effects on pur ects on purchase intention chase intenti. *Journal of Global Bussiness Insights*, *4*(2), 141–155.

Fitzgerald, M., Kruschwitz, N., Bonnet, D., & Welch, M. (2013). Embracing digital technology: A new strategic imperative. *MIT Sloan Management Review*. https://www.capgemini.com/se-en/wp-content/uploads/sites/29/2017/07/embracing_digital_technology_a_new_strategic_imperative.pdf

Flavián, C., Gurrea, R., & Orús, C. (2016). Choice confidence in the webrooming purchase process: The impact of online positive reviews and the motivation to touch. *Journal of Consumer Behaviour*, *15*(5), 459–476. doi:10.1002/cb.1585

Flavián, C., Gurrea, R., & Orús, C. (2019a). Feeling Confident and Smart with Webrooming: Understanding the Consumer's Path to Satisfaction. *Journal of Interactive Marketing*, *47*, 1–15. doi:10.1016/j.intmar.2019.02.002

Flavián, C., Gurrea, R., & Orus, C. (2020). Combining Channels to Make Smart Purchases: The Role of Webrooming and Showrooming. *Journal of Retailing and Consumer Services*, *52*, 101923. doi:10.1016/j.jretconser.2019.101923

Flavian, C., Gurrea, R., & Orús, C. (2021). Mobile word of mouth (m-WOM): Analysing its negative impact on webrooming in omnichannel retailing. *International Journal of Retail & Distribution Management*, *49*(3), 394–420. doi:10.1108/IJRDM-05-2020-0169

Flavián, C., Ibáñez-Sánchez, S., & Orús, C. (2019). The Impact of Virtual, Augmented and Mixed Reality Technologies on the Customer Experience. *Journal of Business Research*, *100*, 547–560. doi:10.1016/j.jbusres.2018.10.050

Fletcher, R. (2001). A holistic approach to internationalisation. *International Business Review*, *10*(1), 25–49. doi:10.1016/S0969-5931(00)00039-1

Følstad, A., & Kvale, K. (2017). Customer journeys: A systematic literature review. *Journal of Service Theory and Practice*, *28*(2), 196–227. doi:10.1108/JSTP-11-2014-0261

Følstad, A., Nordheim, C. B., & Bjørkli, C. A. (2018). What makes users trust a chatbot for customer service? An exploratory interview study. In S. S. Bodrunova (Ed.), *Internet Science* (pp. 194–208). Springer. doi:10.1007/978-3-030-01437-7_16

Fonseca, J. L. C. (2015). *Fatores de satisfação e fidelização no processo de compra online – Um estudo de caso na Universidade Lusófona do Porto* [Master Theses, Lusófona do Porto]. Repositório Científico Lusófona. http://recil.grupolusofona.pt/handle/10437/6956

Fonseca, J. (2002). *Metodologia da pesquisa científica* [Scientific Research Methodology]. Apostila. Editora Fortaleza.

Fournier, S. (1998). Consumers and their brands: Developing relationship theory in consumer research. *The Journal of Consumer Research*, *24*(4), 343–353. doi:10.1086/209515

Frasquet-Deltoro, M., Molla-Descals, A., & Miquel-Romero, M. J. (2021). Omnichannel retailer brand experience: Conceptualisation and proposal of a comprehensive scale. *Journal of Brand Management*, *28*(4), 388–401. doi:10.105741262-021-00233-x

Freitas, C. (2017). https://www.jm-madeira.pt/opinioes/ver/741/O_poder_da_familiaridade_na_preferencia_musical

Frith, S. (2003). Music and Everyday Life. In M. Clayton, T. Herbert, & R. Middleton (Eds.), *The Cultural Study of Music* (pp. 92–101). Routledge.

Gabriel, M. (2010). *Marketing na Era Digital: Conceptos.* Platforms e Estratégias.

Gabrielsson, M., & Gabrielsson, P. (2011). Internet-based sales channel strategies of born global firms. *International Business Review*, *20*(1), 88–99. doi:10.1016/j.ibusrev.2010.05.001

Gadalla, E., Keeling, K., & Abosag, I. (2013). Metaverse-retail service quality: A future framework for retail service quality in the 3D internet. *Journal of Marketing Management*, *29*(13–14), 1493–1517. doi:10.1080/0267257X.2013.835742

Galan, J. P. (2009). Music and responses to advertising: The effects of musical characteristics, likeability and congruency. *Recherche et Applications en Marketing*, *24*(4), 3–22. doi.org/10.1177/205157090240040l

Galko, L., Porubän, J., & Senko, J. (2018). Improving the user experience of electronic university enrolment. *2018 16th International Conference on Emerging eLearning Technologies and Applications (ICETA),* 179–184. 10.1109/ICETA.2018.8572054

Gallemard, J. (2022). *Understanding the Metaverse and its Potential for Customer Experience.* https://blog.smart-tribune.com/en/metaverse-customer-experience

Gamification-Research.org. (2021). http://http://gamification-research.org

Gao, L., Melero, I., & Sese, F. J. (2020). Multichannel Integration along the Customer Journey: A Systematic Review and Research Agenda. *Service Industries Journal*, *40*(15-16), 1087–1118. doi:10.1080/02642069.2019.1652600

Gao, M., & Huang, L. (2021). Quality of channel integration and customer loyalty in omnichannel retailing: The mediating role of customer engagement and relationship program receptiveness. *Journal of Retailing and Consumer Services*, *63*, 102688. doi:10.1016/j.jretconser.2021.102688

Gao, W., Fan, H., Li, W., & Wang, H. (2021). Crafting The Customer Experience in Omnichannel Contexts: The Role of Channel Integration. *Journal of Business Research*, *126*, 12–22. doi:10.1016/j.jbusres.2020.12.056

Garg, P., Gupta, B., Chauhan, A. K., Sivarajah, U., Gupta, S., & Modgil, S. (2021). Measuring the perceived benefits of implementing blockchain technology in the banking sector. *Technological Forecasting and Social Change*, *163*, 163. doi:10.1016/j.techfore.2020.120407

GaronJ. (2022). Legal Implications of a Ubiquitous Metaverse and a Web3 Future. *Available at* SSRN, 4002551. doi:10.2139/ssrn.4002551

Gartner. (2022a). *Metaverse Hype to Transition into New Business Models that Extend Digital Business*. https://www.gartner.com/en/newsroom/press-releases/2022-02-07-gartner-predicts-25-percent-of-people-will-spend-at-least-one-hour-per-day-in-the-metaverse-by-2026

Gartner. (2022b). *What Is a Metaverse?* https://www.gartner.com/en/articles/what-is-a-metaverse

Gasparin, I., Panina, E., Becker, L., Yrjölä, M., Jaakkola, E., & Pizzutti, C. (2022). Challenging the "Integration Imperative": A Customer Perspective on Omnichannel Journeys. *Journal of Retailing and Consumer Services*, *64*, 102829. doi:10.1016/j.jretconser.2021.102829

Gelter, J., Lexhagen, M., & Fuchs, M. (2020, December 15). A Meta-Narrative Analysis of Smart Tourism Destinations: Implications for Tourism Destination Management. *Current Issues in Tourism*, *24*(20), 2860–2874. doi:10.1080/13683500.2020.1849048

Gerber, Barnard, & van der Merwe. (2007). *Towards a semantic web layered architecture*. The 25th Conference on IASTED International Multi-Conference.

Ghaderi, Z. (2021, November 4). Factors Affecting International Tourists' Intention to Use Local Mobile Apps in Online Purchase. *Asia Pacific Journal of Tourism Research*, *26*(12), 1285–1301. doi:10.1080/10941665.2021.1983626

Gill, M., & VanBoskirk, S. (2016). *The Digital Maturity Model 4.0 Benchmarks: Digital Business Transformation Playbook*. Forrester Research, Inc. Available: https://dixital.cec.es/wp-content/uploads/presentacions/presentacion06.pdf

Global Webindex. (2019). https://www.gwi.com/reports/social-2019

Gonzalez Crespo, R., Ferro Escobar, R., Joyanes Aguilar, L., Velazco, S., & Castillo Sanz, A. G. (2013). Use of ARIMA mathematical analysis to model the implementation of expert system courses by means of free software OpenSim and Sloodle platforms in virtual university campuses. *Expert Systems with Applications*, *40*(18), 7381–7390. doi:10.1016/j.eswa.2013.06.054

González-Reverté. (2019, December 3). Building Sustainable Smart Destinations: An Approach Based on the Development of Spanish Smart Tourism Plans. *Sustainability*, *11*(23), 6874. doi:10.3390u11236874

Goode, W., & Hatt, P. (1973). Métodos. In *Pesquisa Social*. Editora Nacional.

GoogleAdsHelp. (2021a). https://support.google.com/google-ads/answer/6052202?hl=pt-BR

GoogleAdsHelp. (2021b). https://support.google.com/google-ads/answer/9035882

GoogleAdsHelp. (2021c). https://support.google.com/youtube/answer/2375464?hl=pt

GoogleAdsHelp. (2021d). https://support.google.com/youtube/answer/2454017?hl=pt

Gorn, G. J. (1982). The effects of music in advertising on choice behavior: A classical conditioning approach. *Journal of Marketing*, *46*(1), 94–101. doi.org/10.1177/002224298204600109

Grand View Research. (2022). *Chatbot Market Size Worth $3.99 Billion By 2030 | CAGR: 25.7%*. https://www.grandviewresearch.com/press-release/global-chatbot-market

Graves, D., & Simon, F. M. (2019). *Pay models for online news in the US and Europe: 2019 update*. Reuters Institute for the Study of Journalism. https://www.digitalnewsreport.org/publications/2019/pay-models-2019-update/

Greenberg, Sutton, & Campbell. (2003). Metadata: A Fundamental Component of the Semantic Web. Bulletin of the American Society for Information Science and Technology, 29(4), 16–18.

Gretzel, U. (2019). *From Smart Destinations to Smart Tourism Regions Privacy View Project Responsible Technology View Project*. Academic Press.

Gretzel, U. (2018). From Smart Destinations to Smart Tourism regions. *Investigaciones Regionales*, *2018*(42), 171–184.

Gretzel, U., & Scarpino-Johns, M. (2018). Destination Resilience and Smart Tourism Destinations. *Tourism Review International*, *22*(3), 263–276. doi:10.3727/154427218X15369305779065

Gretzel, U., Sigala, M., Xiang, Z., & Koo, C. (2015a). Smart tourism: Foundations and developments. *Electronic Markets*, *25*(3), 179–188. doi:10.100712525-015-0196-8

Gretzel, U., Werthner, H., Koo, C., & Lamsfus, C. (2015b). Conceptual foundations for understanding smart tourism ecosystems. *Computers in Human Behavior*, *50*, 558–563. doi:10.1016/j.chb.2015.03.043

Grewal, D., Noble, S. M., Roggeveen, A. L., & Nordfalt, J. (2020). The future of in-store technology. *Journal of the Academy of Marketing Science*, *48*(1), 96–113. doi:10.100711747-019-00697-z

Grewal, D., & Roggeveen, A. L. (2020). Understanding retail experiences and customer journey management. *Journal of Retailing*, *96*(1), 3–8. doi:10.1016/j.jretai.2020.02.002

Grider, D., & Maximo, M. (2021). *Metaverse Report: Web 3.0 and Virtual Cloud Economy*. https://en.0xzx.com/metaverse-report-web-3-0-and-virtual-cloud-economy/

Groh, F. (2012). *Gamification: State of the art definition and utilization*. Institute of Media Informatics Ulm University.

Grove, W. (2020). How will COVID-19 Impact Digital Transformation? *Skybox Security*. https://blog.skyboxsecurity.com/how-will-covid-19-impact-digital-transformation/

Grudin, J., & Jacques, R. (2019). Chatbots, humbots, and the quest for artificial general intelligence. *Proceedings of the 2019 CHI Conference on Human Factors in Computing Systems*, 1–11. 10.1145/3290605.3300439

Guimarães, N., Miranda, F., & Figueira, A. (2019). Identifying journalistically relevant social media texts using human and automatic methodologies. *International Journal of Grid and Utility Computing*, *11*(1), 72–83. doi:10.1504/IJGUC.2020.103971

Gu, J.-C., Lee, S.-C., & Suh, Y.-H. (2009). Determinants of behavioral intention to mobile banking. *Expert Systems with Applications*, *36*(9), 11605–11616. doi:10.1016/j.eswa.2009.03.024

Guo, Y., Zhang, M., & Wang, V. L. (2021). Webrooming or showrooming? The moderating effect of product attributes. *Journal of Research in Interactive Marketing*. Advance online publication. doi:10.1108/JRIM-08-2020-0161

Guzmán, J. R. S. (1993). *Teoría de la publicidad*. Tecnos.

Hackl, C. (2020, August 2). Now Is The Time To Talk About Ethics And Privacy In The Metaverse. *Forbes*. https://www.forbes.com/sites/cathyhackl/2020/08/02/now-is-the-time-to-talk-about-ethics--privacy-in-the-metaverse/?sh=36b6ca3fae6c

Hagel, J. III. (2016). We need to foster our definition of Entrepreneurship. *Harvard Business Review*, (September), 1–5.

Hahn, M., & Hwang, I. (1999). Effects of tempo and familiarity of background music on message processing in TV advertising: A resource-matching perspective. *Psychology and Marketing*, *16*(8), 659–675. doi.org/10.1002/(SICI)1520-6793

Hajdas, M., Radomska, J., & Silva, S. C. (2020). The omni-channel approach: A utopia for companies? *Journal of Retailing and Consumer Services*, *102131*. Advance online publication. doi:10.1016/j.jretconser.2020.102131

Hakulinen, Auvinen, & Korhonen. (2013). Empirical Study on the Effect of Achievement Badges in TRAKLA2 Online Learning Environment. *Learning and Teaching in Computing and Engineering (LaTiCE)*, 21-24.

Halbheer, D., Stahl, F., Koenigsberg, O., & Lehmann, D. (2014). Choosing a digital content strategy: How much should be free? *International Journal of Research in Marketing*, *31*(2), 192–206. doi:10.1016/j.ijresmar.2013.10.004

Hamari, J., & Eranti, V. (2011). Framework for designing and evaluating game achievements. *Proceedings of DiGRA 2011: Think Design Play*, 115-134.

Hamari, J., & Koivisto, J. (2013). Social motivations to use gamification: An empirical study of gamifying exercise. *Proceedings of the 21st European Conference on Information Systems.*

Hamblen, M. (2014). *Mobile Data Traffic Is Expected to Explode 11-Fold by 2018.* www.computerworld.com/article/2487327/wireless-networking/mobile-data-traffic-is expectedto-explode-11-fold-by-2018.html

Hamouda, M. (2019). Omni-channel banking integration quality and perceived value as drivers of consumers' satisfaction and loyalty. *Journal of Enterprise Information Management*, *32*(4), 608–625. doi:10.1108/JEIM-12-2018-0279

Hanafizadeh, P., Behboudi, M., & Koshksaray, A. A. (2012). Mobile-banking adoption by Iranian clients. *Telematics and Informatics*, *31*(1), 62–78. doi:10.1016/j.tele.2012.11.001

Handley, L. (2017). https://www.cnbc.com/2017/02/16/sixty-five-percent-of-people-skip-online-video-ads-heres-what-to-do.html

Hanebrink, J., Kunze, H. M., Mix, T., Löffler, J., Felten, D., & Leppien, K. (2015). *How to Navigate digitization of-manufacturing sector.* McKinsey Industry 4.0 GlobalExpert Survey. Available online: https://www.mckinsey.com/~/media/McKinsey/Business%20Functions/Operations/Our%20Insights/Industry%2040%20How%20to%20navigate%20digitization%20of%20the%20manufacturing%20sector/Industry-40-How-to-navigate-digitization-of-the-manufacturing-sector.ashx

Harrisson-Boudreau, J.-P., & Bellemare, J. (2022). Going Above and Beyond eCommerce in the Future Highly Virtualized World and Increasingly Digital Ecosystem. *Lecture Notes in Mechanical Engineering*, 789–797. doi:10.1007/978-3-030-90700-6_90

Hasija, S., Shen, Z. J. M., & Teo, C. P. (2020). Smart City operations: Modeling challenges and opportunities. *Manufacturing & Service Operations Management*, *22*(1), 203–213. doi:10.1287/msom.2019.0823

Hassouneh, D., & Brengman, M. (2011). Virtual worlds: A gateway for SMEs toward internationalization. *Journal of Brand Management*, *19*(1), 72–90. doi:10.1057/bm.2011.24

Hassouneh, D., & Brengman, M. (2015). Retailing in social virtual worlds: Developing a typology of virtual store atmospherics. *Journal of Electronic Commerce Research*, *16*(3), 218–241. https://www.scopus.com/inward/record.uri?eid=2-s2.0-84940050446&partnerID=40&md5=0c56284bbc6c664d43606bf0dcfd3d79

Haytham, Koutb, & Suoror. (2009). Toward An Agreement on Semantic Web Architecture. *Proceedings of World Academy of Science, Engineering And Technology, 37.*

Hecker, S. (1984). Music for advertising effect. *Psychology and Marketing*, *1*(3-4), 3–8. doi.org/10.1002/mar.4220010303

Hemnath. (2010). *Web 4.0 - A New Web Technology.* http://website-quality.blogspot.com/2010/01/web-40-new-web-technology.html/

Hendrickx, K. (2021). The rise of social journalism: An explorative case study of a youth-oriented Instagram news account. *Journalism Practice*, 1–16. Advance online publication. doi:10.1080/17512786.2021.2012500

Herget, A. K., Schramm, H., & Breves, P. (2018). Development and testing of an instrument to determine Musical Fit in audio-visual advertising. *Musicae Scientiae*, *22*(3), 362–376. doi.org/10.1177/1029864917731127

Hermida, A. (2012). Social journalism: exploring how social media is shaping journalism. In E. Siapera & A. Veglis (Eds.), *The Handbook of Global Online Journalism* (1st ed., pp. 309–328). John Wiley & Sons, Inc. doi:10.1002/9781118313978.ch17

Herrero-Crespo, A., Viejo-Fernández, N., Collado-Agudo, J., & Pérez, M. J. S. (2021). Webrooming or showrooming, that is the question: Explaining omnichannel behavioural intention through the technology acceptance model and exploratory behaviour. *Journal of Fashion Marketing and Management*, *26*(3), 401–419. doi:10.1108/JFMM-05-2020-0091

Hervé, A., Schmitt, C., & Baldegger, R. (2020). Internationalization and Digitalization: Applying digital technologies to the internationalization process of small and medium sized enterprises. *Technology Innovation Management Review*, *10*(7), 28–40. doi:10.22215/timreview/1373

Hickman, E., Kharouf, H., & Sekhon, H. (2020). An omnichannel approach to retailing: Demystifying and identifying the factors influencing an omnichannel experience. *International Review of Retail, Distribution and Consumer Research*, *30*(3), 266–288. doi:10.1080/09593969.2019.1694562

Hilken, T., Chylinski, M., Keeling, D. I., Heller, J., de Ruyter, K., & Mahr, D. (2022). How to strategically choose or combine augmented and virtual reality for improved online experiential retailing. *Psychology & Marketing, 39*(3), 495-507. doi: 10.1002/

Hilken, T., Heller, J., Chylinski, M., Keeling, D. I., Mahr, D., & de Ruyter, K. (2018). Making Omnichannel an Augmented Reality: The Current and Future State of the Art. *Journal of Research in Interactive Marketing*, *12*(4), 509–523. doi:10.1108/JRIM-01-2018-0023

Hill, J., Ford, W. R., & Farreras, I. G. (2015). Real conversations with artificial intelligence: A comparison between human–human online conversations and human–chatbot conversations. *Computers in Human Behavior*, *49*, 245–250. doi:10.1016/j.chb.2015.02.026

Hollensen, S. (2011). *Global Marketing: A Decision Oriented Approach* (5th ed.). Pearson Education Limited.

Hollensen, S., Kotler, P., & Opresnik, M. O. (2022). Metaverse – the new marketing universe. *The Journal of Business Strategy*. Advance online publication. doi:10.1108/JBS-01-2022-0014

Hossain, T. M. T., Akter, S., Kattiyapornpong, U., & Dwivedi, Y. (2020). Reconceptualizing integration quality dynamics for omnichannel marketing. *Industrial Marketing Management, 87*, 225-241. doi:10.1016/j.indmarman.2019.12.006

Hossain, M., Azam, M. S., & Quaddus, M. (2021). Small firm entry to e-marketplace for market expansion and internationalization: A theoretical perspective. *Journal of International Entrepreneurship*, *19*(4), 560–590. doi:10.100710843-021-00297-5

Hsieh, H.-F., & Shannon, S. E. (2005). Three approaches to qualitative content analysis. *Qualitative Health Research*, *15*(9), 1277–1288. doi:10.1177/1049732305276687 PMID:16204405

Huang, W. H. Y., & Soman, D. (2013). *Gamification of Education. Research Report Series: Behavioural Economics in Action*. University of Toronto.

Hunicke, R., LeBlanc, M., & Zubek, R. (2004). MDA: A formal approach to game design and game research. *Proceedings of the AAAI Workshop on Challenges in Game AI,* 4-5.

Huotari, K., & Hamari, J. (2011). Gamification from the perspective of service marketing. *Proceedings of ACM CHI 2011 Workshop on Gamification.*

Huré, E., Picot-Coupey, K., & Ackermann, C. L. (2017). Understanding omni-channel shopping value: A mixed-method study. *Journal of Retailing and Consumer Services*, *39*, 314–330. doi:10.1016/j.jretconser.2017.08.011

Hüseyinoğlu, I. Ö. Y., Galipoğlu, E., & Kotzab, H. (2017). Social, local and mobile commerce practices in omni-channel retailing: Insights from Germany and Turkey. *International Journal of Retail & Distribution Management*, *45*(7-8), 711–729. doi:10.1108/IJRDM-09-2016-0151

Hyche, J. (2016). *The Future of Branch Banking*. Louisiana Banker Association. https://www.lba.org/files/John%20 Hyche,%20The%20Future%20of%20Branch%20%20Banking.pdf

IBM. (2019). *IaaS, PaaS and SaaS – IBM Cloud service models*. IBM. https://www.ibm.com/cloud/learn/iaas-paas-saas

Ieva, M., & Ziliani, C. (2018). Mapping touchpoint exposure in retailing: Implications for developing an omnichannel customer experience. *International Journal of Retail & Distribution Management*, *46*(3), 304–322. doi:10.1108/ IJRDM-04-2017-0097

INE. (2020). *Inquérito à Utilização de Tecnologias da Informação e da Comunicação pelas Famílias 2020*. INE.

Inman, J. J., & Nikolova, H. (2017). Shopper-facing retail technology: A retailer adoption decision framework incorporating shopper attitudes and privacy concerns. *Journal of Retailing*, *93*(1), 7–28. doi:10.1016/j.jretai.2016.12.006

Insider. (2022). *Former Google CEO says Facebook's metaverse is "not necessarily the best thing for human society" and expresses concerns about safety of artificial intelligence*. https://www.businessinsider.com/former-google-ceo-expresses-concerns-about-facebook-metaverse-

Inversini, A., Cantoni, L., & De Pietro, M. (2014). Destination Online Communication: Why Less is Sometimes More. A Study of Online Communications of English Destinations. *Journal of Travel & Tourism Marketing*, *31*(5), 563–575. doi:10.1080/10548408.2014.883949

IPSOS Connect, . (2020). *Ipsos/Google Advertising Attention Research:Report*. E-book.

Iqbal, M. S., Hassan, M. U., & Habibah, U. (2018). *Impact of self-service technology (SST) service quality on customer loyalty and behavioral intention: The mediating role of customer satisfaction*. Cogent Business & Management.

i-SCOOP. (2022). *Industry 4.0 and the fourth industrial revolution explained*. Available online: https://www.i-scoop.eu/industry-4-0/

Ishfaq, N., & Mengxing, H. (2021). Consumer usage behavior of internet-based services (IBS) in Pakistan during COVID-19 crisis from the perspective of technology acceptance model. *Environmental Science and Pollution Research International*, 1–16. doi:10.100711356-021-15868-1 PMID:34550524

Iyer, S. P., & Aggarwal, A. (2019). The Role of Music in Brand Recall - A Study of Select Two-Wheeler Advertisements. *International Journal of Management, Technology And Engineering*, *4*(1), 481–503.

Jacob, F., Pez, V., & Volle, P. (2021). Shifting to phygital experience management with design science: A six-step method to manage customer journeys. *Journal of Strategic Marketing*, •••, 1–22. doi:10.1080/0965254X.2021.2016894

Jacobs, S. (2011). *Serious Play Conference 2011: Microsoft's 'Productivity Games'*. http://www.gamasutra.com/view/ news/36824/Serious_Play_Conference_2011_Mic rosofts_Productivity_Games.php

Jadil, Y., Rana, N. P., & Dwivedi, Y. K. (2021). A meta-analysis of the UTAUT model in the mobile banking literature: The moderating role of sample size and culture. *Journal of Business Research*, *132*, 354–372. doi:10.1016/j.jbusres.2021.04.052

Jain, S., Schweiss, T., Bender, S., & Werth, D. (2021). Omnichannel Retail Customer Experience with Mixed-Reality Shopping Assistant Systems. In *Advances in Visual Computing. International Symposium on Visual Computing 2021. Lecture Notes in Computer Science* (vol. 13017). Springer. 10.1007/978-3-030-90439-5_40

Jain, M., Kumar, P., Kota, R., & Patel, S. N. (2018). Evaluating and informing the design of chatbots. *Proceedings of the 2018 Designing Interactive Systems Conference*, 895–906. 10.1145/3196709.3196735

Jeon, J.-E. (2021). The Effects of User Experience-Based Design Innovativeness on User– Metaverse Platform Channel Relationships in South Korea. *Journal of Distribution Science*, *19*(11), 81–90. doi:10.15722/jds.19.11.202111.81

Johanson, J., & Mattson, L. (1988). *Internationalization in Industrial Systems: A Network Aproach*. Croom Helm.

Johanson, J., & Vahlne, J. (1977). The Internationalization Process of the Firm - A model of knowledge development and increasing foreign market commitment. *Journal of International Business Studies*, *8*(1), 23–32. doi:10.1057/palgrave.jibs.8490676

Johanson, J., & Vahlne, J. (2009). The Uppsala Internationalization Process Model Revisited: From liability of foreignness to liability of outsidership. *Journal of International Business Studies*, *40*(9), 1411–1431. doi:10.1057/jibs.2009.24

Jones, A. L., Miller, J. W., Griffis, S. E., Whipple, J. M., & Voorhees, C. M. (2021). An examination of the effects of omni-channel service offerings on retailer performance. *International Journal of Physical Distribution & Logistics Management*, *52*(2), 150–169. doi:10.1108/IJPDLM-06-2020-0175

Joo, H., & Lee, E. J. (2016). A study on the user acceptance model of omni channel service based on Unified Theory of Acceptance and Use of Technology (UTAUT). *Family and Environment Research*, *54*(4), 405–414. doi:10.6115/fer.2016.031

Jornal de Negócios. (2022). *Está fechada a primeira aquisição de casa com criptomoedas em Portugal* [The first home Purchase with Cryptocrrencies in Portugal is closed]. https://www.jornaldenegocios.pt/mercados/criptoativos/detalhe/esta-fechada-a-primeira-aquisicao-de-casa-com-criptomoedas-em-portugal

Juaneda-Ayensa, E., Mosquera, A., & Sierra Murillo, Y. (2016). Omnichannel customer behavior: Key drivers of technology acceptance and use and their effects on purchase intention. *Frontiers in Psychology*, *7*, 1117. doi:10.3389/fpsyg.2016.01117 PMID:27516749

Julio, K. B. (2019). https://www.meioemensagem.com.br/home/midia/2019/10/08/publicidade-digital-deve-se-igualar-a-tradicional-ate-2023.html

Kaabachi, S. B., Mrad, S. B., Fiedler, A., & Kpossa, M. R. (2021). The Role of Personal Innovativeness in French Omnichannel Banking. *Journal of Euromarketing*, *30*, 72–94. doi:10.9768/jem.v30i3and4.35

Kanematsu, H., Kobayashi, T., Ogawa, N., Fukumura, Y., Barry, D. M., & Nagai, H. (2012). Nuclear Energy Safety Project in Metaverse. *Smart Innovation. Systems and Technologies*, *14*, 411–418. doi:10.1007/978-3-642-29934-6_39

Kang, J. Y. M. (2018). Showrooming, Webrooming, and User-Generated Content Creation in the Omnichannel Era. *Journal of Internet Commerce*, *17*(2), 145–169. doi:10.1080/15332861.2018.1433907

Kang, J. Y. M. (2019). What drives omnichannel shopping behaviors? Fashion lifestyle of social-local-mobile consumers. *Journal of Fashion Marketing and Management*, *23*(2), 224–238. doi:10.1108/JFMM-07-2018-0088

Kaplan, A. M., & Haenlein, M. (2010). Users of the world, unite! The challenges and opportunities of Social Media. *Business Horizons*, *53*(1), 53–68. doi:10.1016/j.bushor.2009.09.003

Karimi, J., & Walter, Z. (2015). The role of dynamic capabilities in responding to digital disruption: A factor-based study of the newspaper industry. *Journal of Management Information Systems*, *32*(1), 39–81. doi:10.1080/07421222.2015.1029380

Karlsson, E., & Rundcrantz, S. (2017). *'Surfing on the wave of Digitalisation' - Digital technology facilitating an SME's internationalisation process in the Fashion Industry - A single case study of Trendy AB* (Master Degree). University of Gothenburg. Retrieved from http://hdl.handle.net/2077/53149

Kaur, K., Bakar, E. A., & Singh, J. (2020). Theoretical framework development on users adoption of omni-channel retailing of fashion apparels based on UTAUT2 and the role of personal innovativeness, brand image and fashion involvement. In *20th Kuala Lumpur International Business, Economics and Law Conference* (pp. 67-81). Academic Press.

Kazancoglu, I., & Aydin, H. (2018). An investigation of consumers' purchase intentions towards omni-channel shopping: A qualitative exploratory study. *International Journal of Retail & Distribution Management*, *46*(10), 959–976. doi:10.1108/IJRDM-04-2018-0074

Kelleher, T. (2009). Conversational voice, communicated commitment, and public relations outcomes in interactive online communication. *Journal of Communication*, *59*(1), 172–188. doi:10.1111/j.1460-2466.2008.01410.x

Khanna, A., Jain, M., Kumar, T., Singh, D., Pandey, B., & Jha, V. (2015). Anatomy and utilities of an artificial intelligence conversational entity. *2015 International Conference on Computational Intelligence and Communication Networks (CICN)*, 594–597. 10.1109/CICN.2015.122

Kietzmann, J. H., Hermkens, K., McCarthy, I. P., & Silvestre, B. S. (2011). Social media? Get serious! Understanding the functional building blocks of social media. *Business Horizons*, *54*(3), 241–251. doi:10.1016/j.bushor.2011.01.005

Kim, B., & Park, M. J. (2019). An examination of the effect of consumer personal traits on show rooming behavior: The effect of product type. *Journal of Internet Commerce*, *18*(1), 24–44. doi:10.1080/15332861.2018.1547569

Kim, J. (2021). Advertising in the Metaverse: Research Agenda. *Journal of Interactive Advertising*, *21*(3), 141–144. doi:10.1080/15252019.2021.2001273

Kim, J. H., Kim, M., Yoo, J., & Park, M. (2020). Consumer decision-making in a retail store: The role of mental imagery and gender difference. *International Journal of Retail & Distribution Management*, *49*(3), 421–445. doi:10.1108/IJRDM-10-2019-0353

Kitchenham, B., & Charters, S. (2007). *Guidelines for performing systematic literature reviews in software engineering.* doi:10.1145/1134285.1134500

Kleinlercher, K., Linzmajer, M., Verhoef, P. C., & Rudolph, T. (2020). Antecedents of webrooming in omnichannel retailing. *Frontiers in Psychology*, *3342*, 606798. Advance online publication. doi:10.3389/fpsyg.2020.606798 PMID:33329282

Kongmanila, X., & Takahashi, Y. (2009). Innovation, export performance and profitability of Lao garment exporters. *International Journal of Economics and Management*, *3*(2), 225–236. http://www.ijem.upm.edu.my/vol3no2/bab01.pdf

Kontogianni, A., & Alepis, E. (2020). Smart tourism: State of the art and literature review for the last six years. *Array*, *6*(January), 100020. doi:10.1016/j.array.2020.100020

Koo, C., Shin, S., Gretzel, U., Hunter, W. C., & Chung, N. (2016). Conceptualization of Smart Tourism Destination Competitiveness. *Asia Pacific Journal of Information Systems*, *26*(4), 561–576. doi:10.14329/apjis.2016.26.4.561

Kotler, P., Kartajaya, H., & Setiawan, I. (2021). *Marketing 5.0: Technology for Humanity*. John Wiley & Sons Inc.

Kozinets, R. V. (2022). Immersive netnography: A novel method for service experience research in virtual reality, augmented reality and metaverse contexts. *Journal of Service Management*. Advance online publication. doi:10.1108/JOSM-12-2021-0481

Kraus, S., Durst, S., Ferreira, J. J., Veiga, P., Kailer, N., & Weinmann, A. (2022). Digital transformation in business and management research: An overview of the current status quo. *International Journal of Information Management*, *63*, 102466.

Kraus, S., Kanbach, D. K., Krysta, P. M., Steinhoff, M. M., & Tomini, N. (2022). Facebook and the creation of the metaverse: Radical business model innovation or incremental transformation? *International Journal of Entrepreneurial Behaviour & Research*, *28*(9), 52–77. doi:10.1108/IJEBR-12-2021-0984

Kroeze, C., & Olivier, M. S. (2012). *Gamifying authentication. Information Security for South Africa.* ISSA.

Kshetri, N. (2017). Potential roles of blockchain in fighting poverty and reducing financial exclusion in the global south. *Journal of Global Information Technology Management*, *20*(4), 201–204. doi:10.1080/1097198X.2017.1391370

Lamas, L. B., & Lamas, M. (2022). New insight into Business Innovation and Sustainable Development: The role of Education. In C. Sarmento (Ed.), *Concepts and Dialogues across Shifting Spaces in Intercultural Business* (pp. 15–31). Cambridge Scholars Publishing.

Larke, R., Kilgour, M., & O'Connor, H. (2018). Build touchpoints and they will come: transitioning to omnichannel retailing. *International Journal of Physical Distribution & Logistics Management, 48*(4), 465-483. doi: 10.1108/IJP-DLM-09-2016-0276/

Latta, J. N., & Oberg, D. J. (1994). A conceptual virtual reality model. *IEEE Computer Graphics and Applications*, *14*(1), 23–29. doi:10.1109/38.250915

Lawry, C. A., & Bhappu, A. D. (2021). Measuring consumer engagement in omnichannel retailing: the mobile in-store experience (MIX) index. *Frontiers in Psychology, 12*. doi: 10.3389/fpsyg.2021.661503

Lazaris, C., Sarantopoulos, P., Vrechopoulos, A., & Doukidis, G. (2021). Effects of Increased Omnichannel Integration on Customer Satisfaction and Loyalty Intentions. *International Journal of Electronic Commerce*, *25*(4), 440–468. doi:10.1080/10864415.2021.1967005

Lazaris, C., Vrechopoulos, A., Sarantopoulos, P., & Doukidis, G. (2022). Additive omnichannel atmospheric cues: The mediating effects of cognitive and affective responses on purchase intention. *Journal of Retailing and Consumer Services*, *64*, 102731. doi:10.1016/j.jretconser.2021.102731

Le, A. N. H., & Le, X. D. N. (2020). A moderated mediating mechanism of omnichannel customer experiences. *International Journal of Retail & Distribution Management*, *49*(5), 595–615. doi:10.1108/IJRDM-02-2020-0054

Lee, L.-H., Braud, T., Zhou, P., Wang, L., Xu, D., Lin, Z., Kumar, A., Bermejo, C., & Hui, P. (2021). *All one needs to know about metaverse: A complete survey on technological singularity, virtual ecosystem, and research agenda.* ArXiv Preprint ArXiv:2110.05352.

Lee, A. M., Lewis, S. C., & Powers, M. (2012). Audience clicks and news placement: A study of time- lagged influence in online journalism. *Communication Research*, *41*(4), 1–26. doi:10.1177/0093650212467031

Lee, H.-J., & Lyu, J. (2016). Personal values as determinants of intentions to use self-service technology in retailing. *Computers in Human Behavior*, *60*, 322–332. doi:10.1016/j.chb.2016.02.051

Lee, K. C., & Chung, N. (2009). Understanding factors affecting trust in and satisfaction with mobile banking in Korea: A modified DeLone and McLean's model perspective. *Interacting with Computers*, *21*(5-6), 385–392. doi:10.1016/j.intcom.2009.06.004

Lee, P., Hunter, W. C., & Chung, N. (2020). Smart tourism city: Developments and transformations. *Sustainability (Switzerland)*, *12*(10), 1–15. doi:10.3390u12103958

Lee, S.-G., Trimi, S., Byun, W. K., & Kang, M. (2011). Innovation and imitation effects in Metaverse service adoption. *Service Business*, *5*(2), 155–172. doi:10.100711628-011-0108-8

Lee, Y., Falahat, M., & Sia, B. (2019). Impact of Digitalization on the Speed of Internationalization. *International Business Research*, *12*(4), 1. doi:10.5539/ibr.v12n4p1

Lee, Z. W., Chan, T. K., Chong, A. Y. L., & Thadani, D. R. (2019). Customer engagement through omnichannel retailing: The effects of channel integration quality. *Industrial Marketing Management*, *77*, 90–101. doi:10.1016/j.indmarman.2018.12.004

Lemon, K. N., & Verhoef, P. C. (2016). Understanding Customer Experience Throughout the Customer Journey. *Journal of Marketing*, *80*(6), 69–96. doi:10.1509/jm.15.0420

Lew, Z., Walther, J. B., Pang, A., & Shin, W. (2018). Interactivity in online chat: Conversational contingency and response latency in computer-mediated communication. *Journal of Computer-Mediated Communication*, *23*(4), 201–221. doi:10.1093/jcmc/zmy009

Liao, Y., Deschamps, F., Loures, E. D. F. R., & Ramos, L. F. P. (2017). Past, present and future of Industry 4.0-a systematic literature review and research agenda proposal. *International Journal of Production Research*, *55*(12), 3609–3629. doi:10.1080/00207543.2017.1308576

Lim, X. J., Cheah, J. H., Dwivedi, Y. K., & Richard, J. E. (2022). Does retail type matter? Consumer responses to channel integration in omni-channel retailing. *Journal of Retailing and Consumer Services*, *67*, 102992. doi:10.1016/j.jretconser.2022.102992

Lischka, J. A., & Garz, M. (2021). Clickbait news and algorithmic curation: A game theory framework of the relation between journalism, users, and platforms. *New Media & Society*. Advance online publication. doi:10.1177/14614448211027174

Liu, D. R., Liao, Y. S., & Lu, J. Y. (2019). Online news recommendations based on topic modeling and online interest adjustment. *Industrial Management & Data Systems*, *119*(8), 1802–1818. doi:10.1108/IMDS-04-2019-0251

Liu, F., Li, J., Mizerski, D., & Soh, H. (2012). Self-congruity, brand attitude, and brand loyalty: A study on luxury brands. *European Journal of Marketing*, *46*(7), 922–937. doi.org/10.1108/03090561211230098

Li, W., Grossman, T., & Fitzmaurice, G. (2012). GamiCAD: a gamified tutorial system for first time AutoCAD users. *Proceedings of the 25th annual ACM symposium on User interface software and technology*, 103-112. 10.1145/2380116.2380131

Loane, S., & Bell, J. (2006). Rapid internationalisation among entrepreneurial firms in Australia, Canada, Ireland and New Zealand: An extension to the network approach. *International Marketing Review*, *23*(5), 467–485. doi:10.1108/02651330610703409

Lockwood, J. (2017). An analysis of web-chat in an outsourced customer service account in the Philippines. *English for Specific Purposes*, *47*, 26–39. doi:10.1016/j.esp.2017.04.001

Lopes, I. M., & Oliveira, P. (2017). Can a small city be considered a Smart City? *Procedia Computer Science*, *121*, 617–624. doi:10.1016/j.procs.2017.11.081

López De Á vila, A., Lancis, E., García, S., Alcantud, A., García, B., & Muñoz, N. (2015, September). *Smart tourism destinations report: Building the future* [Informe destinos turísticos inteligentes: Construyendo el future]. Madrid, Spain: SEGITTUR. https://www.thinktur.org/media/Libro-Blanco-Destinos-Tursticos-Inteligentes-construyendo-el-futuro.pdf

López de Avila, A. (2015). *Smart destinations: XXI century tourism*. ENTER2015 Conference on Information and Communication TechNologies in Tourism, LugaNo, Switzerland.

Loureiro, B. M. P. (2017). *Smart Cities na Política Europeia e o Smart Living em Portugal*. Academic Press.

Lovelock, C. H., & Wirtz, J. (2016). Services Marketing: people, technology, strategy. In C. H. Lovelock & J. Wirtz (Eds.), *Services Marketing: people, technology, strategy* (pp. 470–480). World Scientific Publishing Co. Inc.

Luger, E., & Sellen, A. (2016). "Like having a really bad pA": The gulf between user expectation and experience of conversational agents. *Proceedings of the 2016 CHI Conference on Human Factors in Computing Systems*, 5286–5297. 10.1145/2858036.2858288

Lu, L., McDonald, C., Kelleher, T., Lee, S., Chung, Y. J., Mueller, S., Vielledent, M., & Yue, C. A. (2022). Measuring consumer-perceived humanness of online organizational agents. *Computers in Human Behavior*, *128*, 107092–107092. doi:10.1016/j.chb.2021.107092

Lytras, M. D., & Serban, A. C. (2020). E-Government Insights to Smart Cities Research: European Union (EU) Study and the Role of Regulations. *IEEE Access: Practical Innovations, Open Solutions*, *8*, 65313–65326. doi:10.1109/ACCESS.2020.2982737

Machado, A., Secinaro, S., Calandra, D., & Lanzalonga, F. (2022). Knowledge management and digital transformation for Industry 4.0: A structured literature review. *Knowledge Management Research and Practice*, 1–19. doi:10.1080/14778238.2021.2015261

Madsen, D. O. (2019). The Emergence and Rise of Industry 4.0 Viewed through the Lens of Management Fashion Theory. *Administrative Sciences*, *9*(3), 71. doi:10.3390/admsci9030071

Malbos, A. N. A., Rochadel, W., De Lima, J. P. C., & Da Silva, J. B. (2014). Aplicação da Realidade Aumentada para simulação de experimentos físicos em dispositivos móveis [Application of Augmented Reality for simulation of physical experiments on Mobile Devices]. *2014 11th International Conference on Remote Engineering and Virtual Instrumentation (REV)*, 231–235.

Malhotra, N. (2019). *Marketing Research: An Applied Orientation*. Person Education.

Malhotra, N. K. (2012). *Marketing Research: an applied orientation* (Vol. 6). Prentice Hall.

Malone, T. W. (1982). Heuristics for designing enjoyable user interfaces: Lessons from computer games. *Proceedings of the 1982 conference on Human factors in computing systems*, 63-68. 10.1145/800049.801756

Marcus, C. (2008). *Web 1.0, Web 2.0, Web 3.0 and Web 4.0 explained*. http://www.marcuscake.com/economic-development/internet-evolution/

Marktest. (2018, November 6). *Quase dois em cada três portugueses leem notícias online* [Almost two out of three Portuguese Read News Online]. https://www.marktest.com/wap/a/n/id~244d.aspx

Marktest. (2020). *7,2 milhões de portugueses usam smartphone*. Retrieved September 26, 2021, from https://www.marktest.com/wap/a/n/id~2700.aspx

Marktest. (2021). *Bareme Internet 2021*. Grupo Marktest.

Marous, J. (2020). *COVID-19 Provides Opportunity For Digital Transformation*. The Financial Brand. https://thefinancialbrand.com/94700/covid-19-digital-banking-transformation-opportunities/

Marques, R. P., Melo, A. I., Natário, M. M., & Biscaia, R. (2021). *Impact of Tourist Activities on Low-Density Territories*. Springer International Publishing. doi:10.1007/978-3-030-65524-2

Martins, F. (2012). Os Caminhos da Autonomia no Online [Dissertação de Mestrado, Ciências da Comunicação, Porto]. Faculdade de Letras da Universidade do Porto.

Masferrer, J. Á. R., Sánchez, F. E., & Hernández, D. F.-O. (2014). Experiences complementing classroom teaching with distance seminars in metaverses and videos. *Journal of Cases on Information Technology, 16*(4), 1–12. doi:10.4018/jcit.2014100101

MathiasL. (2016). https://mindminers.com/blog/pesquisa-qualitativa-quantitativa/

Matyas, S., Matyas, C., Schlieder, C., Kiefer, P., Mitarai, H., & Kamata, M. (2008). Designing location-based mobile games with a purpose: collecting geospatial data with CityExplorer. Proceedings of the 2008 International Conference on Advances in Computer Entertainment Technology, 244-247. doi:10.1145/1501750.150180610.1145/1501750.1501806

MazurM. (2021). Non-Fungible Tokens (NFT). The Analysis of Risk and Return. Available at SSRN 3953535. doi:10.2139/ssrn.3953535

McIntosh, M. J., & Morse, J. M. (2015). Situating and constructing diversity in semi-structured interviews. *Global Qualitative Nursing Research, 2*, 1–12. doi:10.1177/2333393615597674 PMID:28462313

Mead, G. (1967). *Mind, Self, & Society: From the Standpoint of a Social Behaviorist* (Vol. 1). University of Chicago Press. doi:10.7208/chicago/9780226516608.001.0001

Meffert, J., & Mendonça, P. (2017). *Digital@Scale*. Planeta.

Mehrabian, A., & Russell, J. A. (1974). *An Approach to Environmental Psychology*. The MIT Press.

Meirinhos, M., & Osório, A. (2016). O estudo de caso como estratégia de investigação em educação. *Eduser - Revista De Educação, 2*(2). https://www.eduser.ipb.pt/index.php/eduser/article/view/24

Mele, C., Russo-Spena, T., Tregua, M., & Amitrano, C. C. (2021). The millennial customer journey: A Phygital mapping of emotional, behavioural, and social experiences. *Journal of Consumer Marketing, 38*(4), 420–433. doi:10.1108/JCM-03-2020-3701

Melero, I., Sese, F. J., & Verhoef, P. C. (2016). Recasting the customer experience in today's omni-channel environment. *Universia Business Review*, (50), 18–37. doi:10.3232/UBR.2016.V13.N2.01

Mendes, B. (2019). *The Problem of Low Density Territories: Four Cases Studies* [A Problemática dos Territórios de Baixa Densidade : Quatro Estudos de Caso]. Academic Press.

Merton, R. K., & Kendall, P. L. (1946). The Focused Interview. *American Journal of Sociology, 51*(6), 541–557. doi:10.1086/219886

Meuter, M. L., Ostrom, A. L., Roundtree, R. I., & Bitner, M. J. (2000). Self-Service Technologies: Understanding Customer Satisfaction with Technology-Based Service Encounters. *Journal of Marketing, 64*(3), 50–64. doi:10.1509/jmkg.64.3.50.18024

Mick, D. G., & Fournier, S. (1998). Paradoxes of technology: Consumer cognizance, emotions, and coping strategies. *The Journal of Consumer Research, 25*(2), 123–143. doi:10.1086/209531

Milgram, P., & Kishino, F. (1994). A taxonomy of mixed reality visual displays. *IEICE Transactions on Information and Systems, 77*(12), 1321–1329.

Mimoun, M. S. B., Miltgen, C. L., & Slama, B. (2022). Is the shopper always the king/queen? Study of omnichannel retail technology use and shopping orientations. *Journal of Retailing and Consumer Services, 65*, 102844. doi:10.1016/j.jretconser.2021.102844

Miniard, P. W., Dickson, P. R., & Lord, K. R. (1988). Some central and peripheral thoughts on the routes to persuasion. *ACR North American Advances, 13*(3), 237-249. doi:10.1016/0167-8116(96)00005-5

Miranda, M. (2018). Youtube, Facebook and Instagram Ads: Get Traffic, Leads and Increase Revenue for your Business. In Marketing Digital & E-Commerce #2. Viseu: Psicosoma.

Mishra, R., Singh, R. K., & Koles, B. (2021). Consumer Decision-Making in Omnichannel Retailing: Literature Review and Future Research Agenda. *International Journal of Consumer Studies*, *45*(2), 147–174. doi:10.1111/ijcs.12617

Mishra, S., Malhotra, G., Arora, V., & Mukhopadhyay, S. (2021b). Omnichannel retailing: Does it empower consumers and influence patronage? *International Journal of Retail & Distribution Management*, *50*(2), 229–250. doi:10.1108/IJRDM-04-2021-0199

Mishra, S., Malhotra, G., Chatterjee, R., & Shukla, Y. (2021a). Consumer retention through phygital experience in omnichannel retailing: Role of consumer empowerment and satisfaction. *Journal of Strategic Marketing*, •••, 1–18. doi:10.1080/0965254X.2021.1985594

Misischia, C. V., Poecze, F., & Strauss, C. (2022). Chatbots in customer service: Their relevance and impact on service quality. *The 13th International Conference on Ambient Systems, Networks and Technologies (ANT) / The 5th International Conference on Emerging Data and Industry 4.0 (EDI40)*, *201*, 421–428. 10.1016/j.procs.2022.03.055

Miskalo Cruz, M., & Gândara, J. (2016). Indicators for monitoring and managing smart tourist destinations [Indicadores para monitoramento e gestão de destinos turísticos inteligentes]. *CULTUR - Revista de Cultura e Turismo, 10*(2), 3–22.

Mittal, S., Romero, D., & Wuest, T. (2018). Towards a smart manufacturing toolkit for SMEs. In *IFIP International Conference on Product Lifecycle Management* (pp. 476–487). Springer. doi:10.1007/978-3-030-01614-2_44

Mohamed Hashim, M. A., Tlemsani, I., & Matthews, R. (2022). Higher education strategy in digital transformation. *Education and Information Technologies*, *27*(3), 3171–3195.

Mohamed Shaffril, H. A., Samsuddin, S. F., & Abu Samah, A. (2021). The ABC of systematic literature review: The basic methodological guidance for beginners. *Quality & Quantity*, *55*(4), 1319–1346. doi:10.100711135-020-01059-6

Moher, D., Liberati, A., Tetzlaff, J., & Altman, D. G. (2009). Preferred Reporting Items for Systematic Reviews and Meta Analyses: The PRISMA Statement. *PLoS Medicine*, *6*(7), e1000097. doi:10.1371/journal.pmed.1000097 PMID:19621072

Montola, M., Nummenmaa, T., Lucero, A., Boberg, M., & Korhonen, H. (2009). Applying game achievement systems to enhance user experience in a photo sharing service. *Proceedings of the 13th International MindTrek Conference: Everyday Life in the Ubiquitous Era*, 94-97. 10.1145/1621841.1621859

Moody's. (2021). *Rating Action: Moody's takes rating actions on six Portuguese banks.* Retrieved September 24, 2021, from https://www.moodys.com/research/Moodys-takes-rating-actions-on-six-Portuguese-banks--PR_454388

Moody's Investors Service. (2021). *Rating Symbols and Definitions.* Author.

Mora, L., & Bolici, R. (2015). The development process of Smart City strategies: the case of Barcelona. *1st International City Regeneration Congress.*

Moreira, M., Au-Yong-Oliveira, M., Goncalves, R., & Costa, C. (2017). *Digital transformation – Opportunities and threats for more consistent competitiveness.* Ed. Syllables & Challenges.

Morgan Chase, J. P. (2022). *Opportunities in the metaverse. How businesses can explore the metaverse and navigate the hype vs. reality.* https://www.jpmorgan.com/content/dam/jpm/treasury-services/documents/opportunities-in-the-metaverse.pdf

Morgan, R. M., & Hunt, S. D. (1994). The commitment-trust theory of relationship marketing. *Journal of Marketing*, *58*(3), 20–38. doi:10.1177/002224299405800302

Morris, T. (2022). *Unsure about the metaverse? Here's what you (might) need to know*. https://blog.gwi.com/chart-of-the-week/understanding-the-metaverse/

Morse, J. M., & Field, P. A. (1995). *Qualitative Research Methods for Health Professionals*. SAGE Publications.

Mosquera, A., Olarte Pascual, C. & Juaneda Ayensa, E. (2017): Understanding the customer experience in the age of omni-channel shopping, *Icono 14*, *15*(2), 166-185. doi:10.7195/ri14.v15i2.1070

Moşteanu, N., Faccia, A. C., & Bhatia, S. (2020). Digital Technologies' Implementation within Financial and Banking System During Socio Distancing Restrictions - Back to the Future. *International Journal of Advanced Research in Engineering and Technology*, *11*(6), 307–315.

Muntean, C. I. (2011). Raising engagement in e-learning through gamification. *Proc. 6th International Conference on Virtual Learning ICVL*, 323-329.

Murfield, M., Boone, C. A., Rutner, P., & Thomas, R. (2017). Investigating logistics service quality in omni-channel retailing. *International Journal of Physical Distribution & Logistics Management*, *47*(4), 263–296. doi:10.1108/IJPDLM-06-2016-0161

Murphy, P., Pritchard, M. P., & Smith, B. (2000). The destination product and its impact on traveller perceptions. *Tourism Management*, *21*(1), 43–52. doi:10.1016/S0261-5177(99)00080-1

Murugesan. (2007). Understanding Web 2.0. *Journal IT Professional*.

Mystakidis, S. (2022). Metaverse. *Encyclopedia*, *2*(1), 486–497. doi:10.3390/encyclopedia2010031

Narayanan, A., Bonneau, J., Felten, E., Miller, A., & Goldfeder, S. (2016). *Bitcoin and cryptocurrency technologies: a comprehensive introduction*. Princeton University Press.

Needle, F. (2021). https://blog.hubspot.com/blog/tabid/6307/bid/23865/13-mind-bending-social-media-marketing-statistics.aspx

Neeli, B. K. (2012). A Method to Engage Employees Using Gamification in BPO Industry. *Services in Emerging Markets (ICSEM), 2012 Third International Conference*, 12-15. 10.1109/ICSEM.2012.27

Nejadirani, F., Behravesh, M., & Rasouli, R. (2011). Developing countries and electronic commerce the case of SMEs. *World Applied Sciences Journal*, *15*(5), 756–764. https://idosi.org/wasj/wasj15(5)11/22.pdf

Neslin, S. A. (2022). The Omnichannel Continuum: Integrating Online and Offline Channels Along the Customer Journey. *Journal of Retailing*, *98*(1), 111–132. doi:10.1016/j.jretai.2022.02.003

Neslin, S. A., Grewal, D., Leghorn, R., Shankar, V., Teerling, M. L., Thomas, J. S., & Verhoef, P. C. (2006). Challenges and Opportunities in Multichannel Customer Management. *Journal of Service Research*, *9*(2), 95–112. doi:10.1177/1094670506293559

Neslin, S. A., & Shankar, V. (2009). Key Issues in Multichannel Customer Management: Current Knowledge and Future Directions. *Journal of Interactive Marketing*, *23*(1), 70–81. doi:10.1016/j.intmar.2008.10.005

Neuberger, C., Nuernbergk, C., & Langenohl, S. (2018). Journalism as multichannel communication. *Journalism Studies*, *20*(9), 1–21. doi:10.1080/1461670X.2018.1507685

Neubert, M. (2018). The Impact of Digitalization on the Speed of Internationalization of Lean Global Startups. *Technology Innovation Management Review*, *8*(5), 44–54. doi:10.22215/timreview/1158

News Times. (2021). *A virtual yacht sells in the Sandbox metaverse for $ 650,000.* https://www.newstimes.com/business/article/A-virtual-yacht-sells-in-the-Sandbox-metaverse-16666199.php

Ng, P. Y. (2011). *An analysis on the effectiveness of music in advertising towards consumer's buying behavior* (Doctoral dissertation). UTAR.

Nguyen, A., McClelland, R., Hoang Thuan, N., & Hoang, T. G. (2022). Omnichannel Marketing: Structured Review, Synthesis, and Future Directions. *International Review of Retail, Distribution and Consumer Research*, *32*(3), 1–45. doi:10.1080/09593969.2022.2034125

Nicolls, D. (2020). COVID-19: A Call for Digital Transformation. *Forbes*. https://www.forbes.com/sites/jumio/2020/04/01/covid-19-a-call-for-digital-transformation/?sh=79aa37434f11

Obercom. (2014, June 22). *A Internet em Portugal – Sociedade em Rede 2014*. https://obercom.pt/a-internet-em-portugal-sociedade-em-rede-2014/

Ochoa, C. (2015). https://www.netquest.com/blog/br/blog/br/amostra-conveniencia

Ödemiş, M. (2022). Smart Tourism Destinations: A Literature Review on Applications in Turkey's Touristic Destinations. *Optimizing Digital Solutions for Hyper-Personalization in Tourism and Hospitality*, 131-153.

OECD. (2007). *Participative web and user-created content: Web 2.0 wikis and social networking. Organization for Economic Cooperation and Development*. OECD.

OECD. (2019). *OECD SME and Entrepreneurship Outlook 2019*. OECD Publishing. https://www.oecd-ilibrary.org/industry-and-services/oecd-sme-and-entrepreneurship-outlook-2019_34907e9c-en

Oktie, H. (2008). *Introduction to Semantic Web Technologies & Linked Data*. http://www.cs.toronto.edu/~oktie/slides/web-of-data-intro.pdf

Oliveira, T., Faria, M., Thomas, M. A., & Popovic, A. (2014). Extending the understanding of mobile banking adoption: When UTAUT meets TTF and ITM. *International Journal of Information Management*, *34*(5), 689–703. doi:10.1016/j.ijinfomgt.2014.06.004

Olsen, R. K., Kalsnes, B., & Barland, J. (2021). *Do Small Streams Make a Big River? Detailing the Diversification of Revenue Streams in Newspapers' Transition to Digital Journalism Businesses*. Digital Journalism. doi:10.1080/21670811.2021.1973905

ONU. (2015). *Sustainable Development Goals* [Objetivos de Desenvolvimento Sustentável]. https://unric.org/pt/objetivos-de-desenvolvimento-sustentavel/

Orejon-Sanchez, R. D., Crespo-Garcia, D., Andres-Diaz, J. R., & Gago-Calderon, A. (2022). Smart Cities' Development in Spain: A Comparison of Technical and Social Indicators with Reference to European Cities. *Sustainable Cities and Society*, *81*(June), 103828. Retrieved May 5, 2022, from. doi:10.1016/j.scs.2022.103828

Ortiz-Ospina. (2019). *The rise of social media*. https://ourworldindata.org/rise-of-social-media

Ossi, N. (2003). *Semantic Web: Definition*. https://www.w3c.tut.fi/talks/2003/0331umedia-on/slide6-0.html

Osterwalder, A., & Pigneur, Y. (2010). *Business Model Generation*. John Wiley & Sons, Inc.

Otowicz, M. (2022). Dimensions of Smart Tourism and Its Levels: An Integrative Literature Review. *Journal of Smart Tourism*, *2*(1), 5–19. doi:10.52255marttourism

Pais, I. (2017). *Territorial Brand: The residents' view of their city (Cidade da Guarda)* [Master's dissertation]. Faculty of Economics of Porto.

Palazón, M., López, M., Sicilia, M., & López, I. (2022). The customer journey: A proposal of indicators to evaluate integration and customer orientation. *Journal of Marketing Communications, 28*(5), 1–32. doi:10.1080/13527266.2022.2051584

Pallant, J. (2016). *SPSS Survival Manual: A Step By Step Guide to Data Analysis Using SPSS Program* (6th ed.). McGraw-Hill Education.

Palmer. (2001). *The Semantic Web: An Introduction.* http://infomesh.net/2001/swintro/

Pangarkar, A., Arora, V., & Shukla, Y. (2022). Exploring phygital omnichannel luxury retailing for immersive customer experience: The role of rapport and social engagement. *Journal of Retailing and Consumer Services, 68*, 103001. doi:10.1016/j.jretconser.2022.103001

Pantano, E., & Viassone, M. (2015). Engaging consumers on new integrated multichannel retail settings: Challenges for retailers. *Journal of Retailing and Consumer Services, 25*, 106–114. doi:10.1016/j.jretconser.2015.04.003

Papagiannidis, S., Bourlakis, M., & Li, F. (2008). Making real money in virtual worlds: MMORPGs and emerging business opportunities, challenges and ethical implications in metaverses. *Technological Forecasting and Social Change, 75*(5), 610–622. doi:10.1016/j.techfore.2007.04.007

Papagiannidis, S., Pantano, E., See-To, E. W. K., & Bourlakis, M. (2013). Modelling the determinants of a simulated experience in a virtual retail store and users' product purchasing intentions. *Journal of Marketing Management, 29*(13–14), 1462–1492. doi:10.1080/0267257X.2013.821150

Pardo, A., & Etayo, C. (2014). Movies and screens: The Spanish audience's choice. *Communicatio Socialis, 27*(4), 131–145. https://dadun.unav.edu/bitstream/10171/37938/1/20141027124907.pdf

PareraE. (2020). https://postcron.com/pt/blog/como-criar-anuncios-no-facebook-ads/

Parida, V., Westerberg, M. & Frishammar, J. (2012). Inbound open innovation activities in high-tech SMEs: the impact on innovation performance. *Journal of Small Business Management, 50*(2), 283-309. doi:10.1111/j.1540-627X.2012.00354.x

Park, J., & Kim, R. B. (2022). Importance of offline service quality in building loyalty of OC service brand. *Journal of Retailing and Consumer Services, 65*, 102493. doi:10.1016/j.jretconser.2021.102493

Park, J., & Stoel, L. (2005). Effect of brand familiarity, experience and information on online apparel purchase. *International Journal of Retail & Distribution Management, 33*(2), 148–160. doi:10.1108/09590550510581476

Park, S., & Kim, S. (2022a). Identifying World Types to Deliver Gameful Experiences for Sustainable Learning in the Metaverse. *Sustainability, 14*(3), 1361. Advance online publication. doi:10.3390u14031361

Park, S., & Kim, Y. G. (2022b). A Metaverse: Taxonomy, Components, Applications, and Open Challenges. *IEEE Access: Practical Innovations, Open Solutions, 10*, 4209–4251. doi:10.1109/ACCESS.2021.3140175

Park, S., Min, K., & Kim, S. (2021). Differences in learning motivation among bartle's player types and measures for the delivery of sustainable gameful experiences. *Sustainability, 13*(16), 9121. Advance online publication. doi:10.3390u13169121

Parry, E., & Urwin, P. (2011). Generational Differences in Work Values: A Review of Theory and Evidence. *International Journal of Management Reviews, 13*(1), 79–96. doi:10.1111/j.1468-2370.2010.00285.x

Patterson, T. E. (2000). *Doing Well and Doing Good: How Soft News and Critical Journalism Are Shrinking the News Audience and Weakening Democracy – And What News Outlets Can Do About It.* Harvard University, John F. Kennedy School of Government. https://shorensteincenter.org/how-soft-news-critical-journalism-are-shrinking-news-audience/

Pedro, A., & Marques, S. (2010). *From Space Construction to Territory Construction* [Da Construção do Espaço à Construção do Território]. Academic Press.

Perdigoto, M., & Picoto, W. (2012). Analysing Intention and Action in Mobile Banking Services. *Portuguese Journal of Management Studies*, *17*(2), 133–152.

Perdomo, R. S. (2019). *Framework das capacidades dinâmicas para impulsionar a transformação digital: O Caso da Empresa de Construção Civil de Florianópolis* [Master Thesis]. Universidade Estadual de Santa Catarina, Brasil. https://sistemabu.udesc.br/pergamumweb/vinculos/000070/00007078.pdf

Pereira, A. (2021). *A Transformação Digital nas PME Portuguesas: Análise das Empresas da Região Centro* [Master Thesis]. Instituto Superior de Entre Douro e Vouga (ISVOUGA), Santa Maria da Feira, Portugal. https://comum.rcaap.pt/handle/10400.26/36356

Pereira, P., & Becker, J. (2020). Smart design engineering: A literature review of the impact of the 4th industrial revolution on product design and development. *Research in Engineering Design*, *31*(2), 175–195. doi:10.100700163-020-00330-z

Pereira, R., & Pinto, M. (2018). Estratégias de internacionalização – Estudo de caso para empresas portuguesas. *Empresas E Empreendedores Na CPLP*, *3*(2), 9–33. doi:10.29073/e3.v3i2.45

Pestana, M., & Gageiro, J. (2014). *Data Analysis for Social Sciences: Complementarity of SPSS*. Ed. Syllabus.

Pew Research Center. (2019, April 23). *The role of social media in news.* https://www.journalism.org/2019/04/23/the-role-of-social-media-in-news/

Phillips, P., & Moutinho, L. (2014). Critical review of strategic planning research in hospitality and tourism. *Annals of Tourism Research*, *48*, 96–120. doi:10.1016/j.annals.2014.05.013

Picot-Coupey, K., Huré, E., & Piveteau, L. (2016). Channel design to enrich customers' shopping experiences: Synchronizing clicks with bricks in an omni-channel perspective - the Direct Optic case. *International Journal of Retail & Distribution Management*, *44*(3), 336–368. doi:10.1108/IJRDM-04-2015-0056

Pînzaru, F., & Mitan, A. (2012). Social media and marketing of the "popcorn" music wave: The success of romanian commercial musicians analysed through their perceived image on facebook and youtube. *Economia e Sociologia*, *5*(2), 125–138. https://www.economics-sociology.eu/files/14_MOD_Paper_Pinzaru%20_Mitan_proofr_03.02.13E&S.pdf

PORDATA. (2019). *Your municipality in numbers!* [O seu município em números!]. Guarda.

Pordata. (2021). *Assinantes / equipamentos de utilizadores do serviço móvel.* Retrieved October 9, 2021, from https://www.pordata.pt/Portugal/Assinantes+++equipamentos+de+utilizadores+do+servi%c3%a7o+m%c3%b3vel-1180

Pordata. (2021). *Indicadores de envelhecimento.* Retrieved October 9, 2021, from https://www.pordata.pt/Portugal/Indicadores+de+envelhecimento-526

Pordata. (2021). *População residente, média anual: total e por grupo etário.* Retrieved October 10, 2021, from https://www.pordata.pt/Portugal/Popula%c3%a7%c3%a3o+residente++m%c3%a9dia+anual+total+e+por+grupo+et%c3%a1rio-10

Portal da Queixa. (2019). *Portal da Queixa internacionaliza-se e cria marca global: Consumers Trust.* https://portaldaqueixa.com/news/portal-da-queixa-internacionaliza-se-e-cria-marca-global-consumers-trust

Porter, M. E. (2001). Strategy and the Internet. *Harvard Business Review*. https://hbr.org/2001/03/strategy-and-the-internet PMID:11246925

Porter, M. E., & Heppelmann, J. E. (2015). How smart, connected products are transforming companies. *Harvard Business Review*, *93*(10), 96–114. https://hbr.org/2015/10/how-smart-connected-products-are-transforming-companies

Powers, M., & Benson, R. (2014). Is the Internet homogenizing or diversifying the news? External pluralism in the U.S., Danish, and French press. *The International Journal of Press/Politics*, *19*(2), 246–265. doi:10.1177/1940161213519680

Powers, M., & Vera-Zambrano, S. (2019). Endure, Invest, ignore: How French and American journalists react to economic constraints and technological transformations. *Journal of Communication*, *69*(3), 320–343. doi:10.1093/joc/jqz015

Prasad, R. (2022). *The transformation of customer experience within the Metaverse*. https://customerthink.com/the-transformation-of-customer-experience-within-the-metaverse/

Priano, F. H., Armas, R. L., & Guerra, C. F. (2016). A model for the smart development of island territories. In Y. Kim, & M. Liu (Eds.), *Proceedings of the 17th International Digital Government Research Conference on Digital Government Research* (pp. 465–474). New York: Association for Computing Machinery. 10.1145/2912160.2912187

Priberam. (2021). https://dicionario.priberam.org/informatividade

Priebatsch, S. (2010). Welcome to the Decade of Games. *Harvard Business Review*. http://blogs.hbr.org/cs/2010/09/welcome_to_the_decade_of_games.html

Proença, J. F., & Rodrigues, M. A. (2021). A comparison of users and non-users of banking self-service technology in Portugal. *Journal of Service Theory and Practice*, 192–210.

Proença, J. F., & Silva, M. M. (2008). Os efeitos da desmaterialização das relações bancárias devido ao uso do Internet Banking e do Marketing Relacional. *Revista Portuguesa e Brasileira de Gestão*, 88–102.

Proença, M. Jr, & Duenhas, R. A. (2020). Smart cities and sustainable cities: Convergence of actions mere advertising? [Cidades inteligentes e cidades sustentáveis: convergência de ações ou mera publicidade?]. *Revista Brasileira de Planejamento e Desenvolvimento*, *9*(2), 317. doi:10.3895/rbpd.v9n2.10234

Público. (2021). *Portugal entre os países da UE com mais dificuldades no acesso à Net*. Retrieved October 20, 2021, from https://www.publico.pt/2021/02/20/infografia/transicao-digital-uniao-europeia-585

Puccinelli, N. M., Wilcox, K., & Grewal, D. (2015). Consumers' response to commercials: When the energy level in the commercial conflicts with the media context. *Journal of Marketing*, *79*(2), 1–18. doi.org/10.1509/jm.13.0026

Purdy, M. (2022). How the Metaverse Could Change Work. *Harvard Business Review - Digital Articles*. https://hbr.org/2022/04/how-the-metaverse-could-change-work

Puschmann, T. (2017). Fintech. *Business & Information Systems Engineering*, *59*(1), 69–76. doi:10.100712599-017-0464-6

PwC. (2020). *How Virtual Reality Is Redefining Soft Skills Training*. PwC.

Qin, H., Peak, D. A., & Prybutok, V. (2021). A virtual market in your pocket: How does mobile augmented reality (MAR) influence consumer decision making? *Journal of Retailing and Consumer Services*, *58*, 102337. doi:10.1016/j.jretconser.2020.102337

Quach, S., Barari, M., Moudrý, D. V., & Quach, K. (2022). Service Integration in Omnichannel Retailing and its Impact on Customer Experience. *Journal of Retailing and Consumer Services*, *102267*, 102267. Advance online publication. doi:10.1016/j.jretconser.2020.102267

Radfahrer, L. (2022). *Entenda o que é o metaverso e por que ele pode não estar tão distante de você* [Understand what the Metaverse is and Why it might nit be so far away from you]. https://www.cnnbrasil.com.br/business/entenda-o-que-e-o-metaverso-e-por-que-ele-pode-nao-estar-tao-distante-de-voce/

Rafael, L. D., Jaione, G. E., Cristina, L., & Ibon, S. L. (2020). An Industry 4.0 maturity model for machine tool companies. *Technological Forecasting and Social Change*, *159*(120203), 1–13. doi:10.1016/j.techfore.2020.120203

Rahman, S. M., Carlson, J., Gudergan, S. P., Wetzels, M., & Grewal, D. (2022). Perceived Omnichannel Customer Experience (OCX): Concept, measurement, and impact. *Journal of Retailing*. Advance online publication. doi:10.1016/j.jretai.2022.03.003

Raja, M. W., Anand, S., & Kumar, I. (2018). Multi-item scale construction to measure consumers' attitude toward advertising music. *Journal of Marketing Communications*, *26*(3), 314–327. doi.org/10.1080/13527266.2018.1471615

Ramos, G. M. de A. (2015). *Institutional Innovation, Tourism and Development in Low Density Territories: the case of Naturtejo Geopark* [Inovação Institucional, Turismo e Desenvolvimento em Territórios de Baixa Densidade: o caso do Geopark Naturtejo]. Academic Press.

Rapp, A., Curti, L., & Boldi, A. (2021). The human side of human-chatbot interaction: A systematic literature review of ten years of research on text-based chatbots. *International Journal of Human-Computer Studies*, *151*, 102630–102630. doi:10.1016/j.ijhcs.2021.102630

Raskino, M., & Waller, G. (2015). *Digital to the Core: Remastering Leadership for Your Industry, Your Enterprise, and Yourself.* Routledge.

Rauschnabel, P. A., Babin, B. J., tom Dieck, M. C., Krey, N., & Jung, T. (2022). What is augmented reality marketing? Its definition, complexity, and future. *Journal of Business Research*, *142*, 1140–1150. doi:10.1016/j.jbusres.2021.12.084

Rebelo, A. (2020). *The impact of digitalization on SMEs international strategies: A Bibliometric Analysis* [Master Thesis]. Faculdade de Economia Universidade do Porto.

Reddy, S., & Reinartz, W. (2017). Digital Transformation - From academic research to practical use. *GfK Verein*, *9*(1), 11-17. https://www.nim.org/sites/default/files/medien/2327/dokumente/2017_gfk_mir_digital_transformation_eng_0.pdf

Rego, F. (2021) Digital transformation in business: a systematic study from the web of science and scopus, *11th International Symposium on technological innovation*, 11(1), 1610-1619. 10.51722/S2318-3403202100011364

Reimers, V., Chao, C.-W., & Gorman, S. (2016). Permission email marketing and its influence on online shopping. *Asia Pacific Journal of Marketing and Logistics*, *28*(2). Advance online publication. doi:10.1108/APJML-03-2015-0037

Reis, F. L. dos. (2018). *Scientific Research and Academic Works - Practical Guide* [Investigação Científica e Trabalhos Académicos - Guia Prático]. Academic Press.

Reis, P., & Baltazar, M. da S. (2019). Low-density rural territories as a space for leisure and tourism. The tourist destination Historic Villages of Portugal [Os territórios rurais de baixa densidade como espaço de lazer e de turismo. O destino turístico Aldeias Históricas de Portugal]. *Sociologia on Line*, *21*, 141–166. doi:10.30553ociologiaonline.2019.21.6

Rentfrow, P. J., Goldberg, L. R., Stillwell, D. J., Kosinski, M., Gosling, S. D., & Levitin, D. J. (2012). The song remains the same: A replication and extension of the MUSIC model. *Music Perception*, *30*(2), 161–185. doi.org/10.1525/mp.2012.30.2.161

Republic Realm. (2021). *The 2021 Metaverse Real Estate Report.* https://www.republicrealm.com/post/the-2021-metaverse-real-estate-report

Research Councils UK. (2017). *Rise of the Machines*. Research Councils UK. https://blogs.rcuk.ac.uk/author/bena/page/2/

Riar, M., Xi, N., Korbel, J. J., Zarnekow, R., & Hamari, J. (2022). Using augmented reality for shopping: A framework for AR induced consumer behavior, literature review and future agenda. *Internet Research*. Advance online publication. doi:10.1108/INTR-08-2021-0611

Rigby, D. (2011). The future of shopping. *Harvard Business Review*, *89*(12), 65–76. https://hbr.org/2011/12/the-future-of-shopping

Ritchie, J. R. B., & Crouch, G. I. (2003). *The Competitive Destination: A Sustainable Tourism Perspective*. https://books.google.pt/books?hl=pt-PT&lr=&id=yvydAwAAQBAJ&oi=fnd&pg=PR5&dq=Ritchie,+J.+R.+B.,+%26+Crouch,+G.+I.+(2003).+The+competitive+destination:+a+sustainable+tourism+perspective.+Wallingford,+Reino+Unido:+CAB+International.&ots=lSIZ6ph9SO&sig=qvdIW5

Rodrigues, Fonseca, & Garcia. (2022). The Use of CRM in Marketing and Communication Strategies in Portuguese Non-Profit Organizations. Navigating Digital Communication and Challenges for Organizations - Advances in E-Business Research, 223-244.

Rodríguez-Torrico, P., Cabezudo, R. S. J., San-Martín, S., & Apadula, L. T. (2021). Let it flow: The role of seamlessness and the optimal experience on consumer word of mouth in omnichannel marketing. *Journal of Research in Interactive Marketing*. Advance online publication. doi:10.1108/JRIM-06-2021-0154

Rodríguez-Torrico, P., Trabold Apadula, L., San-Martín, S., & San Jose Cabezudo, R. (2020). Have an omnichannel seamless interaction experience! Dimensions and effect on consumer satisfaction. *Journal of Marketing Management*, *36*(17-18), 1731–1761. doi:10.1080/0267257X.2020.1801798

Rogers, D. L. (2016). *The digital transformation playbook*. Columbia University Press. doi:10.7312/roge17544

Rogers, E. M. (1995). *The Diffusion of Innovations*. Free Press.

Rogers, J., & Sparviero, S. (2011). Same tune, different words: The creative destruction of the music. *The Observatory Journal*, *5*(4), 1–30. https://pdfs.semanticscholar.org/bec0/5cc20dfc50c9c1eb5208a69eaff73c6c71fd.pdf

Rookes, P., & Willson, J. (2000). *Perception: Theory, Development and Organisation* (1st ed.). Routledge.

Roos, J. M., & Kazemi, A. (2021). Personality traits and Internet usage across generation cohorts: Insights from a nationally representative study. *Current Psychology (New Brunswick, N.J.)*, *40*(3), 1287–1297. doi:10.100712144-018-0033-2

Rupeika-Apoga, R., Larisa, B., & Kristine, P. (2022). Digital Transformation of Small and Medium Enterprises: Aspects of Public Support. *Journal of Risk and Financial Management*, *15*(45), 45. Advance online publication. doi:10.3390/jrfm15020045

Russell, C. A. (2007). *Advertainment: Fusing Advertising and Entertainment*. University of Michigan: Yaffe Center.

Rüßmann, M., Lorenz, M., Gerbert, P., Waldner, M., Justus, J., Engel, P., & Harnisch, M. (2015). *Industry 4.0: The future of productivity and growth in manufacturing industries*. Boston Consulting Group. https://image-src.bcg.com/Images/Industry_40_Future_of_Productivity_April_2015_tcm9-61694.pdf

Saghiri, S., Wilding, R., Mena, C., & Bourlakis, M. (2017). Toward a three-dimensional framework for omni-channel. *Journal of Business Research*, *77*, 53–67. doi:10.1016/j.jbusres.2017.03.025

Saha, R., Manna, R., & Geetha, G. (2012). CAPTCHINO-A Gamification of Image-based CAPTCHAs to Evaluate Usability Issues. *Computing Sciences (ICCS), 2012 International Conference*, 95-99. 10.1109/ICCS.2012.18

Sahni, N. S., Stanford, G. S. B., & Zhang, C. (2020). *Are Consumers Averse to Sponsored Messages? The Role of Search Advertising in Information Discovery* [Master's thesis]. Standford University Graduate School of Bussiness.

Sahu, K. C., Naved Khan, M., & Gupta, K. D. (2021). Determinants of webrooming and showrooming behavior: A systematic literature review. *Journal of Internet Commerce*, *20*(2), 137–166. doi:10.1080/15332861.2020.1863041

Salkin, C., Oner, M., Ustundag, A., & Cevikcan, E. (2018). A Conceptual Framework for Industry 4.0. In A. Ustundag & E. Cevikcan (Eds.), *Industry 4.0: Managing the Digital Transformation* (pp. 3–23). Springer. doi:10.1007/978-3-319-57870-5_1

Salvietti, G., Ziliani, C., Teller, C., Ieva, M., & Ranfagni, S. (2022). Omnichannel retailing and post-pandemic recovery: Building a research agenda. *International Journal of Retail & Distribution Management*, *50*(8/9), 1156–1181. Advance online publication. doi:10.1108/IJRDM-10-2021-0485

Sandroni, A. G. (2015). *Breve Historia y Origen del Internet*. https://web.archive.org/web/20150505102148/http://www.academia.edu/5489717/BREVE_HISTORIA_Y_ORIGEN_DEL_INTERNET

Sanseverino, G. G., & Santos, M. F. L. (2021). Experimenting with user-generated content in journalistic practices: Adopting a user-centric storytelling approach during the covid-19 pandemic coverage in Latin America. *Brazilian Journalism Research*, *17*(2), 244–279. doi:10.25200/BJR.v17n2.2021.1369

Santos-Júnior, A. (2019). Entendiendo La Gobernanza de Los Destinos Turísticos Inteligentes: El Caso de Florianópolis -Brasil [Understanding Smart Tourism Destinations' Governance: The Case of Florianópolis – Brazil]. *International Journal of Information Systems and Tourism (IJIST).*

Sarstedt, M., & Mooi, E. (2019). *A Concise Guide to Market Research: The Process, Data, and Methods* (3rd ed.). Springer. doi:10.1007/978-3-662-56707-4

Saunders, M., Lewis, P., & Thornhill, A. (2016). *Research Methods for Business Students* (Vol. 7). Pearson.

Savić, J., & Pavlović, G. (2018). Analysis of Factors of Smart Tourism Development in Serbia. *Menadzment U Hotelijerstvu I Turizmu*, *6*(1), 81–91. doi:10.5937/menhottur1801081S

Scarle, S., Arnab, S., Dunwell, I., Petridis, P., Protopsaltis, A., & de Freitas, S. (2012). E-commerce transactions in a virtual environment: Virtual transactions. *Electronic Commerce Research*, *12*(3), 379–407. doi:10.100710660-012-9098-4

Schenk, B., Mateusz, D., Schwabe, C., & Schwabe, G. (2021). What Citizens Experience and How Omni-Channel Could Help-Insights from a Building Permit Case. *Information Technology & People*, *2021*, 277–299. doi:10.1108/ITP-06-2020-0374

Scherer, A., Wünderlich, N. V., & von Wangenheim, F. (2015). The value of self-service: Long-term effects of technology-based self-service usage on customer retention. *Management Information Systems Quarterly*, *39*(1), 177–200. doi:10.25300/MISQ/2015/39.1.08

Schmitt, C., & Baldegger, R. (2020). Editorial: Digitalization and Internationalization. *Technology Innovation Management Review*, *10*(4), 3–4. doi:10.22215/timreview/1342

Schober, P., Boer, C., & Schwarte, L. A. (2018). Correlation Coefficients: Appropriate Use and Interpretation. *Anesthesia and Analgesia*, *126*(5), 1763–1768. doi:10.1213/ANE.0000000000002864 PMID:29481436

Schoneville, S. (2007). *This just in: Analysis of factors influencing online newspaper reading behaviour* (Publication No. 140784811) [Masters dissertation, Universiteit Twente]. SemanticScholar.

Schram, W. (1954). *The Process and Effects of Mass Communication*. University of Illinois Press.

Schwertner, K. (2017). Digital transformation of business. *Trakia Journal of Sciences*, *15*(Suppl. 1), 388–393. doi:10.15547/tjs.2017.s.01.065

SEGITTUR. (2015). *Smart Destinations Report: building the future*. https://www.segittur.es/opencms/export/sites/segitur/.content/galerias/descargas/documentos/Libro-Destinos-Inteligentes-en-Ingls.pdf

Selnes, F., & Hansen, H. (2001). The potential hazard of self-service in developing customer loyalty. *Journal of Service Research*, *4*(2), 79–90. doi:10.1177/109467050142001

Shah, D., Rust, R. T., Parasuraman, A., Staelin, R., & Day, G. S. (2006). The Path to Customer Centricity. *Journal of Service Research*, *9*(2), 113–124. doi:10.1177/1094670506294666

Shankar, A., Gupta, M., Tiwari, A. K., & Behl, A. (2021). How does convenience impact showrooming intention? Omnichannel retail strategies to manage global retail apocalypse. *Journal of Strategic Marketing*, 1–22. doi:10.1080/0965254X.2021.1965189

Shannon, C. E. (1948). A Mathematical Theory of Communication. *The Bell System Technical Journal*, *15*(3), 379–423. doi:10.1002/j.1538-7305.1948.tb01338.x

Sheehan, B., Jin, H. S., & Gottlieb, U. (2020). Customer service chatbots: Anthropomorphism and adoption. *Journal of Business Research*, *115*, 14–24. doi:10.1016/j.jbusres.2020.04.030

Sheth, J. N. (2021). Future of brick and mortar retailing: How will it survive and thrive? *Journal of Strategic Marketing*, *29*(7), 598–607. doi:10.1080/0965254X.2021.1891128

Shih, N. J., Diao, P. H., & Chen, Y. (2019). ARTS, an AR tourism system, for the integration of 3D scanning and smartphone AR in cultural heritage tourism and pedagogy. *Sensors (Switzerland)*, *19*(17), 3725. Advance online publication. doi:10.339019173725 PMID:31466354

Shi, S., Wang, Y., Chen, X., & Zhang, Q. (2020). Conceptualization of omnichannel customer experience and its impact on shopping intention: A mixed-method approach. *International Journal of Information Management*, *50*, 325–336. doi:10.1016/j.ijinfomgt.2019.09.001

Shi, W., Wang, X., Oh, Y. J., Zhang, J., Sahay, S., & Yu, Z. (2020). Effects of persuasive dialogues: testing bot identities and inquiry strategies. In *Proceedings of the 2020 CHI Conference on Human Factors in Computing Systems* (pp. 1–13). Association for Computing Machinery. 10.1145/3313831.3376843

Shon, M., Shin, J., Hwang, J., & Lee, D. (2020). Free contents vs. inconvenience costs: Two faces of online video advertising. *Telematics and Informatics*, 101476. doi:10.1016/j.tele.2020.101476

Siachou, E., Vrontis, D., & Trichina, R. (2021). Can traditional organizations be digitally transformed by themselves? The moderating role of absorptive capacity and strategic interdependence. *Journal of Business Research*, *124*, 408–421. doi:10.1016/j.jbusres.2020.11.011

Sibs. (2021). *Multibanco*. Retrieved October 9, 2021, from https://www.sibs.com/marcas/multibanco/

Sicart, M. (2008). Defining game mechanics. *Game Studies*, *8*(2).

Silva, S. C. E., Martins, C. C., & Sousa, J. M. D. (2018). Omnichannel approach: Factors affecting consumer acceptance. *Journal of Marketing Channels*, *25*(1-2), 73–84. doi:10.1080/1046669X.2019.1647910

Simmonds, L., Bellman, S., Kennedy, R., Nenycz-Thiel, M., & Bogomolova, S. (2020). Moderating effects of prior brand usage on visual attention to video advertising and recall: An eye-tracking investigation. *Journal of Business Research*, *111*, 241–248. doi.org/10.1016/j.jbusres.2019.02.062

Simões, V. C. (1997). *Estratégias de Internacionalização das Empresas Portuguesas em Comércio e Investimento Internacional*. ICEP - Investimentos, Comércio e Turismo de Portugal.

Singh, S., & Srivastava, R. K. (2020). Understanding the intention to use mobile banking by existing online banking customers: An empirical study. *Journal of Financial Services Marketing*, *25*(3-4), 86–96. doi:10.105741264-020-00074-w

Sjøbakk, B. (2018). The Strategic Landscape of Industry 4.0. In I. Moon, G. M. Lee, J. Park, D. Kiritsis, & G. von Cieminski (Eds.), *Advances in Production Management Systems: Smart Manufacturing for Industry 4.0* (pp. 122–127). Springer. doi:10.1007/978-3-319-99707-0_16

Skjeret, F., Steen, F., & Wyndham, G. A. (2019). Paywalls and the demand for online news. *NHH Department of Economics Discussion Paper*, 7. doi:10.2139/ssrn.3364834

Smart, J., Cascio, J., & Paffendorf, J. (2007). Metaverse roadmap: pathways to the 3D web. *Metaverse: A Cross-Industry Public Foresight Project.*

Smith, J. (2017). *Transforming Travel: Realising the Potential of Sustainable Tourism*. https://books.google.pt/books?id=eQJDDwAAQBAJ&printsec=frontcover&dq=Transforming+Travel+Realising+the+potential+of+sustainable+tourism&hl=pt-PT&sa=X&ved=2ahUKEwjR4NWE0qXvAhXF8OAKHZiPApoQ6AEwAHoECAMQAg#v=onepage&q=Transforming%20Travel%20Realising%20the%20potenti&f=false

SproutSocial. (2021). https://sproutsocial.com/insights/social-media-statistics/

Stalmachova, K., Chinoracky, R., & Strenitzerova, M. (2021). Changes in Business Models Caused by Digital Transformation and the COVID-19 Pandemic and Possibilities of Their Measurement—Case Study. *Sustainability*, *14*(1), 127. Advance online publication. doi:10.3390u14010127

Statista. (2021). *Internet usage in India*. https://www.statista.com/topics/2157/internet-usage-in-india/

Statista. (2021). *Online banking penetration in Portugal from 2005 to 2018*. Retrieved May 29, 2022 from https://www.statista.com/statistics/380873/online-banking-penetration-in-portugal/

Statista. (2021a). https://www.statista.com/outlook/216/147/digital-advertising/portugal

Statista. (2021b). https://www.statista.com/statistics/617136/digital-population-worldwide/

Statista. (2021c). https://www.statista.com/statistics/272014/global-social-networks-ranked-by-number-of-users/

Statista. (2022). *Forecast of internet user numbers in Portugal from 2017 to 2026*. Retrieved May 29, 2022, from https://www.statista.com/map/europe/portugal/internet

Stephen, A. T., & Toubia, O. (2010). Deriving value from social commerce networks. *JMR, Journal of Marketing Research*, *47*(2), 215–228. doi.org/10.1509/jmkr.47.2.215

Stephenson, N. (2003). *Snow crash: A novel*. Spectra.

Stjepić, A.M., Ivančić, L., & Vugec, D.S. (2020). Mastering digital transformation through business process management: Investigating alignments, goals, orchestration, and roles. *Journal of Entrepreneurship, Management and Innovation*, *16*(1), 41-74.

Stolterman, E., & Fors, A. C. (2004). Information Technology and the Good Life. International Federation for Information Processing Digital Library. *Information Systems Research*, *143*, 687–692. https://www8.informatik.umu.se/~acroon/Publikationer%20Anna/Stolterman.pdf

Strange, R., & Zucchella, A. (2017). Industry 4.0, global value chains and international business. *Multinational Business Review*, *25*(3), 174–184. doi:10.1108/MBR-05-2017-0028

Sudhir, B. (2006). *AJAX - Asynchronous Java Script and XML*. ITS - Information Technology and Systems Management.

Sukhodolov, Y. A. (2019). The Notion, essence and peculiarities of Industry 4.0 as a sphere of industry. In *Industry 4.0: industrial revolution of the 21st century*. Springer. doi:10.1007/978-3-319-94310-7_1

Sunaga, T., Meng, Y., & Zhuang, X. (2020). Interactions between instrumental timbre and consumers' regulatory focus. *Journal of Business Research*, *121*, 1–12. doi.org/10.1016/j.jbusres.2020.08.014

Susanti, R., Soetomo, S., Buchori, I., & Brotosunaryo, P. M. (2016). Smart Growth, Smart City and Density: In Search of The Appropriate Indicator for Residential Density in Indonesia. Procedia - Social and Behavioral Sciences, 227, 194–201. doi:10.1016/j.sbspro.2016.06.062

Susi, T., Johannesson, M., & Backlund, P. (2007). *Serious Games: An Overview. Technical Report*. School of Humanities and Informatics, University of Skövde.

Sweetser, P., & Wyeth, P. (2005). GameFlow: A model for evaluating player enjoyment in games. *Computers in Entertainment*, *3*(3), 3–3. doi:10.1145/1077246.1077253

Synergy Research Group. (2022). *As Quarterly Cloud Spending Jumps to Over $50B, Microsoft Looms Larger in Amazon's Rear Mirror*. https://www.srgresearch.com/articles/as-quarterly-cloud-spending-jumps-to-over-50b-microsoft-looms-larger-in-amazons-rear-mirror

Tandoc, E. Jr, & Vos, T. (2015). The Journalist is Marketing the News. *Journalism Practice*, *10*(8), 950–966. doi:10.1080/17512786.2015.1087811

Tandyonomanu, D. (2018, October). Ads on *YouTube*: Skip or Watch? In *1st International Conference on Social Sciences (ICSS 2018)*. Atlantis Press. .66 doi:10.2991/icss-18.2018

Tan, T. M., Makkonen, H., Kaur, P., & Salo, J. (2022). How do ethical consumers utilize sharing economy platforms as part of their sustainable resale behavior? The role of consumers' green consumption values. *Technological Forecasting and Social Change*, *176*, 121432. Advance online publication. doi:10.1016/j.techfore.2021.121432

Tan, T. M., & Salo, J. (2021). Ethical Marketing in the Blockchain-Based Sharing Economy: Theoretical Integration and Guiding Insights. *Journal of Business Ethics*. Advance online publication. doi:10.100710551-021-05015-8

Tavares, E. (2021). Lucros da banca crescem 49 milhões para quase 540 milhões no primeiro semestre. *Dinheiro Vivo*. Retrieved September 29, 2021, from https://www.dinheirovivo.pt/empresas/bancos-lucram-mais-49-milhoes-ate-junho-13993544.html

Teece, D. J. (2010). Business Models, Business Strategy and Innovation. *Long Range Planning*, *43*(2-3), 172–174. doi:10.1016/j.lrp.2009.07.003

Teece, D. J. (2020). Hand in Glove: Open Innovation and the Dynamic Capabilities Framework. *Strategic Management Review*, *1*(2), 233–253. doi:10.1561/111.00000010

Teece, D. J., & Linden, G. (2017). Business models, value capture, and the digital enterprise. *Journal of Organization Design*, *6*(1), 8. doi:10.118641469-017-0018-x

Thakor, A. V. (2020). Fintech and banking: What do we know? *Journal of Financial Intermediation*, *41*, 1–13. doi:10.1016/j.jfi.2019.100833

That Ad Music Blog. (2014). http://aquelamusicanuncio.blogspot.com/?view=classic

The New York Times. (2022). *Crypto Is Cool. Now Get on the Yacht*. https://www.nytimes.com/2021/11/05/technology/nft-nyc-metaverse.html

ThinkWithGoogle. (2015). https://www.thinkwithgoogle.com/intl/pt-br/estrategias-de-marketing/video/youtube-insights-segmentacao/

Thoben, K.-D., Wiesner, S., & Wuest, T. (2017). "Industrie 4.0" and Smart Manufacturing – A Review of Research Issues and Application Examples. *International Journal of Automotive Technology*, *11*(1), 4–19. doi:10.20965/ijat.2017.p0004

Thurman, N., & Fletcher, R. (2019). Has digital distribution rejuvenated readership? *Journalism Studies*, *20*(4), 542–562. doi:10.1080/1461670X.2017.1397532

Tiessen, J. H., Wright, R. W., & Turner, I. (2001). A model of e-commerce use by internationalizing SMEs. *Journal of International Management*, *7*(3), 211–233. doi:10.1016/S1075-4253(01)00045-X

Tillström, J. (2012). *Gamification in automotive marketing: A conceptual framework for implementation* [Bachelor's Thesis]. Metropolia Ammattikorkeakoulu.

Tim, B.-L., Christian, B., Tom, H., & Kingsley, I. (2008). Linked Data on the Web. *17th International World Wide Web Conference*.

Tim, B.-L., James, H., & Ora, L. (2001). The Semantic Web. *Scientific American*, *5*(1).

Tintarev, N., O'donovan, J., & Felfernig, A. (2016). Introduction to the special issue on human interaction with artificial advice givers. *ACM Transactions on Interactive Intelligent Systems*, *6*(4), 1–12. doi:10.1145/3014432

Toader, D. C., Boca, G., Toader, R., Măcelaru, M., Toader, C., Ighian, D., & Rădulescu, A. T. (2020). The effect of social presence and chatbot errors on trust. *Sustainability*, *12*(1), 256. Advance online publication. doi:10.3390u12010256

Tueanrat, Y., Papagiannidis, S., & Alamanos, E. (2021). A Conceptual Framework of the Antecedents of Customer Journey Satisfaction in Omnichannel Retailing. *Journal of Retailing and Consumer Services*, *61*, 102550. doi:10.1016/j.jretconser.2021.102550

Tunebat. (2021a). https://tunebat.com/Info/Closer-Lemaitre-Jennie-A-/3h5IIiL9vK5aR0DZO6jD7D

Tunebat. (2021b). https://tunebat.com/Info/Don-t-Start-Now-Dua-Lipa/6WrI0LAC5M1Rw2MnX2ZvEg

Tunebat. (2021c.) https://tunebat.com/Info/Nuvole-Bianche-Rousseau/7EoBV6PM8qlIkoDms2RthA

Tyrväinen, O., Karjaluoto, H., & Saarijärvi, H. (2020). Personalization and hedonic motivation in creating customer experiences and loyalty in omnichannel retail. *Journal of Retailing and Consumer Services*, *57*, 102233. doi:10.1016/j.jretconser.2020.102233

Uhl, A., & MacGillavry, K. (2016). Customer Centricity. In A. Uhl & L. A. Gollenia (Eds.), *Digital Enterprise Transformation* (pp. 149–171). Routledge. doi:10.4324/9781315577166-6

UNTWO. (2020b). *Digital Transformation*. UNWTO. https://www.unwto.org/digital-transformation

UOL. (2022). *Com tema de arte NFT, hamburgueria recebe pagamento em criptomoedas* [NFT-Themed Art, Hamburgueria receives payment in Cryptocureencies]. UOL.

Upadhyay, A. K., & Khandelwal, K. (2022). Metaverse: The future of immersive training. *Strategic HR Review*, *21*(3), 83–86. doi:10.1108/SHR-02-2022-0009

Ürgüplü, Ö., & Yumurtacı Hüseyinoğlu, I. Ö. (2021). The mediating effect of consumer empowerment in omni-channel retailing. *International Journal of Retail & Distribution Management*, *49*(11), 1481–1496. doi:10.1108/IJRDM-10-2020-0403

Valdellós, A. M. S. (2011). La música en los comerciales televisivos: el arma secreto. *Sinfonía Virtual: Revista de Música Clásica y Reflexión Musical, 1*(18), 3.

Van Nguyen, A. T., McClelland, R., & Thuan, N. H. (2022). Exploring Customer Experience During Channel Switching in Omnichannel Retailing Context: A Qualitative Assessment. *Journal of Retailing and Consumer Services, 64*, 102803. doi:10.1016/j.jretconser.2021.102803

Vannucci, V., & Pantano, E. (2019). Digital or human touchpoints? Insights from consumer-facing in-store services. *Information Technology & People, 33*(1), 296-310. doi: 10.1108/ITP-02-2018-0113/

Vargas Sánchez, A. (2016). Exploring the concept of smart tourist destination. *Enlightening Tourism: A Pathmaking Journal, 6*(2), 178–196. doi:10.33776/et.v6i2.2913

Vasconcellos-Guedes, L., & Guedes, L. F. (2007). *E-surveys: Advantages and limitations of electronic questionnaires via internet in the context of scientific research.* X SemeAd-Seminário em Administração FEA/USP.

Vaskal, S., Massarol, M., Bagarottol, E. M., & Mas, F. D. (2021). The Digital Transformation of Business Model Innovation: A Structured Literature Review. *Frontiers in Psychology*, 1–18. https://www.frontiersin.org/articles/10.3389/fpsyg.2020.539363/full PMID:33584396

Velazco, C. (2012). *AT&T's Wireless Data Traffic Doubles Every Year, but Throttling Is Not the Solution.* http://techcrunch.com/ 2012/02/14/atts-wireless-data-traffic-doubles-everyyear-but-throttling-is-not-the-solution

Venkatesh, V., & Davis, F. D. (1996). A Model of the Antecedents of Perceived Ease of Use: Development and Test. *Decision Sciences, 27*(3), 451–481. doi:10.1111/j.1540-5915.1996.tb01822.x

Venkatesh, V., & Davis, F. D. (2000). A Theoretical Extension of the Technology Acceptance Model: Four Longitudinal Field Studies. *Management Science, 46*(2), 186–204. doi:10.1287/mnsc.46.2.186.11926

Venkatesh, V., Morris, M. G., Davis, G. B., & Davis, F. D. (2003). User Acceptance of Information Technology: Toward a Unified View. *Management Information Systems Research Center, University of Minnesota, 27*(3), 425–478. doi:10.2307/30036540

Venkatesh, V., Thong, J. Y., & Xu, X. (2012). Consumer Acceptance and Use of Information Technology: Extending the Unified Theory of Acceptance and Use of Technology. *Management Information Systems Quarterly, 36*(1), 157–178. doi:10.2307/41410412

Verhoef, P. C., Kannan, P. K., & Inman, J. J. (2015). From Multi-Channel Retailing to Omni-Channel Retailing: Introduction to the Special Issue on Multi-Channel Retailing. *Journal of Retailing, 91*(2), 174–181. doi:10.1016/j.jretai.2015.02.005

Verhoef, P. C., Neslin, S. A., & Vroomen, B. (2007). Multichannel Customer Management: Understanding the Research-Shopper Phenomenon. *International Journal of Research in Marketing, 24*(2), 129–148. doi:10.1016/j.ijresmar.2006.11.002

Verhoef, P. C., Stephen, A. T., Kannan, P. K., Luo, X., Abhishek, V., Andrews, M., Bart, Y., Datta, H., Fong, N., Hoffman, D. L., Hu, M. M., Novak, T., Rand, W., & Zhang, Y. (2017). Consumer Connectivity in A Complex, Technology-Enabled, and Mobile-Oriented World with Smart Products. *Journal of Interactive Marketing, 40*(1), 1–8. doi:10.1016/j.intmar.2017.06.001

Verma, A., & Shukla, V. (2019). Analyzing the Influence of IoT in Tourism Industry. SSRN *Electronic Journal.* doi:10.2139/ssrn.3358168

Vermeulen, I., & Beukeboom, C. J. (2016). Effects of music in advertising: Three experiments replicating single-exposure musical conditioning of consumer choice (Gorn 1982) in an individual setting. *Journal of Advertising, 45*(1), 53-61. .1088809 doi:10.1080/00913367.2015

Viana, C., & Hortinha, J. (2005). *Marketing Internacional, 2ª Edição*. Edições Sílabo.

Vidal-Tomás, D. (2022). The new crypto niche: NFTs, play-to-earn, and metaverse tokens. *Finance Research Letters, 47*, 102742. Advance online publication. doi:10.1016/j.frl.2022.102742

Viejo-Fernandez, N., Sanzo-Pérez, M. J., & Vázquez-Casielles, R. (2020). Is showrooming really so terrible? Start understanding showroomers. *Journal of Retailing and Consumer Services, 54*, 102048. doi:10.1016/j.jretconser.2020.102048

Virtudes, A., Abbara, A., & Sá, J. (2017). Dubai: A Pioneer Smart City in the Arabian Territory. *IOP Conference Series. Materials Science and Engineering, 245*(5), 052071. Advance online publication. doi:10.1088/1757-899X/245/5/052071

Vives, X. (2017). The Impact of Fintech on Banking. *European Economy – Banks, Regulation, and the Real Sector*, 97-105.

Vives, X. (2019). Digital Disruption in Banking. *Annual Review of Financial Economics, 11*(1), 243–272. doi:10.1146/annurev-financial-100719-120854

Volgger, Erschbamer, G., & Pechlaner, H. (2021, January). Destination Design: New Perspectives for Tourism Destination Development. *Journal of Destination Marketing & Management, 19*, 100561. doi:10.1016/j.jdmm.2021.100561

Von Ahn, L., & Dabbish, L. (2008). Designing games with a purpose. *Communications of the ACM, 51*(8), 58–67. doi:10.1145/1378704.1378719

Von Briel, F. (2018). The future of omnichannel retail: A four-stage Delphi study. *Technological Forecasting and Social Change, 132*, 217–229. doi:10.1016/j.techfore.2018.02.004

W3C. (1999). *Resource Description Framework (RDF) Model and Syntax Specification*. https://www.w3.org/TR/1999/REC-rdf-syntax-19990222/

W3C. (2004). *The Unicode Consortium*. http://www.unicode.org/

Waddell, T. F. (2020). The authentic (and angry) audience. *Digital Journalism, 8*(2), 249–266. doi:10.1080/21670811.2018.1490656

Wade, M. (2015). *Digital Business Transformation. A Conceptual Framework*. Global Center for Digital Business Transformations. https://www.imd.org/contentassets/d0a4d992d38a41ff85de509156475caa/framework

Wallace, M. (2006). Virtual worlds, virtual lives. *PC World, 24*(11), 133–136. https://www.scopus.com/inward/record.uri?eid=2-s2.0-33750551484&partnerID=40&md5=fbc08e8b2d59ea37bbbfd49f23c16f57

Wang, D., Li, X., & Li, Y. (2013). China's "Smart Tourism Destination" initiative: A taste of the service-dominant logic. *Journal of Destination Marketing & Management, 2*(2), 59–61. doi:10.1016/j.jdmm.2013.05.004

Wang, J., & Wang, S. (2022). Revisiting the showrooming effect on online and offline retailers: The strategic role of in-store service. *Journal of Retailing and Consumer Services, 66*, 102884. doi:10.1016/j.jretconser.2021.102884

Wang, J., Wang, T., Shi, Y., Xu, D., Chen, Y., & Wu, J. (2022). Metaverse, SED Model, and New Theory of Value. *Complexity, 2022*, 4771516. doi:10.1155/2022/4771516

Wang, Li, X. R., Zhen, F., & Zhang, J. H. (2016). How Smart Is Your Tourist Attraction?: Measuring Tourist Preferences of Smart Tourism Attractions via a FCEM-AHP and IPA Approach. *Tourism Management, 54*(June), 309–320. doi:10.1016/j.tourman.2015.12.003

Wang, M., Zhou, T., & Wang, D. (2020). Tracking the evolution processes of Smart Cities in China by assessing performance and efficiency. *Technology in Society, 63*(June), 101353. doi:10.1016/j.techsoc.2020.101353

Watson, G. IV, Weaven, S., Perkins, H., Sardana, D., & Palmatier, R. (2018). International Market Entry Strategies: Relational, Digital, and Hybrid Approaches. *Journal of International Marketing*, *26*(1), 30–60. doi:10.1509/jim.17.0034

Weekes, S. (2019). *Insight Report Smart waste, an opportunity cities should not trash*. Academic Press.

Wei, M.-F., Luh, Y.-H., Huang, Y.-H., & Chang, Y.-C. (2021). Young Generation's Mobile Payment Adoption Behavior: Analysis Based on an Extended UTAUT Model. *Theoretical and Applied Electronic Commerce Research*, 618-637.

Westerlund, M. (2020). Digitalization, Internationalization and Scaling of Online SMEs. *Technology Innovation Management Review*, *10*(4), 48–57. doi:10.22215/timreview/1346

Westerman, G., Bonnet, D., & McAfee, A. (2015). *Liderando na Era Digital*. M. Books.

Weston, M. J. (2022). Strategic Planning for a Very Different Nursing Workforce. *Nurse Leader*, *20*(2), 152–160. doi:10.1016/j.mnl.2021.12.021 PMID:35075352

Wikipedia. (2020). https://pt.wikipedia.org/wiki/Jingle

Wikipedia. (2021). https://en.wikipedia.org/wiki/Music_in_advertising

Wilcox, K., & Stephen, A. T. (2013). Are close friends the enemy? Online social networks, self-esteem, and self-control. *The Journal of Consumer Research*, *40*(1), 90–103. doi:10.1086/668794

Wittgenstein, L. (1953). *Philosophical investigations* (4th ed.). John Wiley & Sons.

Wójcik, D. (2020). Financial geography II: The impacts of FinTech - Financial sector and centres, regulation and stability, inclusion and governance. *Progress in Human Geography*, *20*(10), 1–12.

WulffM.MorgenthalerA. (2020). *Wumo*. https://wumo.com/wumo/2020/07/16

Xavier, A. (2015). *Storytelling: stories that leave a mark* (5th ed.). Best Seller.

Xiang, Z. (2021, March 31). A New Platform to Support and Define an Emerging Field. *Journal of Smart Tourism*, *1*(1), 1–2. doi:10.52255marttourism.2021.1.1.1

Xiaopu, S., Runtong, Z., Xiaomin, Z., & Quan, Z. (2016). Design theory, modelling and the application for the Internet of Things service. *Enterprise Information Systems*, *10*(3), 249–267. doi:10.1080/17517575.2015.1075592

Xu, K., & Lombard, M. (2017). Persuasive computing: Feeling peer pressure from multiple computer agents. *Computers in Human Behavior*, *74*, 152–162. doi:10.1016/j.chb.2017.04.043

Yang, K. C., Huang, C. H., Yang, C., & Yang, S. Y. (2017). Consumer attitudes toward online video advertisement: YouTube as a platform. *Kybernetes*, *46*(5), 840–853. doi.org/10.1108/K-03-2016-0038

Ye, Ye, H., & Law, R. (2020, April 22). Systematic Review of Smart Tourism Research. *Sustainability*, *12*(8), 3401. doi:10.3390u12083401

Yin, R. K. (2014). Case Study Research: Design and Methods (5th ed.). Academic Press.

Yin, R. K. (2015). Estudo de caso. Planejamento e métodos. Tradução de Daniel Grassi (5th ed.). Bookman.

Youn, S., & Jin, S. (2021). "In A.I. We Trust?" The effects of parasocial interaction and technopian versus luddite ideological views on chatbot-based customer relationship management in the emerging "feeling economy". *Computers in Human Behavior*, *119*, 106721. doi:10.1016/j.chb.2021.106721

YouTubeHelp. (2021a). https://support.google.com/youtube/answer/188038?hl=pt

YouTubeHelp. (2021b). https://support.google.com/youtube/answer/2454017?hl=pt

YouTubeOfficialBlog. (2021). https://blog.youtube/press/

Yrjölä, M., Spence, M. T., & Saarijärvi, H. (2018). Omni-channel retailing: Propositions, examples and solutions. *International Review of Retail, Distribution and Consumer Research*, *28*(3), 259–276. doi:10.1080/09593969.2018.1445657

Zajonc, R. B. (1980). Feeling and thinking: Preferences need no inferences. *The American Psychologist*, *35*(2), 151. doi.org/10.1037/0003-066X.35.2.151

Zander, M. F., Apaolaza-Ibáñez, V., & Hartmann, P. (2010). Music in Advertising: Effects on Brand and Endorser Perception. Advances in Advertising Research, 1, 127-140.

Zen, F., & Faizin, M. A. (2021). An Overview of Digitalization and Internationalization Msmes: A Conceptual Paper. Proceedings of the BISTIC Business Innovation Sustainability and Technology International Conference (BISTIC 2021). *Advances in Economics, Business and Management Research*, *193*, 195-200. https://www.atlantis-press.com/proceedings/bistic-21/125963905

Zhang, M., Li, Y., Sun, L., & Moustapha, F. A. (2022). Integrated store service quality measurement scale in omni-channel retailing. *International Journal of Retail & Distribution Management*, *50*(7), 839–859. Advance online publication. doi:10.1108/IJRDM-02-2021-0056

Zhang, M., Ren, C., Wang, G. A., & He, Z. (2018). The impact of channel integration on consumer responses in omni-channel retailing: The mediating effect of consumer empowerment. *Electronic Commerce Research and Applications*, *28*, 181–193. doi:10.1016/j.elerap.2018.02.002

Zhang, T., Lu, C., & Kizildag, M. (2018). Banking "on-the-go": Examining consumer's adoption of mobile banking services. *International Journal of Quality and Service Sciences*, *10*(3), 279–295. doi:10.1108/IJQSS-07-2017-0067

Zhou, M., Leenders, M., & Cong, L. M. (2018). Ownership in the virtual world and the implications for long-term user innovation success. *Technovation*, *78*(June), 56–65. doi:10.1016/j.technovation.2018.06.002

Zhou, R., Khemmarat, S., Gao, L., Wan, J., & Zhang, J. (2016). How YouTube videos are discovered and its impact on video views. *Multimedia Tools and Applications*, *75*(10), 6035–6058. doi.org/10.1007/s11042-015-3206-0

Zhu, K., Dong, S., Xu, S. X., & Kraemer, K. L. (2006). Innovation diffusion in global contexts: Determinants of post-adoption digital transformation of European companies. *European Journal of Information Systems*, *15*(6), 601–616. doi:10.1057/palgrave.ejis.3000650

Zoghaib, A. (2019). Typology of advertising music components and consumers' responses to the brand. [English Edition]. *Recherche et Applications en Marketing*, *34*(3), 47–82. doi.org/10.1177/2051570718828893

Zubiaga, A., Heravi, B., An, J., & Kwak, H. (2019). Social media mining for journalism. *Online Information Review*, *43*(1), 2–6. doi:10.1108/OIR-02-2019-395

Zúñiga, H. G., Weeks, B., & Ardèvol-Abreu, A. (2017). Effects of the news-finds-me perception in communication: Social media use implications for news seeking and learning about politics. *Journal of Computer-Mediated Communication*, *22*(3), 105–123. doi:10.1111/jcc4.12185

Zygiaris, S. (2013). Smart City Reference Model: Assisting Planners to Conceptualize the Building of Smart City Innovation Ecosystems. *Journal of the Knowledge Economy*, *4*(2), 217–231. doi:10.100713132-012-0089-4

About the Contributors

Jorge Remondes is a professor at Instituto Superior de Entre Douro e Vouga and at Instituto Politécnico do Porto. Researcher at CEOS - Center for Organizational and Social Studies. Editor-in-chief at International Journal of Marketing, Communication and New Media. Author. Trainer of executives and at the Ministry of Foreign Affairs of Portugal.

Sandrina Teixeira is an adjunct professor at School of Accounting and Administration of Porto, Polytechnic Institute of Porto, Portugal. Since 2013 has Coordinator of the Digital Marketing Masters and of the Degree in Business Communication. Between 2013-2014 has Coordinator of the Degree in Marketing. She is responsible by various courses, such as: unit of Marketing and Advertising Communication, Development of Multimedia Content, Seminars in the Digital Marketing Masters. She got a PhD at University Vigo of Communication, Advertising and Public Relation, in 2011. Her research interests are in digital communication, transparency in public administration and new consumers. Her supervised several master degree dissertations, was arguer in various proofs masters in various institutions of higher education and presided several master juries degrees.

* * *

Carolina Afonso has a PhD in Marketing (Universidad Complutense de Madrid). Invited Assistant Professor of Marketing & Strategy at Lisbon School of Economics and Management (ISEG), University of Lisbon. Reseacher at Advance.

Lídia Aguiar is a Master in Tourism and Business Development. PhD in Tourism by the University of Girona. She has focused her activity on research projects, in several areas taking place at the Center for Interdisciplinary Research and Community Intervention (CIIIC), at ISCET. She is dedicated to the area of research applied to tourist and cultural itineraries and to research on tourism sustainability in low-density territories. Participates in several National and International Congresses. It has several articles published. She is currently coordinating professor at the Higher Institute of Business and Tourism Sciences where she teaches several Curricular Units in the field of Tourism.

Belem Barbosa received her PhD in Business and Management Studies – Specialisation in Marketing and Strategy from the University of Porto, Portugal. She is Assistant Professor at the School of Economics and Management of the University of Porto. She is full researcher at GOVCOPP, the Research Unit of Governance, Competitiveness, and Public Policies, and Invited Researcher at cef.up Center for

Economics and Finance at the University of Porto. Her research interests lie primarily in the areas of digital marketing and consumer behavior.

Julien Diogo is CCO of the ICN Agency, neuromarketing agency, CCO of PsicoSoma, publisher and training center and CCO of expertMind, international LMS platform. He is Adjunct Professor at the Polytechnic Institute of Viseu, at the School of Education of Viseu, in UC such as Market Analysis, Consumer Behavior, Strategic Communication, or Innovation/Creativity and Entrepreneurship, and Guest professor at ISAG - Instituto Superior de Administração e Gestão in the Executive MBA and in the Specialization Course in Neuromarketing. He is a consultant in the areas of Applied Cognitive Neuroscience, Strategy and Marketing, with national and international companies Professional Trainer and also consultant in professional e-training techniques, resources and e-learning. He is author of several scientific articles in the areas of consumer, marketing and cognitive neuroscience, as well as books on cognitive neuroscience applied to communication, design and territories, with participation in books such as "Principles of NeuroMarketing"; "Consumer behavior"; "Why Marketing is Sexy and Smart", "Digital Marketing & Ecommerce", "NeuroCities, a Neuroscience process applied to Place Branding" or even "Neurobranding".

Manuel José Fonseca is a PhD in Information Sciences, Manuel José Serra da Fonseca was born in Lisbon in 1971. He played professional activities such as marketing communications technician and professor of higher education. He is currently Adjunct Professor at the Polytechnic Institute of Viana do Castelo (Portugal) and Coordinator of Marketing Master's degree at its Business School (ESCE-IPVC). He is a researcher at UNIAG (Applied Management Research Unit, since 2014). He participated in 3 national projects I&D and supervised about 40 undergraduate and masterworks in the areas of Communication Sciences and Business Sciences.

Jorge Esparteiro Garcia is an adjunct professor at Polytechnic Institute of Viana do Castelo (IPVC) since 2005/2006. He is an Afilliated Researcher at INESC TEC in the Software Engineering area since 2015. He has a PhD in Informatics Engineering from Faculty of Engineering of University of Porto (FEUP) with a thesis titled "Requirements Change Management based on Web Usage Mining". He received his master degree in Informatics Engineering from the Faculty of Engineering of University of Porto (FEUP) and his bachelor's degree in Computer Science from Faculty of Sciences of University of Porto (FCUP). His research interests are in the field of software engineering, web usage mining, requirements management and digital marketing.

Chaitra Katti is an Independent Researcher and worked as a marketing professional for SMEs and Large Corporations. Her Interest lies in Digital Marketing, Consumer Ethics and Consumer emotions

Marco Lamas is a Post-Doc in Entrepreneurship Education, PhD in Education and Entrepreneurship and bachelor's degree in International Relations. Phdcandidate in Business and Economics . Senior Professor at ISLA – Instituto Politécnico de Gestão e Tecnologia (ISLA - IPGT) and Invited Senior Professor in Entrepreneurship and Management at ISCAP-P.Porto (Porto Polytechnic), where he is a member of the Board and sub-coordinator of Porto Executive Academy, and coordinates the following Executive programs: Business Creation and Development; Science, Entrepreneurship and Innovation; Lean Innovation; and Smart Cities Management. He collaborates with other Higher Education Institutions. Research member of the Center for Intercultural Studies (CEI), the Center for Organizational and Social Studies of the Polytechnic of Porto (CEOS.PP), both from ISCAP-P.Porto. Participates in international scientific conferences, presenting articles and as a member of the scientific committee. He supervises Master's thesis and he is, on a regular basis, member of Master's and Doctoral examination jury. He is Editor and Reviewer in several scientific journals. Business coach, mentor, management consultant and trainer.

Ana Maria Lima has a Master Degree in Digital Marketing from Porto Polytechnic Institute and is a PhD student at Vigo University. At the same time, develops a career in communication, owning a company in digital communication and content marketing. Her research interests lie primarily on digital communication, online media, and online business strategy.

Dulcineia Catarina Moura is a PhD in Economics from the University of Beira Interior and post-graduate in Territorial Marketing. Researcher as an integrated member of NECE – Research Center in Business Sciences (UBI). She is the executive coordinator of a regional development association, which promotes tourism in the region of influence (Territórios do Côa, ADR). Guest Professor at ISCET - Instituto Superior de Ciências Empresariais e Turismo (Porto).

Alecsander Pereira is Master in Business Management from Instituto Superior de Entre Douro e Vouga and postgraduate in Digital Marketing and e-Commerce. Is currently responsible for the information systems area of ISVOUGA, is also responsible for the IT technical support of the companies DIGITALGREEN and ZONAVERDE and is writing IT articles and news for the Admirável Mundo Novo section of the newspaper Correio da Feira and Meio Byte magazine.

Raquel Pereira is a Ph.D. in Economics (Santiago de Compostela University, Spain). Master in Business and Industrial Economy (Minho University) and graduation in Economical and Political International Relations (Minho University). She started her professional activity at former ICEP (Investment and External Commerce of Portugal) now named AICEP (Portuguese Agency for Investment and External Commerce). Full professor at ISCAP - Porto Accounting and Business School (Polytechnic Institute of Porto) in the area of Economics. She is also the program director of the International Commerce degree and a member of the pedagogical board of ISCAP. Currently, she is a member of the scientific boards and a reviewer in scientific international journals and international conferences. Researcher at the Center for Organizational and Social Studies (CEOS.PP) (Polytechnic of Porto - ISCAP), her main research interests include Regional Economic Development, Internationalization, International commerce, and Entrepreneurship.

Pedro Rodrigues was born November 20, 1996 in Viana do Castelo, Portugal. Graduated in Distribution and Logistics Management, Post-Graduated in Digital Marketing and E-Business and, currently, Master in Marketing.

Jiarui Song is an independent researcher. Her research interests include FinTech, Chinese economy and business analytics.

João Teixeira is a master's student in Management at the Universidade de Lisboa and holds a bachelor's degree in Management from the Universidade Católica Portuguesa. Following an internship in the banking sector, he is currently pursuing a career in retail at Sonae MC.

Ahmet Tuğrul Tuğer is an Assistant Professor in Marketing and Logistics at Piri Reis University, Turkey. He got a PhD in Business and Management from Gebze Technical University, Turkey. He teaches several courses such as: Logistics Management, Retailing Management, Services Marketing. He has academic publications in national and international conferences and journals. His research areas are in services marketing, omni-channel marketing, sustainable marketing and logistics management.

Pedro Mota Veiga is graduated in Probabilities and Statistics (University of Lisbon, Portugal), and he holds a PhD in Management (University of Beira Interior, Portugal) and post-graduation in Applied Economics (University of Coimbra, Portugal). He is Assistant Professor at Universidade Católica Portuguesa, Polytechnic Institute of Viseu and where teaches subjects in Marketing, Management and Data Analysis. He is currently the scientific coordinator of the Entrepreneurship, Competitiveness and Innovation research line of the research center NECE - Center for Studies in Business Sciences at the University of Beira Interior and his research is focused on Innovation, Entrepreneurship, Knowledge Management, Digital Transformation and Quantitative Methods. He has published several articles in international journals and book chapters about these research domains. He works as a scientific consultant for several organizations.

Poshan (Sam) Yu is a Lecturer in Accounting and Finance in the International Cooperative Education Program of Soochow University (China). He is also an External Professor of FinTech and Finance at SKEMA Business School (China), a Visiting Professor at Krirk University (Thailand) and a Visiting Researcher at the Australian Studies Centre of Shanghai University (China). Sam leads FasterCapital (Dubai, UAE) as a Regional Partner (China) and serves as a Startup Mentor for AIC RAISE (Coimbatore, India). His research interests include financial technology, regulatory technology, public-private partnerships, mergers and acquisitions, private equity, venture capital, start-ups, intellectual property, art finance, and China's "One Belt One Road" policy.

Index

H

I

J

K

L

M

N

O

P

S

T

U

V

W

Y

www.ingramcontent.com/pod-product-compliance
Lightning Source LLC
LaVergne TN
LVHW060423300926
849162LV00006B/1511
* 9 7 8 1 6 6 8 4 5 5 3 8 8 *